Tube Forming

Yukihisa Kuriyama · Atsushi Shirayori ·
Noah Utsumi · Masaaki Mizumura

Tube Forming

Secondary Processing of Tubes for Weight Reduction and High Functionality

Springer

Yukihisa Kuriyama
The University of Tokyo
Tokyo, Japan

Noah Utsumi
Department of Integrated Mechanical Engineering
Daido University
Nagoya, Japan

Atsushi Shirayori
Department of Fundamental Engineering
Utsunomiya University
Utsunomiya, Japan

Masaaki Mizumura
Department of Mechanical Engineering
National Institute of Technology Kure College
Kure, Japan

ISBN 978-981-95-3420-3 ISBN 978-981-95-3421-0 (eBook)
https://doi.org/10.1007/978-981-95-3421-0

This Springer imprint is published by the registered company Springer Nature Singapore Pte Ltd.
The registered company address is: 152 Beach Road, #21-01/04 Gateway East, Singapore 189721, Singapore

Preface to the English Edition

We are delighted to present this English edition of *Tube Forming* in a more accessible format for international engineers. It is a great pleasure for us to share well-organized and up-to-date information on Japan's technology of plasticity.

This English edition is a translation of the Japanese textbook *Tube Forming*, originally published in 2019 as part of the *New Plastic Forming Technologies Series* by the Japan Society for Technology of Plasticity (JSTP). It was translated by the primary authors of the Japanese edition and members of JSTP's Tube Forming Committee.

We hope this textbook proves useful to engineers around the world.

Tokyo, Japan
September 2025

Yukihisa Kuriyama

Preface to the Japanese Edition

Tube forming is a secondary processing technology for tubes, which involves processing tubes formed by primary processing techniques such as roll forming, piercing-rolling, and plate bending. It includes multiple processing techniques such as cutting, bending, expanding and shrinking the tube, flaring and squeezing the tube end, hole punching, and joining. *Tube Forming*, one of the *New Plastic Forming Technology Series*, systematically describes these secondary processing techniques for tubes. The revised edition, published nearly 25 years after the original version (1992), serves as a comprehensive guide for these techniques.

Looking back from the time of the original publication until 25 years prior, Japan was in a period of rapid economic growth, during which various industrial technologies progressed. The development of materials capable of withstanding processing, coupled with advancements in processing techniques, led to progress in this field. However, since 1995, the working-age population has declined, and Japan has entered a mature phase with economic growth rates declining from the previous population bonus era, where the working-age population was more than twice the dependent population.

To summarize the last 25 years, the significant change in tube forming is that it has expanded from primarily piping to structural application. Lightweight has become essential due to energy and resource conservation efforts, and hollow pipes are ideal for this purpose because they provide both lightweight and rigidity/strength. Tube processing is necessary because straight pipes cannot form structures, and thus, tube forming technology has evolved. Regarding materials, copper pipes have not undergone significant changes, but the application of titanium pipes has expanded, and steel and aluminum pipes with excellent formability have been developed, followed by advancements in high-strength materials. Processing techniques have been supported by advances in computing, such as complex and precise processing process control via computers and high-precision computer simulation of processing conditions and mechanisms. In the field of tube forming, the evolution from hydraulic bulge forming to tube hydroforming, which allows for more complex integral forming, is a significant example of progress.

This book strives to provide universal descriptions that will not become outdated in the next 25 years while incorporating advancements in such technologies. The chapter structure has been simplified, and efforts have been made to systematize the technology. In the *New Plastic Forming Technology Series*, "bending processing" is not included, so the bending of shaped materials is also included in this textbook. In addition, in this revision, we considered technology transfer to be an important task and extracted and archived the knowledge of skilled engineers' processing techniques. We developed a method for this purpose, and we are deeply grateful for the valuable knowledge disclosed by Masanobu Nakamura of the former Tube Forming Company and Masao Okada of Taiyo Corporation, which we were able to include in this textbook with their assistance.

As information storage on the internet continues to increase and access to it becomes easier, textbooks with systematized information, rather than fragmented knowledge on individual topics, are considered to have certain value. As the population further declines and technology transfer becomes a challenge in the future, it would be greatly appreciated if this book could serve as a reference for the next generation of engineers.

This book was written mainly by members of the Tube Forming Committee of the Japan Society for Technology of Plasticity, and I am grateful to all the authors. In particular, I would like to express our gratitude to Prof. Atsushi Shirayori of Utsunomiya University, Prof. Noah Utsumi of Saitama University, and Dr. Masaaki Mizumura of Nippon Steel and Sumitomo Metal Corporation for their discussions and contributions not only in writing and organizing their respective chapters but also in shaping the overall structure and description of this book. Although I was involved in research and development of primary and secondary processing of steel sheets and pipes at Nippon Steel, after moving to the university, I mainly focused on research other than plastic working. Therefore, it was thanks to these individuals that this book was completed and I was able to fulfill my role as a go-between. Finally, I would like to express our gratitude to Prof. Moto Asakawa and Prof. Junichi Endo for their valuable advice throughout the book.

Tokyo, Japan
March 2019

Yukihisa Kuriyama

Competing Interests The authors have no competing interests to declare that are relevant to the content of this manuscript.

Authors of Original Japanese Edition

	Lead Author	Kuriyama Yukihisa
栗山幸久 Kuriyama Yukihisa	The University of Tokyo*	Chapter 1, Chapter 2, Chapter 7, Chapter 9 (9.3)
水越秀雄 Mizukoshi Hideo	UACJ Corporation	Chapter 2 (2.2.2)
内海能亜 Utsumi Noah	Saitama University	Chapter 3 (3.1, 3.5–3.7)
加藤和明 Kato Kazuaki	Sango Co., Ltd.*	Chapter 3 (3.2.1), Chapter 5 (5.2, 5.6)
奥出裕亮 Okude Yusuke	Tokyo Metropolitan Industrial Research Institute	Chapter 3 (3.2.2)
長谷川収 Hasegawa Osamu	Tokyo Metropolitan College of Industrial Technology	Chapter 3 (3.3, 3.4)
水村正昭 Mizumura Masaaki	Nippon Steel and Sumitomo Metal Corporation	Chapter 4
淵澤定克 Fuchizawa Sadakatsu	Utsunomiya University	Chapter 4
福村卓巳 Fukumura Takumi	Yamamoto Suiatsu Kogyosho Co., Ltd.	Chapter 4
白寄篤 Shirayori Atsushi	Utsunomiya University	Chapter 4, Chapter 8 (8.1, 8.4)
久保木孝 Kuboki Takashi	The University of Electro-Comunication	Chapter 5 (5.1, 5.3–5.5)
陣内雄士 Jinnouchi Yuji	Sango Co., Ltd.	Chapter 5 (5.2, 5.6)
北澤君義 Kitazawa Kimiyoshi	Shinshu University	Chapter 6
寺前俊哉 Teramae Toshiya	Hitachi, Ltd.	Chapter 7
山内浩行 Yamauchi Hiroyuki	Sugino Machine Ltd.	Chapter 8 (8.2.1–8.2.3)
山藤博文 Sando Hirofumi	BENKAN Corporation	Chapter 8 (8.2.4)
村上碩哉 Murakami Hiroya	Tokyo Institute of Technology*	Chapter 8 (8.3)
吉原正一郎 Yoshihara Syoichiro	Shibaura Institute of Technology	Chapter 9 (9.1)
古島剛 Furushima Tsuyoshi	The University of Tokyo	Chapter 9 (9.2)

* former affiliation as of 2019 February

Authors of the English Edition

English translation and adaptation by Yukihisa Kuriyama, Atsushi Shirayori, Noah Utsumi, and Masaaki Mizumura.

Figures were retouched to adjust for the differences in space required by English and Japanese words and texts. Some figures were modified for clarity and readability.

Contents

Chapter 1
Introduction

1.1 Overview

The secondary forming of tube materials is referred to as tube forming, but broadly speaking, it often includes various processing techniques such as separation and joining in addition to forming. As shown in Fig. 1.1, there are various processing techniques for forming, separating, and joining tube materials, and tube forming is a comprehensive processing technique for processing products using tubes as materials, rather than a single processing method. Furthermore, tube forming is also applied to subsequent processing steps such as deep drawing products with similar shapes to tube materials.

Fig. 1.1 Classification of tube forming by processing

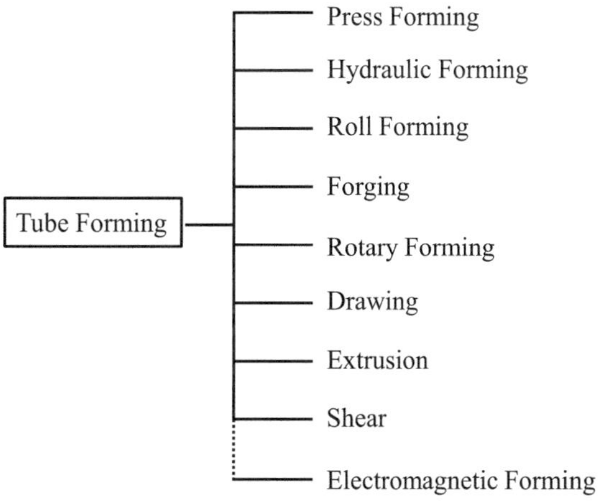

Y. Kuriyama et al., *Tube Forming*, https://doi.org/10.1007/978-981-95-3421-0_1

Tube materials have been widely used for various fluid transportation purposes since ancient times, and because of this, bend processing for piping, processing to form pipe connections and supports, as well as manufacturing techniques for joints from tube materials have been developed. Furthermore, as the pipe manufacturing technology progressed to form higher strength materials with higher dimensional accuracy, tube materials began to be widely used as structural components for buildings and automobiles. Changes from the previous edition of this textbook include the practical use of hydroforming in the late 1990s, which led to the use of tube materials in automobile structural components, and advances in processing technologies such as bending of high-strength tube materials. Such technological advances have been made possible by improvements in control technology and equipment, and eccentric and inclined spinning have also become practical due to the progress in control technology. Additionally, research and development of in-process control has progressed, which involves controlling the processing conditions by monitoring the deformation state using sensors to avoid fracture by proper distribution of the thickness of the tube.

Many theoretical and experimental formulas based on the theory of plasticity have been derived and proposed to evaluate the necessary processing force, mechanical properties required for forming, and formability assessment in tube forming. The development of FEM (Finite Element Method) technology has enabled detailed analysis of deformation behavior during forming, but tube forming is a difficult process to handle with the FEM due to issues such as contact and friction with tools, and comparison with analytical solutions is also important. Studies have been conducted on evaluation methods of deformation characteristics, which are important inputs for the FEM, and standard methods for evaluating both axial and circumferential properties have been proposed. By using such accurate material properties, the deformation behavior of tube materials in tube forming can be accurately determined using the FEM. On the other hand, the forming limit, especially the fracture limit, depends on the deformation ability of the material. In sheet metal forming, the forming limit can be practically determined by using the occurrence of necking, which leads to fracture shortly after its occurrence, as the forming limit. This necking limit is obtained from bifurcation theory. Similar theoretical methods for determining the forming limit in tube materials have also been investigated, and experimental methods for evaluating the forming limit and forming margin have been proposed, which allow an advanced formability assessment based on their evaluations.

1.2 Processing Method Classification

When classifying the processing methods used in tube forming according to the processing style, as shown in Fig. 1.1, various processing techniques in the titles of other textbooks in the New Plastic Forming Technology series are listed, indicating tube forming is a comprehensive processing technology system related to tubes. Tube forming processes are classified into three major categories—forming, joining, and separation—and each processing method is further classified under these broad categories, as shown in Fig. 1.2.

There are various forming methods used in tube forming, including bending and twisting of the entire tube, shaping and thickness variation at the tube end, as well as bulging, shrinking, and variable thickness processing in the tube body.

Various welding methods and adhesives are used for joining, and plastic deformation-based joining methods, such as crimping, rotary expanders, and can seaming, are also applied for joining by shaping the end of the tube.

For separation, there are many methods involving removal processing such as cutting and grinding using saws, grinding stones, and specialized pipe cutting machines represented by the bite cutting method. When production efficiency is the primary concern over cutting surface accuracy, cutting using press dies and cutting with wedge-shaped rolls are used. In addition, cutting with press dies is useful for piercing holes in the pipe body and cutting the end surface into complex shapes.

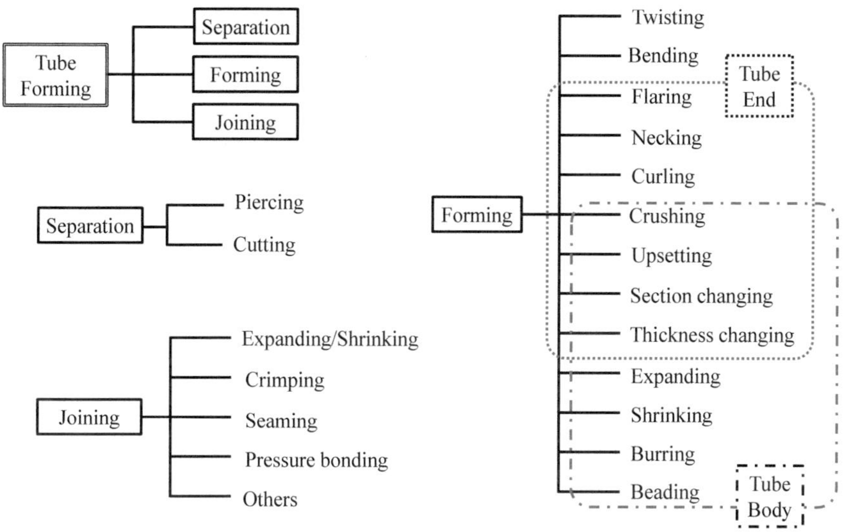

Fig. 1.2 Classification of tube forming by forming method

1.3 Product Design Using Tubes and Process Design for Tube Forming

1.3.1 Examples of Products Made from Tubes

Pipes and tubes have a long history of use in transporting various fluids, with copper pipes used in the water supply pipes of Egyptian temples 4,800 years ago and lead pipes used in the water pipes of Rome 2,000 years ago. Materials for water pipes have evolved from lead, brass, copper, iron, and stainless steel to include resin-lined steel pipes in recent years. In addition, the range of applications has expanded from water pipes to include air conditioning pipes, piping for chemical plants and power generation facilities, and pipelines for oil and gas. These are all metal pipes, and development has been carried out on their joining methods, joint processing, and bending processing.

Pipes and tubes are also used as containers by closing their ends. They are widely used as propane gas containers for household use, LP gas cylinders for taxis, as well as metal liners for compressed natural gas cylinders and hydrogen cylinders for fuel cell vehicles. The use of these cylinders is expected to expand for hydrogen gas transport and storage. In these pressure vessel applications, closing and joining the tube end are the main tube forming processes. The main requirement for these tubes and containers is pressure resistance design, and the forming process is not complex.

There is a different application of pipe materials as structural parts, apart from piping and containers. They are used as functional structural members such as automobile door impact beams, large crane beams, structural members of truss structures covering large spaces, power transmission shafts, and torsion springs. These are made possible because the tubes have a hollow closed cross-section, and their stiffness for bending and twisting is at a high weight-saving ratio. Tube forming makes it possible to process them into shapes that match other parts or into desired dimensions and shapes as structural members. This has led to the expansion of their use in office equipment parts, automobile engine cradles, rear suspension torsion beams, pillar reinforcement members, and other applications.

1.3.2 Product Design Using Tube Materials

When using tubes as piping or pressure vessels, the basic requirement is pressure resistance, and heat resistance may also be required for heat exchangers, boiler tubes, and some plant piping. Pressure resistance can be achieved by selecting appropriate material strength, thickness, and outer diameter. On the other hand, when used as structural parts, weight reduction of the parts is often required. Since the material cost of tubes is higher than that of sheet and round bars, weight reduction is also important from the viewpoint of cost reduction by reducing the amount of material used. When used as structural parts, the following advantages can be obtained by

Table 1.1 Rigidity and section modulus of round bar and thin-walled tube

	Round bar	Thin-walled tube
Flexural rigidity *EI*	$E\frac{\pi}{64}d_s^4$	$E\frac{\pi}{8}D^3t$
Section modulus *Z*	$\frac{\pi}{32}d_s^3$	$\frac{\pi}{4}D^2t$
Cross-sectional area *A*	$\frac{\pi}{4}d_s^2$	πDt
Specific rigidity *EI*/*A*	$\frac{E}{16}d_s^2$	$\frac{E}{8}D^2$
Specific section modulus *Z*/*A*	$\frac{1}{8}d_s$	$\frac{1}{4}D$

replacing round bars with tubes: (1) weight reduction due to hollowing, (2) strength increase due to work-hardening, and (3) easy processing to achieve optimal outer diameter and thickness.

(1) **Weight reduction by hollow cross-section**

Replacing a solid round bar with a hollow tube can increase the stiffness and sectional modulus for the same mass by concentrating material farther from the neutral axis. In strength design (design based on stress limits at the onset of material yielding), sectional modulus is the basic property of a member, whereas in stiffness design (design based on limits of deflection), the area moment of inertia is the basic property. Table 1.1 shows the bending stiffness and sectional modulus for a solid round bar and a thin-walled tube, where d_s is the diameter of the solid round bar, D is the outer diameter of the thin-walled tube, and t is the thickness. The specific stiffness, or stiffness per unit mass, and specific sectional modulus, or sectional modulus per unit mass, are both twice as large for the thin-walled tube compared to the solid round bar, indicating the lightweighting effect achieved by replacing the solid bar with a tube.

Detailed investigation of lightweighting using thick-walled tubes has also been conducted [1, 2]. However, while the effect of lightweighting is shown in the literature, when replacing solid rods with tubes, flattening and yielding of the cross-section due to bending become issues. Therefore, the ratio of tube wall thickness to outer diameter, which serves as a guide for cross-section flattening and yield considerations, is shown in Fig. 1.3 along with the lightweighting effect (mass ratio). Empirically, when the thickness-to-outer- diameter ratio is less than about 0.03, cross-section flattening and yield considerations become important. Therefore, when the outer diameter ratio between the hollow material and solid material (rod) is 1.5 or greater for the same strength or rigidity, caution is required for hollowing. Here, the outer diameter of the solid material (rod) is d_s, the outer diameter of the hollow material (tube) with the same strength or rigidity is d_o, the wall thickness is t, and the mass ratio between the hollow material and the solid material is α.

(2) **Increased strength due to work-hardening**

In tube forming, materials can be work-hardened by large cold forming such as ironing and hydroforming. This is advantageous in strength design because the yield

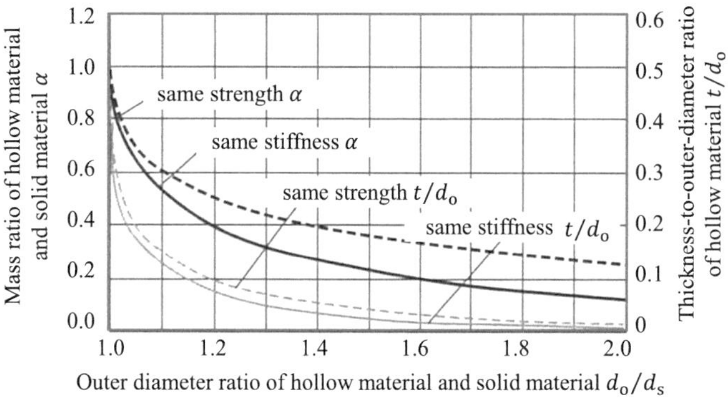

Fig. 1.3 Changes in equivalent mass and wall-thickness-to-outer-diameter ratio by replacing solid material with hollow material

point after forming is a concern. STKM13A, JIS carbon steel tube for machine structures, is expected to have a strength increase up to around 600 MPa with 40% strain applied.

(3) **Easy to achieve optimal outer diameter and thickness**

In the case of solid bars, it is difficult to achieve optimal properties and mass since the entire section needs to be modified. However, in the case of tubular materials, it is easier to modify the outer diameter and wall thickness separately, or partially change the cross-sectional shape, resulting in a high degree of design flexibility for parts. Furthermore, it is possible to reduce the stress acting on the parts over its entire length.

1.3.3 Process Design for Tube Forming

Tube forming is an excellent forming technology that can be used to shape complex shapes by combining various forming techniques using tubes as materials. However, the design of intermediate forming shapes and the selection of forming methods, tools, and machines for each intermediate forming shape and final shape are important for obtaining the desired final shape.

Various tools and media, such as dies, rolls or spatulas, hydraulic or rubber and heating and cooling instead of die, are used for various tube-forming processes. The selection of these tools and processing machines is also an important factor for the success of tube forming, considering the size, strength, forming amount, and productivity of the workpiece.

This process design heavily relies on the experience of skilled engineers, and is essentially tacit knowledge. In the Tube Forming Research Committee, this tacit knowledge was extracted and structured to describe and pass on the techniques. Section 9.3 provides examples of experienced process design.

References

1. Manabe, K.: Material-saving and weight-reduction effects of tubular hydroformed components. In: Proceedings of International Workshop of Environment & Economic Issues in Metal processing (ICEM-98), 119–125 (1998)
2. The Japan Society for Technology of Plasticity: "Tube Hydroforming" (2015), pp.11–12, Morikita Publishing

Chapter 2
Materials for Tube Forming

2.1 Overview

A material used for tube forming is a plastically deformable material with a circular tube shape. This tube is categorized based on the material and whether it has seams or not. Materials used for tube forming are mainly ductile metal materials with high elongation to allow for plastic deformation, among which steel (carbon steel, stainless steel), aluminum, and copper are widely used. In recent years, titanium has also been increasingly applied due to its excellent corrosion resistance and lightweight properties. Magnesium is also gaining attention as a lightweight material, but it has not yet been applied for mass production. Tungsten, zirconium alloy, and other materials are limited to special uses such as nuclear power generation and reactor materials.

Tubes for tube forming are classified into those with longitudinal seams and those without seams depending on the manufacturing process. Seamless tubes are produced by piercing or extruding cylindrical materials and then elongating them. Tubes with seams are produced by bending sheet materials in the circumferential direction and joining the two edges. Although these tubes have seams in the axial direction due to the tube manufacturing process, there is a wider variety of materials available for selection compared to seamless tubes. In addition, there are two methods for forming in the circumferential direction: one is to form a coil-shaped sheet and then cut it to a fixed length after joining, and the other is to form a rectangular sheet or plate that has been cut to a fixed length by pressing or bending and then joining. The roll forming method has high productivity because it is a continuous process, but it requires many rolls for each outer diameter and has dimensional constraints. On the other hand, the short sheet or plate bending method has low productivity because it is a batch process, but it can produce various dimensions with simple preparation of forming tools. There are several methods of joining, such as electric resistance welding, arc welding (mainly TIG welding), laser welding, and forge welding. Electric resistance welding and forge welding are combined with roll forming due to the relationship

Y. Kuriyama et al., *Tube Forming*, https://doi.org/10.1007/978-981-95-3421-0_2

Table 2.1 Typical metal tubes in JIS

Material	Standard name	Material symbols	Standard number
Steel	Carbon steel tubes for machine structure	STKM 11–20	JIS G 3445
	Electric resistance welded carbon steel tubes for automobile	STAM 200–500	JIS G 3472
	Carbon steel tubes for general structural purposes	STK 290–540	JIS G 3444
	Carbon steel pipes for ordinary piping	SGP	JIS G 3452
	Stainless steel tubes for machine and structural purposes	SUS*** TKA	JIS G 3446
	Stainless steel pipes	SUS**** TPY	JIS G3459
Aluminum	Aluminum and aluminum alloy extruded tubes and cold-drawn tubes	A**** TE, TES, TD, TDS	JIS H 4080
	Aluminum and aluminum alloy welded pipes and tubes	A**** TW, TWA, TWS, …	JIS H 4090
	Aluminum and aluminum alloy extruded profiles	A**** S, SS	JIS H 4100
Copper	Copper and copper alloy seamless pipes and tubes	C**** T, TS	JIS H 3300
	Copper and copper alloy welded pipes and tubes	C**** TW, TWS	JIS H 3320
Titanium	Titanium and titanium alloys-seamless pipes	TTP*** H, C, W, WC	JIS H 4630
	Titanium and titanium alloys-welded pipes	TTP*** W, WC, Pd, …	JIS H 4635
Magnesium	Magnesium alloy seamless pipes and tubes	MT*	JIS H 4202
	Magnesium alloy extruded profiles	MS*	JIS H 4204

between the appropriate joining speed and the forming speed, and these methods have high productivity. Table 2.1 shows a list of metal tubes defined in JIS as a reference for tube-forming materials. Steel and titanium tubes have many types, so we have selected those that are relatively commonly used for tube forming.

2.2 Various Pipes and Tubes

2.2.1 Steel Pipes and Tubes

Iron and steel materials are widely used as tube-forming materials because they have high strength and excellent formability, and are cheaper than non-ferrous metal materials. The method of manufacturing tubes with dimensions suitable for tube forming is represented by seamlessness steel tubes, which are produced by Mannesmann piercing and mandrel rolling, and ERW steel tubes, which are produced with a combination of roll forming and electric resistance welding. Carbon steel tubes for machine structures (STKM) are widely used, but electric resistance welded carbon steel tubes for automobile structures (STAM), which conform to the standards of automotive steel sheets, are also used for automobile components.

Figure 2.1 shows a schematic diagram of the general manufacturing process for small-diameter seamless pipes. A cylindrical material is pierced using the Mannesmann method, then rolled with a mandrel and elongated by a stretch reducer (SR) to produce the desired dimensions. All of these steps are performed by hot forming. Seamless pipes have the advantage of having uniform strength, toughness, and corrosion resistance in the circumferential direction because there are no longitudinal seams. However, because the manufacturing process is hot forming and it is difficult to ensure the concentricity of the tool on the inner and outer surfaces of the pipe during piercing and rolling, the wall thickness uniformity and ovality are inferior to those of electric resistance welded pipes. Additionally, the material used is a solution-hardened carbon steel; high formability materials cannot be selected because of the manufacturing process.

Figure 2.2 shows the typical manufacturing process for small-diameter electric resistance welded pipes [1]. A strip is roll-formed, and the edges are joined by high-frequency electric resistance welding. The bead formed at the weld is cut off,

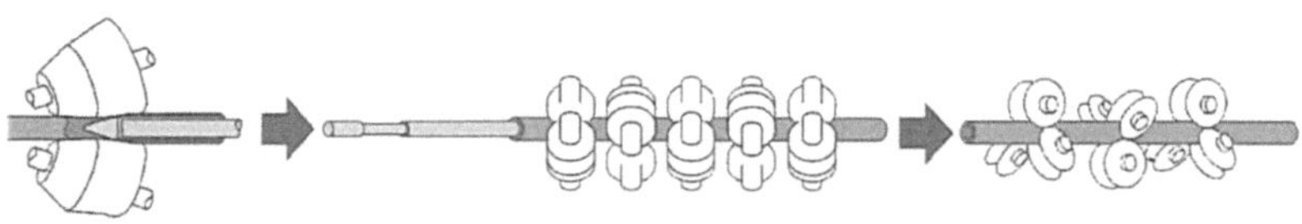

Fig. 2.1 Typical manufacturing process of small diameter seamless tubes

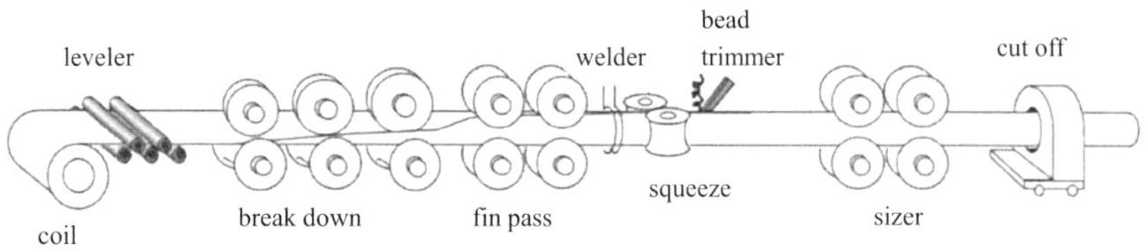

Fig. 2.2 Typical manufacturing process of small-diameter ERW tubes

and the pipe is finished to the required dimensions. These processes are typically all performed at room temperature. Although there have been concerns about the reliability of the electric resistance welded pipes due to the presence of weld seam, monitoring and diagnostic techniques for the welding process have developed, and it is now no longer a problem as a material for tube forming [2, 3]. The welded part is hardened due to rapid cooling, and its deformation ability differs from that of other parts, but the hardness can be made the same as that of the base material by heat-treating the welded part [1]. When large deformation is necessary, it is advisable to use tubes with heat treatment on the weld seams. The thickness accuracy is guaranteed in the sheet rolling process, and because the pipe is roll-formed at room temperature, the accuracy of thickness and roundness is higher than that of seamless pipes.

Steel tubes for tube forming are widely used, with STKM being commonly used and STAM being used for automobile components. STKM is a standard based on the carbon content of the chemical composition, with carbon content increasing from STKM 11 to STKM 15 and manganese content increasing from STKM 16, resulting in increased strength. STAM is a standard based on tensile strength, with its nominal designation indicating the tensile strength. Tables 2.2 and 2.3 show the mechanical properties of JIS standards for STKM and STAM. These are plain carbon steels, but STKM 20 and STAM G are allowed to have composite additions of niobium (Nb) and vanadium (V) up to 0.15%. Plain carbon steels contain carbon, manganese, silicon, phosphorus, and sulfur, which provide strength through solid solution strengthening, but the limit of their tensile strength is around 550 MPa. To achieve higher strength, Nb and V are added for precipitation strengthening and grain refinement. Molybdenum (Mo) is added to prevent softening of the heat-affected zone during electric resistance welding of these steels [4]. To further increase strength or maintain elongation while increasing strength, transformation strengthening is applied by transforming austenite at high temperature, into martensite or other structures through rapid cooling to room temperature.

The representative transformation-strengthened steels are dual-phase (DP) steels that disperse the hard martensite phase into the high formability ferrite phase, and transformation-induced plasticity (TRIP) steels that retain the austenite phase in the ferrite phase and transform it into martensite due to strain during forming. These complex phase steels are widely used in the field of sheet metal forming and are also beginning to be applied in tube forming. Materials that are single phase martensite with faster cooling rates, commonly known as quenched and tempered steels, have been in longer practical use for automobile door impact beams [5] than complex phase

Table 2.2 Mechanical properties of carbon steel tubes for machine structure (STKM JIS G 3445)

Class		Material symbol	Tensile strength [MPa]	Yield strength [MPa]	Elongation [%]	Flattening	Bending	
						Height H	Angle [deg]	Inner R [mm]
11	A	STKM 11 A	≥290	–	≥35	(1/2) D	180	4 D
12	A	STKM 12 A	≥340	≥175	≥35	(2/3) D	90	6 D
	B	STKM 12 B	≥390	≥275	≥35	(2/3) D	90	6 D
	C	STKM 12 C	≥470	≥355	≥25	–	–	–
13	A	STKM 13 A	≥370	≥215	≥30	(2/3) D	90	6 D
	B	STKM 13 B	≥440	≥305	≥20	(3/4) D	90	6 D
	C	STKM 13 C	≥510	≥380	≥15	–	–	–
14	A	STKM 14 A	≥410	≥245	≥25	(3/4) D	90	6 D
	B	STKM 14 B	≥500	≥355	≥15	(7/8) D	90	8 D
	C	STKM 14 C	≥550	≥410	≥15	–	–	–
15	A	STKM 15 A	≥470	≥275	≥22	(3/4) D	90	6 D
	C	STKM 15 C	≥580	≥430	≥12	–	–	–
16	A	STKM 16 A	≥510	≥325	≥20	(7/8) D	90	8 D
	C	STKM 16 C	≥620	≥460	≥12	–	–	–
17	A	STKM 17 A	≥550	≥345	≥20	(7/8) D	90	8 D
	C	STKM 17 C	≥650	≥480	≥10	–	–	–
18	A	STKM 18 A	≥440	≥275	≥25	(7/8) D	90	6 D
	B	STKM 18 B	≥490	≥315	≥23	(7/8) D	90	8 D
	C	STKM 18 C	≥510	≥380	≥15	–	–	–
19	A	STKM 19 A	≥490	≥315	≥23	(7/8) D	90	6 D
	C	STKM 19 C	≥550	≥410	≥15	–	–	–
20	A	STKM 20 A	≥540	≥390	≥23	(7/8) D	90	6 D

* Based on JIS G3445:2010 D: outer diameter of the tube, H: crush height at flattening test

steels. These materials are heated by high-frequency induction and immediately quenched by water to increase their strength to approximately 1.5 GPa, protecting passengers in the event of a side collision.

Furthermore, warm SR rolled ERW steel pipes [6] have been developed to increase strength at the same elongation. The r-value can be improved by developing a microstructure through large deformation at elevated temperatures in warm SR rolling. However, the uneven thickness distribution due to polygonization of the inner surface is a problem in SR rolling. To address this issue, a technology called twist SR [7] has been developed that reduces the thickness unevenness by arranging the SR rolls at different angles, and it has been applied to automotive suspension components. Figure 2.3 shows the steel grades, strength, and representative applications, including DP steel, TRIP steel, and other complex microstructure steels. Although the application of complex phase steels such as DP steel and TRIP steel

Table 2.3 Mechanical properties of electric resistance welded carbon steel tubes for automobiles

Class	Material symbol	Tensile strength [MPa]	Yield strength [MPa]	Axial elongation [%]	Flaring ratio [%]	Summary
G	STAM 290 GA	≥290	≥175	≥40	1.25 *D*	Tubes for general purpose automobile structural parts
	STAM 290 GB	≥290	≥175	≥35	1.20 *D*	
	STAM 340 G	≥340	≥195	≥35	1.20 *D*	
	STAM 390 G	≥390	≥235	≥30	1.20 *D*	
	STAM 440 G	≥440	≥305	≥25	1.15 *D*	
	STAM 470 G	≥470	≥325	≥22	1.15 *D*	
	STAM 500 G	≥500	≥355	≥18	1.15 *D*	
H	STAM 440 H	≥440	≥355	≥20	1.15 *D*	Tubes for the automobile structural parts with special consideration on yield strength
	STAM 470 H	≥470	≥410	≥18	1.10 *D*	
	STAM 500 H	≥500	≥430	≥16	1.10 *D*	
	STAM 540 H	≥540	≥480	≥13	1.05 *D*	

* Based on JIS G3472:2013 *D*: outer diameter of the tube

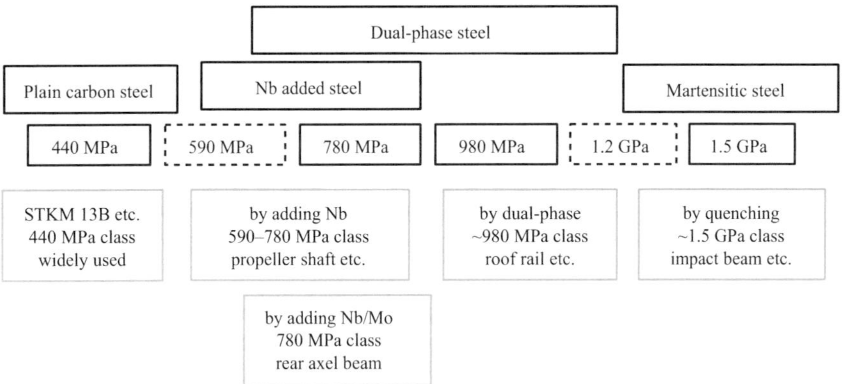

Fig. 2.3 Steel type, strength, and application of steel tubes for tube forming

is currently limited, they are expected to be widely applied in tube forming due to their high strength and high formability.

2.2.2 Aluminum Tubes

There are both seamless and welded aluminum and aluminum alloy tubes available. JIS H 4080 specifies seamless aluminum and aluminum alloy tubes with circular cross-sections, which are classified into 21 types based on their chemical composition

Table 2.4 Chemical composition and mechanical properties of typical aluminum alloys

Alloy	Chemical composition									H.T	YP [MPa]	TS [MPa]	El [%]	*n*	*r*
	Si	Fe	Cu	Mn	Mg	Cr	Zn	Ti	Others						
1050	0.25	0.4	0.05	0.05	0.05	–	0.05	0.03	V	O	30	80	40	0.27	0.69
1100	0.95 (Si + Fe)		0.05 –0.2	0.05	–	–	0.1	–	–	H112	>20	>75	>25	–	–
2014	0.05 –0.2	0.7	3.9 –5.0	0.4 –1.2	0.2 –0.8	0.1	0.25	0.15	–	O	< 125	< 245	>12	–	–
2024	0.5	0.5	3.8 –4.9	0.3 –0.9	1.2 –1.8	0.1	0.25	0.15	–	O	< 125	< 245	>12	–	–
										T4	>295	>390	>12	–	–
3003	0.6	0.7	0.05 –0.2	1.0 –1.5	–	–	0.1	–	–	O	40	110	30	0.24	0.65
										H112	>35	>95	>35	–	–
5052	0.25	0.4	0.1	0.1	2.2 –2.8	0.15 –0.35	0.1	–	–	O	90	195	25	0.28	0.63
5083	0.4	0.4	0.1	0.4 –1.0	4.0 –4.9	0.05 –0.25	0.25	0.15	–	O	145	290	24	0.26	0.74
6063	0.2 –0.6	0.35	0.1	0.1	0.45 –0.9	0.1	0.1	0.1	–	O	40	100	35	0.22	–
										T5	200	240	17	0.14	–
6N01	0.4 –0.9	0.35	0.35	0.5	0.4 –0.8	0.3	0.25	0.1	–	O	50	100	25	–	–
										T5	250	280	12	–	–
6061	0.4 –0.8	0.7	0.15 –0.4	0.15	0.8 –1.2	0.04 –0.35	0.25	0.15	–	O	55	125	25	0.17	–
										T6	270	310	12	–	–
7003	0.3	0.35	0.2	0.3	0.5 –1.0	0.2	0.5 –6.5	0.2	–	T5	>245	>285	>10	–	–
7075	0.4	0.5	1.2 –2.0	0.3	2.1 –2.9	0.18 –0.28	5.1 –6.1	0.2	–	O	105	230	17	0.19	–
										T6	505	570	11	0.12	–
7N01	0.3	0.35	0.2	0.2 –0.7	1.0 = 2.0	0.3	4.0 –5.0	0.2	–	T5	>245	>325	>10	–	–

* H.T.: heat treatment, YP: proof stress, TS: tensile strength, El: elongation

and into 2 classes based on dimensional tolerance grades, and are further classified into extruded and drawn tubes. Drawn tubes have better dimensional accuracy and are thinner and smaller in diameter than extruded tubes. The manufacturing range of tube dimensions varies depending on whether extruded or drawn and on the alloy composition.

JIS H 4090 specifies welded aluminum and aluminum alloy tubes—9 types of high-frequency induction welding tubes and 9 types of arc welding tubes—including inert gas arc welding, and covers a wide range of typical dimensions of welded tubes. Aluminum alloy is a typical non-ferrous metal used for tube forming. Aluminum alloy materials are not only more expensive than steel materials but also have drawbacks such as low Young's modulus and large springback, and low local elongation. However, because the density is about 1/3 that of steel, it is used for lightweight structural members and automobile parts. There are both extruded and cast aluminum

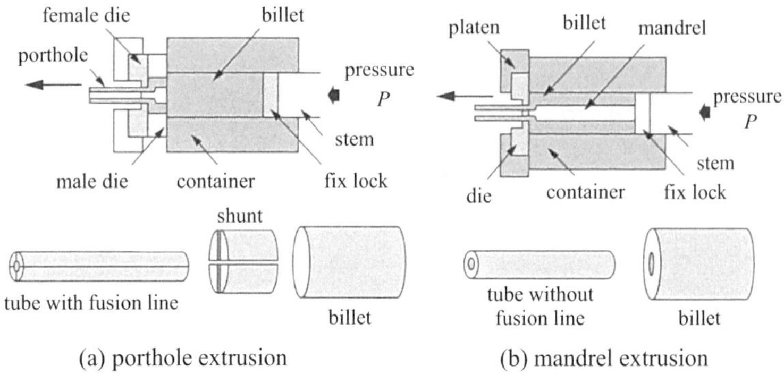

Fig. 2.4 Extrusion process of hollow sections

alloys, each with non-heat-treated and heat-treated types, but from a formability standpoint, extruded alloys are typically used for tube forming.

Table 2.4 shows the chemical composition and mechanical properties [8–10] of representative aluminum alloys. The 1000 (pure Al), 3000 (Al-Mn), 4000 (Al-Si), and 5000 (Al-Mg) series are non-heat-treated type, while the 2000 (Al-Cu), 6000 (Al-Mg-Si), and 7000 (Al-Zn-Mg[-Cu]) series are heat-treated type.

The application of aluminum alloys to automobile parts has been widely studied from the 1000 to the 7000 series [10], but among them, the 6000 series alloys have been widely adopted, followed by the 5000 series alloys. Aluminum alloys have different characteristics depending on the type of alloy, so this textbook describes the specific features and applications of different alloys with examples of parts.

In addition to material properties, there are two types of extrusion methods to obtain a hollow extruded material—port hole extrusion and mandrel extrusion—and their outlines are shown in Fig. 2.4. It is important to note that the applicability of port hole extrusion is determined according to the material and usage. Port hole extrusion can produce complex hollow cross-sectional shapes by diverting and joining the material in the die, but the characteristics of the fusion line are different from those of the base material. The effect of the fusion line by port hole extrusion on the forming limit is shown in Fig. 2.5. The forming limit was clearly lowered due to the influence of the fusion line by port hole [10].

(1) **1000-series aluminum**

Due to their low deformation resistance and good formability, A1050-O material is often used as a research subject. It is also commonly used as a comparison material when investigating the forming limits of other aluminum alloys, as it is readily available. However, due to its low strength after forming, there are few examples of

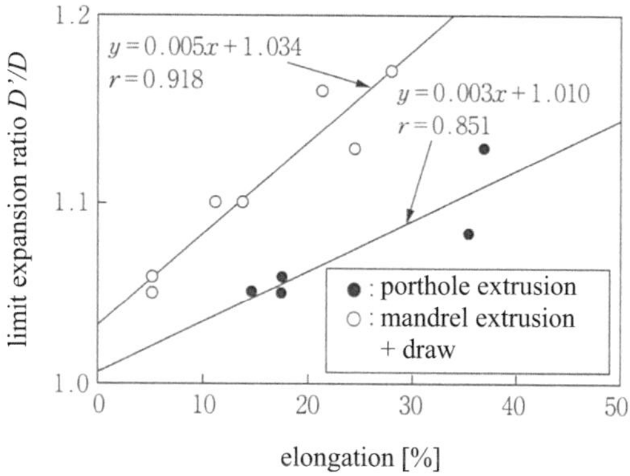

D': *expanded outer diameter*, *D*: original outer diameter

Fig. 2.5 Difference in forming limit due to extrusion method [10]

it being used as a structural member. On the other hand, its corrosion resistance and thermal conductivity are the most excellent among aluminum alloys, making it an attractive material for functional parts that can utilize these properties.

(2) 3000-series aluminum alloys

The 3000 series aluminum alloys, like the 1000 series, are often used as comparison materials for other aluminum alloys [9–12]. A3003-O being a representative alloy of non-heat-treated alloy, dimensional accuracy after processing is not reduced by heat treatment. However, its formability is not very high and its strength after processing is also low, so there are few structural applications. In addition to being widely used as air conditioning piping material, it is also used as a material for non-structural building materials, beverage cans, and photoreceptor drums for printers, where excellent surface properties after cold forming are required.

(3) 5000-series aluminum alloys

The 5000 series alloy, like the 3000 series, is a non-heat-treated type, but has higher strength than the 3000 series, making it suitable for automobile parts. Basic hydroforming properties have been investigated for alloys such as A5052, A5154, and A5083 for this purpose [9–16]. In Europe, the 5000 series aluminum sheets are widely supplied at reasonable price, and electric resistance welded pipes using the excellent weldability of the 5000 series are popular [17, 18]. A5154 has almost the same formability as A5052, which is a representative alloy of the 5000 series. A5154 has approximately 20% higher strength than A5052 and has been applied to suspension subframes by hydroforming. Additionally, A5083 is suitable for automobile frames and suspension components due to its superior strength, weldability, and corrosion resistance. In Japan, almost all 5000 series pipes are produced by extrusion.

Although aluminum alloys have low formability at room temperature, formability can be improved at high temperatures, enabling large plastic deformation. A5754 alloy pipes and A5182 (Al-Mg-3.5 Mn) alloy pipes have been reported to show significantly larger expansion deformation by gas-bulge forming at high temperatures than at room temperature [19, 20]. Aluminum rear and front subframes with complex cross-sectional shapes have been commercialized by the hot bulge forming method (high-temperature hydroforming) using specially developed 5000 series (Al-Mg-Fe alloy) materials [21].

(4) **6000-series aluminum alloys**

A6063, which is widely used as a structural material for automobiles, and A6061, A6N01 (A6005C), A6082, and others have been studied and put into practical use. A6063 has good formability and corrosion resistance, and has been used in space frames, which is the skeleton of the automobile body, as well as in roof racks, door frames, and engine cradles [22]. Additionally, Al-Mg-Si 0.5 aluminum alloy extrusion was applied for strut bearers (suspension parts) of all-aluminum space frame cars using hydroforming [22]. A6061 is a representative alloy of the 6000 series, and is a suitable alloy for bumper reinforcement and suspension components due to its excellent strength and corrosion resistance.

(5) **7000-series aluminum alloys**

7000 series is the strongest among aluminum alloys, and the representative A7075 has been mainly used for aircraft by machining. Among the 7000 series, A7003 and A7N01 (A7204) are used for railway vehicles and ships. They are alloys that can be extruded for port holes, and have good weldability and relatively good formability. In general, the higher the strength of the material, the worse the stress corrosion cracking resistance, so careful attention is needed in the selection of alloys, heat treatment, forming methods, and so on. Recently, the application of high-temperature forming of high-strength aluminum alloys has also been considered using in-die cooling.

2.2.3 *Other Tube Materials*

(1) **Copper tubes and brass tubes**

Copper is a metal with a long history of use, dating back to ancient times. It is widely used due to its excellent electrical and thermal conductivity, good corrosion resistance, and excellent formability. Copper is a precious metal along with gold and silver, which belong to the 11th group in the periodic table of elements. While iron and titanium have electrons involved in metal bonding localized near adjacent atoms, in group 11 elements, electrons involved in metal bonding spread throughout the entire crystal, making them very malleable.

JIS H 3300 and JIS H 3320 define seamless and welded tubes made of copper and copper alloys. Copper is classified based on deoxidation methods: (1) oxygen-free copper, which is dissolved in a reducing atmosphere, (2) tough pitch copper, which is produced by reducing oxygen levels and removing impurities using a shaft furnace, and (3) phosphorus deoxidized copper, which utilizes phosphorus as a deoxidizer.

Copper alloys such as brass (containing approximately 30–40% Zn), red brass (containing approximately 15% Zn), and cupronickel (containing approximately 10–30% Ni) are defined.

About 70% of copper tubes are used in air conditioning systems, and they are also widely used in water supply and hot water supply piping. Internal grooved pipes are mainly used in air conditioning systems, but corrugated tubes and low-fin tubes with increased heat transfer area are also used. Capillary tubes used for refrigerant piping are slender tubes ranging from 4.0 mm/3.0 mm to 1.8 mm/0.5 mm in outer diameter/inner diameter. One-piece capillary tubes with an original tube section, reduced diameter section (about 1/2), and slender tube section have been produced by utilizing copper's high formability. Development of high-strength and high-corrosion-resistant copper alloy tubes is underway to respond to changes in refrigerants and high pressure caused by environmental issues [23].

Copper has excellent formability, allowing for product development without the need for new secondary processing technologies. Therefore, the focus is on developing tubes with expanded heat transfer area and integrated tubes. For brass, development of a forming technology for one-piece water faucets (spigots) from brass tubes is underway [24].

(2) **Titanium tubes**

Titanium and titanium alloys are used in various fields due to their lightweight, high specific strength (about twice that of steel and three times that of aluminum), excellent corrosion resistance, and high biocompatibility [25]. Titanium is a relatively new metal, with industrial production starting in 1948. Gas shielding is required for arc welding, but resistance welding is possible, similar to steel materials. Because of a hexagonal close-packed lattice, titanium and titanium alloys have low formability, but pure titanium (JIS types 1 and 2) can also be deep drawn. Titanium alloys are broadly classified into alpha (α), alpha + beta (α + β), and beta (β) types, by the structure of their phases at room temperature (α phase: hexagonal close-packed lattice, β phase: body-centered cubic lattice). Titanium alloy with higher fraction of β phase, which has many slip systems for plastic deformation, has higher formability. To stabilize the β phase at room temperature, vanadium, molybdenum, chromium, etc. are added. Representative titanium alloys of each type are Ti-5Al-2.5Sn for the α type, Ti-6Al-4 V for the α + β type, and Ti-15 V-3Al-3Cr-3Sn for the β type. The α type has excellent corrosion resistance, heat resistance, and creep resistance, the α + β type has excellent strength and ductility, and the β type has excellent formability and can be highly strengthened by solid solution aging treatment. Titanium and titanium alloys are used in aircraft and sports equipment, taking advantage of their lightweight and high strength characteristics, and in marine structures and bridge piers, taking advantage of their corrosion resistance. Moreover, because of their high biocompatibility, they are used in artificial bones, artificial joints, surgical clips, etc. and because of their lightness and high corrosion resistance, they are widely used as building materials (roofing materials), etc. As tube materials, they are used in piping for thermal power plants and seawater desalination plants, as well as exhaust pipes for automobiles and motorcycles.

(3) **Magnesium tubes**

Magnesium is the lightest among practical metals and has the highest specific strength. Magnesium has a hexagonal close-packed (hcp) lattice, which is different from iron's body-centered cubic lattice and aluminum's face-centered cubic lattice. Because the hcp lattice has few slip systems for plastic deformation, cold forming is difficult and warm forming is necessary to achieve enough deformation [26]. Joining can be achieved through friction stir welding. Commercial production of magnesium began in 1886, similar to aluminum, but development was delayed due to the difficulty of refining it.

The main alloys for extruded and drawn magnesium products are AZ31, which contains 3% Al and 1% Zn, with solid solution strengthening and work-hardening, and is used in the form of sheets, tubes, bars, and shapes. AZ61, which has up to 6% Al added to increase strength and corrosion resistance, is also widely used as an extruded and drawn product. ZK60, which has Zn added instead of Al up to 6%, and a small amount of Zr added for grain refinement and to ensure strength and ductility, is also used. The approximate strengths of these materials are 255 MPa for AZ31, 305 MPa for AZ61, and 340 MPa for ZK60. Lightweight magnesium tubes are used to make walking aids, but the application of extruded and drawn products has mainly been for replacing plastics, such as for laptop, cell phone, and camera casings. They are also used in automobiles, but only to a limited extent as a core material for steering wheels. For use in structural parts for automobiles and railway vehicles, it is necessary to have flame retardancy, and flame-retardant alloys (such as AZ X611 with added calcium) are being developed.

2.3 Formability Tests

Formability tests for tube materials include tests to evaluate the mechanical properties of the material, tests to evaluate the material's forming limits, and tests to evaluate the formability of various types of tube-forming methods, as shown in Table 2.5. Friction is also an important factor in formability, and measurement of the friction coefficient in tube forming has been investigated, but there is no established test method. With the increasing use of FEM analysis to study tube-forming processes, the accuracy of the material properties as input data to the FEM becomes important, and studies have been conducted to evaluate the mechanical properties and formability at secondary processing of tubes [27].

Table 2.5 Forming tests of tubular materials

Type	Test method	Summary/special notes
Material properties	Longitudinal tensile test	Cut out specimen, tubular specimen
	Circumferential tensile test	Ring tensile test
	Bi-axial bulge test	Complex load of internal pressure and axial load
Forming limit	Longitudinal tensile test	Ductile fracture criteria
	Bi-axial bulge test	Forming limit curves
Formability	Bending test	Wrap around grooved die
	Flattening test	Flattening, crushing
	Flaring test	Flaring, flanging, ring expansion
	Bulge test	Free bulge and bulge with die

2.3.1 Material Properties Tests

(1) **Tensile test**

The tensile test is a representative method for examining the properties of tube materials, as well as for sheet and rod materials. It is specified by JIS and ISO standards that only axial force is applied to the specimen during testing to ensure uniaxial stress conditions. The obtained material properties include the "yield stress" corresponding to the onset of plastic deformation, "tensile strength" corresponding to the maximum load in uniaxial tension, "elongation at fracture", and so on. The elongation up to the onset of necking is called uniform elongation, and the elongation from then to fracture is called local elongation. Before the onset of local deformation, deformation occurs under uniform uniaxial stress conditions. The obtained stress–strain characteristics become the equivalent stress–strain characteristics, and thus, this is a fundamental test for evaluating material properties. There are two types of tensile test specimens for tube materials: one is a cylindrical specimen in the form of the tube itself, and the other is an arc-shaped specimen cut from the tube in the axial direction, which is similar to a sheet specimen. Although there is no difference in yield stress and tensile strength between the two types of specimens, the elongation at fracture is considerably larger for cylindrical specimens than for arc-shaped specimens. This difference is due to the fact that the arc-shaped specimen has a geometry that opens in the width direction like a sheet specimen, whereas the cylindrical specimen has a closed shape, resulting in different necking behavior between the two. Both cylindrical and arc-shaped specimens have proportional and standardized specimens, and the thickness of the arc-shaped specimen is the same as the original thickness. Figure 2.6 shows examples of cylindrical and arc-shaped specimens. The gage length L is 50 mm for

Outer diameter D (mm)	Shape	Fixed form	L (mm)	W (mm)	Proportional	L (mm)	W (mm)
small diameter	Tube	11	50	—	14 C	$5.65\sqrt{A}$	—
$D \leqq 50$	Arc	12 A	50	19			
$50 < D \leqq 170$	Arc	12 B	50	25	14 B	$5.65\sqrt{A}$	$\leq 8\,T$
$170 < D$	Arc	12 C	50	38			

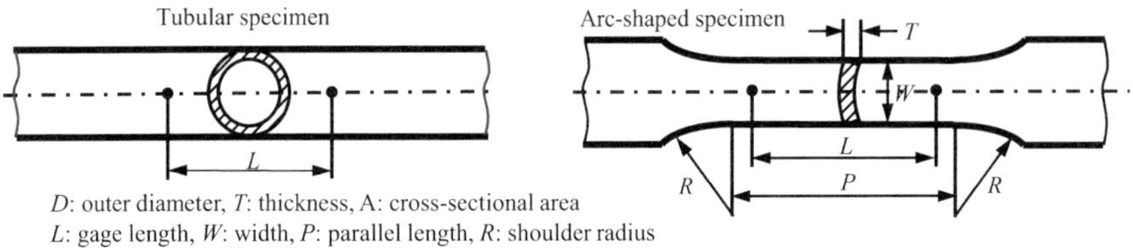

Fig. 2.6 Tensile test specimen of tubular materials

non-proportional test specimens regardless of the classification, and for proportional test specimens, it is 5.65 times the square root of the cross-sectional area A, regardless of the classification. The width W of the test piece is determined by three classifications based on the outer diameter of tube for non-proportional test specimens, and it is limited to 8 times the thickness T for proportional test specimens.

(2) **Under multi-axial stress testing**

Under multi-axial stress testing, it is possible to determine the deformation characteristics, including the circumferential properties. Experimental studies conducted in the early 1930s and mid-1950s involved applying external loads on thin-walled cylindrical tubes through combinations of tension and internal pressure [28, 29], as well as tension and torsion [30, 31] to identify yield surfaces. In recent years, a bulge forming (biaxial bulge forming) method has been developed, where tension and internal pressure loads are applied. In this test method, stress is determined by measuring the curvature of the tube during loading, along with axial force and internal pressure, and strain is measured by strain gage or digital image correlation (DIC), enabling the measurement of material properties in longitudinal and circumferential directions. While a specialized apparatus is required, this technique allows for accurate measurement of mechanical properties in both axial and circumferential directions. Biaxial bulge testing is a testing method where a cylindrical test specimen is subjected to both axial force (T) and internal pressure (P), generating a biaxial stress state at the center of the specimen. By employing feedback control of T and P based on the measured stress during loading, it is possible to obtain stress–strain curves along the desired stress path from uniaxial tension in the axial direction to uniaxial tension in the circumferential direction. Additionally, by loading until failure, it is possible to measure the forming limit strain and stress. The axial stress σ_φ and circumferential stress σ_θ at the center of the cylindrical test specimen can be determined from the equilibrium of forces in the axial and radial directions, as given by the following equation.

$$\sigma_\varphi = \frac{P\pi(D/2 - t)^2 + T}{\pi(D - t)t}, \quad \sigma_\theta = \frac{(R_\varphi - t)(D - 2t)}{(2R_\varphi - t)t}P - \frac{D - t}{2R_\varphi - t}\sigma_\varphi \tag{2.1}$$

Here, D is the outer diameter of the tube, t is the thickness, and R_φ is the curvature radius of in the axial direction. The stress is determined by the values of D, R_φ, t, and the external forces T and P. If we denote the initial outer diameter as D_0, D is given by the following equation based on the measured circumferential strain $\varepsilon_\theta{}^S$ on the surface.

$$D = D_0 \exp\left(\varepsilon_\theta^s\right) \tag{2.2}$$

If we denote the initial thickness as t_0, the thickness t can be determined from the measured value of circumferential strain $\varepsilon_\theta{}^S$ on the surface using the condition of constant volume, as given by the following equation.

$$t = {}^D/_2 - \left\{ \left({}^D/_2\right)^2 - \frac{(D_0 - t_0)t_0}{\exp\left(-\varepsilon_\theta^s\right)} \right\}^{1/2} \tag{2.3}$$

The axial strain ε_φ and circumferential strain ε_θ at the center of the cylindrical test specimen, in the middle of the thickness, can be determined by the following equations.

$$\varepsilon_\varphi = \varepsilon_\varphi^S - \ln\frac{R_\varphi}{R_\varphi - t/2}, \quad \varepsilon_\theta = \ln\frac{D_0 \exp\left(\varepsilon_\theta^s\right) - t}{D_0 - t_0} \tag{2.4}$$

By measuring the axial strain on the surface $\varepsilon_\varphi{}^S$, circumferential strain on the surface $\varepsilon_\theta{}^S$, axial curvature radius R_φ, and applied axial force T and internal pressure P, the stresses and strains in the axial and circumferential directions can be determined. This enables the experimental measurement of stress–strain relationships. Figure 2.7 illustrates the configuration of a servo-controlled biaxial bulge testing machine.

In tube forming such as expanding and flaring, the circumferential properties are important, and a method for evaluating circumferential properties is desired. Therefore, the tensile test of ring-shaped specimens cut from tubes is also specified in ISO 8496. However, since there is extra deformation such as bending until tensile deformation occurs in ring-shaped specimens, the test is not suitable for accurately measuring the uniaxial tensile properties in the circumferential direction but rather for evaluating the circumferential strength.

The measured yield surfaces of aluminum and steel tubes using biaxial bulge testing by Kuwabara et al. are shown in Fig. 2.8 [33, 34]. The aluminum tubes were extruded A5154-H112. The steel tubes were cold-drawn electric resistance welded

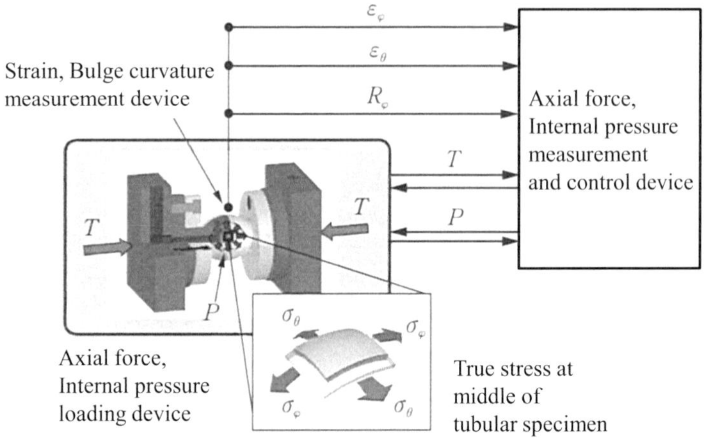

Fig. 2.7 Servo-controlled biaxial bulge test facility [32]

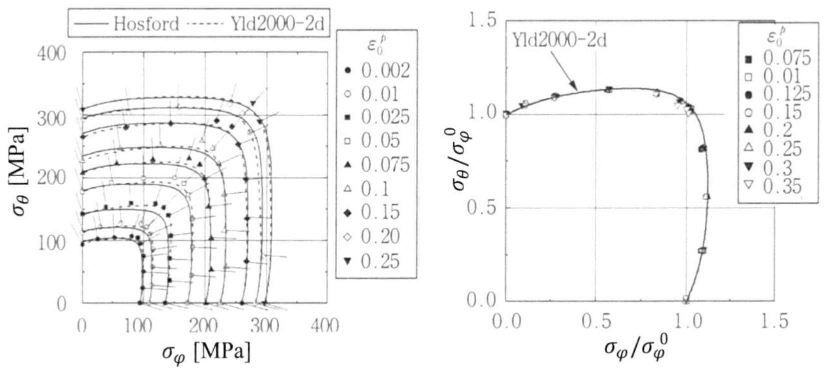

(a) extruded aluminum alloy tube [33] (b) cold-drawn and annealed ERW tube [34]

Fig. 2.8 Measured iso-plastic work loci of tubular materials by biaxial bulge test

tubes to achieve uniform thickness, followed by annealing to have the same material properties at the welded sections as the base material.

All tests were conducted with a linear stress path (proportional loading) maintaining a constant stress ratio $\sigma_\theta : \sigma_\varphi$. The cylindrical test specimens were loaded until failure. The reference for the accumulated plastic work was defined as reaching the reference strain in the longitudinal uniaxial tension test. At each proportional loading level, points corresponding to this accumulated plastic work were plotted to form an iso-plastic-work surface. In the case of steel tubes, the iso-plastic-work surface was normalized by the longitudinal uniaxial tension stress.

The figure includes yield surfaces based on the Yld2000-2d yield function and yield surfaces proposed by Hosford, allowing for a comparison with the yield surface obtained from the biaxial bulge test. Stress–strain measurements in biaxial bulge forming tests assume axisymmetric deformation, so it is important to consider the thickness and strength non-uniformity, such as in the welded seam of the tube. However, it is an effective testing method that allows for the measurement of yield surfaces and work-hardening of the tube material.

2.3.2 *Material Forming Limit Test*

In sheet metal forming, the forming limit diagram is widely used, and there are also ISO standards for it. By drawing grids or circles on the sheet for strain measurement, the forming limit of the formed product can be easily determined by measuring the strain at the fracture location (ISO 12004-1). Additionally, proportional load tests with different strain ratios, such as the Nakajima method, can be conducted to obtain the forming limit curve (FLC) of the material on the strain plane (ISO 12004-2). With established test methods and ease of evaluation, it is widely used in sheet metal forming such as pressing.

In the forming of bulk materials such as forging, the evaluation of forming limits is performed based on the ductile fracture criterion. This criterion assumes that material damage (damage value) is determined by stress and strain, and when the integrated value along the deformation path reaches a critical value, fracture occurs [35]. Various models integrate the maximum principal stress (σ_{max}) or the average normal stress (σ_{m}) along the path. The model by Cockloft & Latham utilizes σ_{max} normalized by the equivalent stress ($\overline{\sigma}$), while the model by Ayada uses the value of σ_{m} divided by the equivalent stress ($\overline{\sigma}$), which is the triaxiality of stress, denoted as $\sigma_{\mathrm{m}}/\overline{\sigma}$. The respective equations are shown below.

$$\int^{\overline{\varepsilon_f}} \frac{\sigma_{\mathrm{max}}}{\overline{\sigma}} d\overline{\varepsilon} = C_{CL} \tag{2.5}$$

$$\int^{\overline{\varepsilon_f}} \frac{\sigma_{\mathrm{m}}}{\overline{\sigma}} d\overline{\varepsilon} = C_A \tag{2.6}$$

There are no standardized testing methods for identifying the parameters of the ductile fracture criterion (critical values of damage, $C_{\mathrm{CL}}, C_{\mathrm{A}}$). However, in the tensile test of round bars, the stress in the necked region can be determined using Bridgman's equation even after necking occurs. By performing strain measurements using digital image correlation or similar methods, the parameters can be identified. For tubular materials as well, testing methods have been proposed for identifying these forming limits and ductile fracture parameters, which will be discussed in detail below.

(1) **Ductile fracture criterion**

A method for identifying the parameters of the ductile fracture criterion has been proposed for tubular materials, similar to the method using the tensile test for round bars [36]. In the tubular tensile test, the outer diameter D and necking curvature

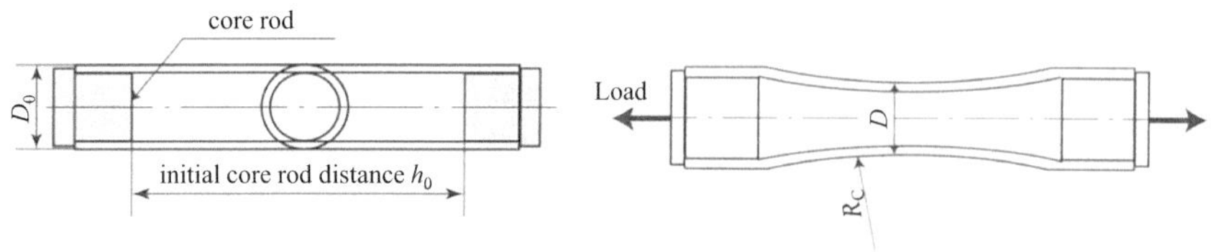

Fig. 2.9 Parameter identification test for ductile fracture criteria of tubular materials

radius R at the central portion are measured from the start of the test until fracture using image analysis. Simultaneously, a grid marked on the tube surface is captured to measure the strain (Fig. 2.9). This method is similar to the one used for ductile fracture parameter identification in the tensile test of round bars. However, unlike round bars, the stress cannot be determined without information about the thickness of the necked portion. Therefore, an additional tensile test of arc-shaped test specimens is necessary to obtain the stress–strain relationship (constitutive equation), and the stress is determined by finite element analysis.

The ductile fracture criterion considers path dependence, making it useful for complex processing in tube forming. However, the parameters of the criterion are not material-specific and are known to depend on the stress triaxiality $\sigma_m/\overline{\sigma}$. Therefore, when applying it to tube forming, considerations such as identifying the ductile fracture parameters under stress triaxiality similar to the intended processing are necessary.

(2) **Forming limit diagram**

In tubular materials, measuring the forming limit curve (FLC) has been very challenging. However, by using the biaxial bulge test and forming until fracture under proportional loading, the FLC can be determined. It should be noted that the FLC can vary if the loading is non-proportional, as observed in sheet metal forming. It has also been demonstrated that the FLC can change when the strain ratio is altered during deformation of tubular materials. Figure 2.10 shows the strain FLC obtained under proportional loading (Fig. 2.10a) and the strain FLC under a combination path with a change in loading path midway (Fig. 2.10b) and (Fig. 2.10c), using an aluminum extruded tube (A5154-H112) [37]. Applying the strain FLC to forming processes with varying deformation paths, such as tube forming, is difficult.

On the other hand, it has been reported that the stress FLC is independent of the deformation path, unlike the conventional strain FLC. The results of obtaining stress FLC under a linear stress path (proportional loading) and a complex stress path using the biaxial bulge test on aluminum extruded tubes and cold-drawn and annealed electric resistance welded tubes are shown in Fig. 2.11 [37]. The fracture limits obtained from the linear stress path and complex stress path are nearly identical. While it has been demonstrated that the FLC exhibits path dependence when strain hardening is path-dependent, the path dependence in stress FLC is much smaller compared to strain FLC.

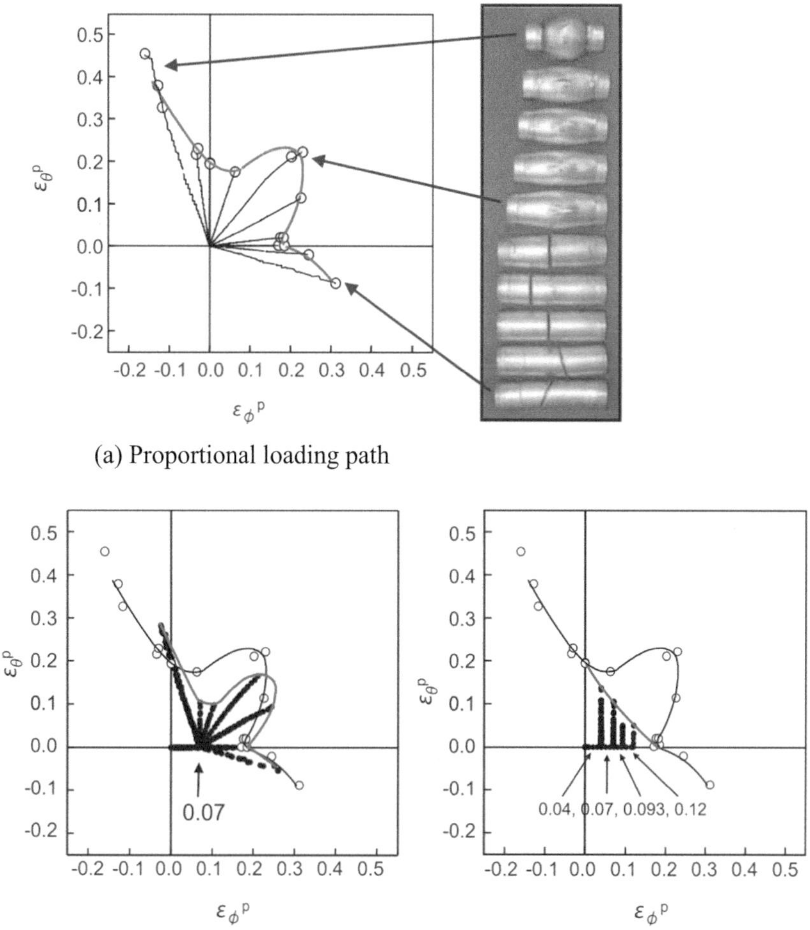

(a) Proportional loading path

(b) Complex loading path 1

(c) Complex loading path 2

Fig. 2.10 Strain FLC of proportional loading path and complex loading paths

Since stress FLC is evaluated based on stress and cannot be directly measured like strain, FEM analysis is required. However, it is a promising method for assessing the feasibility of forming in complex processes such as tube forming.

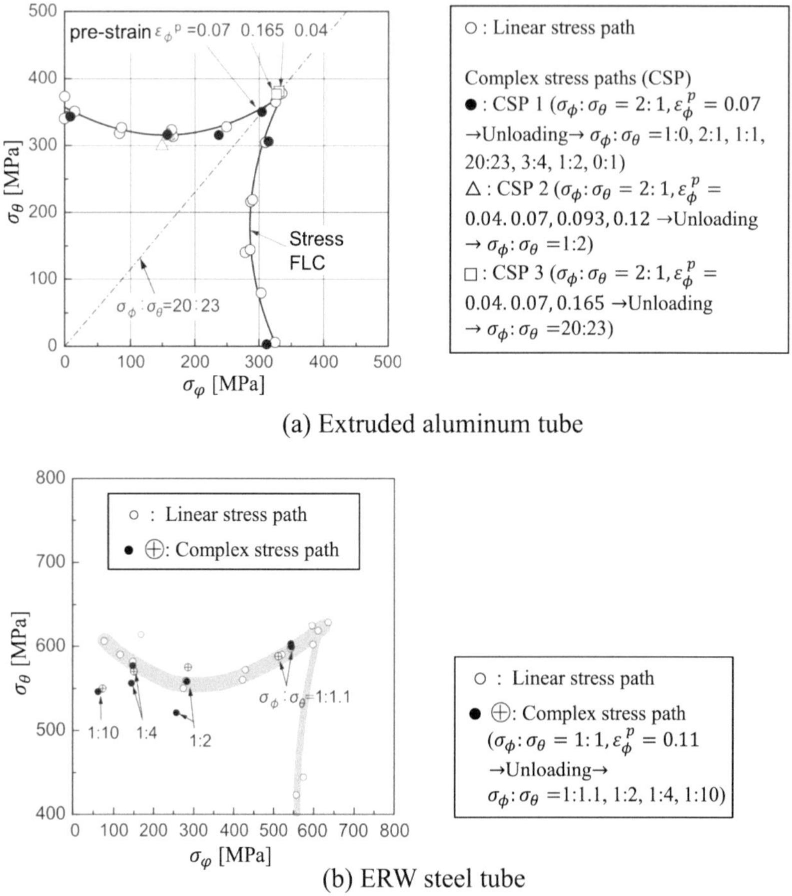

Fig. 2.11 Stress FLC of linear stress path and complex stress path

2.3.3 *Formability Tests*

(1) **Bending test**

The bending test, along with the flattening test, is primarily conducted to inspect for any defects in tubular materials. It is specified in JIS for steel tubes for machine structures, which require bending up to the specified bending radius without any signs of scratches or cracks. ISO 8491 specifies the test method of bending up to the prescribed bending radius and bending angle to examine the plastic deformability. Figure 2.12 illustrates the bending test method defined by ISO. It evaluates the bendability by winding the specimen with a groove corresponding to the pipe diameter D around a bending radius r up to a specified angle α, where t is the thickness of tube. Similar criteria are also applied in JIS.

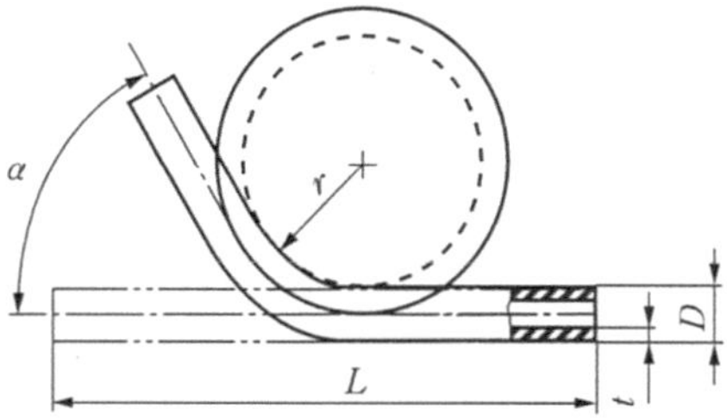

Fig. 2.12 Bend test of tubes (ISO 8491)

In practical bending processes, there are various methods such as rotary bending, push bending, and press bending, each with different constraints, like mandrels. Therefore, there is no standardized test method for evaluating bendability in general, considering the variations in bending processes.

(2) **Flattening test**

The flattening test, like the bending test, is specified in JIS and ISO to inspect for defects in tubular materials. It is a test conducted to determine the presence of scratches or cracks by compressing the specimen with a flat tool up to a specified height *H* or until full contact. Figure 2.13 illustrates the flattening test method defined by ISO 8492. In the figure, *D* represents the pipe diameter, and *t* represents the thickness. In the case of contact flattening, it is specified that the length of the contact area should be more than half of the inner distance *b* of the contact area. For pipes with welded or joined seams, the flattening test is conducted with the seam positioned perpendicular to the compression direction to evaluate the integrity of the seam.

(3) **Expansion test**

End-forming of tubes involves operations such as flaring, flanging, and curling, which are widely used to create shapes that enable the connection of tubes with other components. Various methods, including press-fitting expansion or rotary forming, are employed for tube end-forming. Due to the variety of end-forming methods, it is challenging to establish a universally applicable evaluation method for the end-formability of tubes. However, compared to other tube-forming techniques, the expansion test closely simulates the actual deformation during the forming process. As a result, it is utilized to assess the end-formability of tube materials.

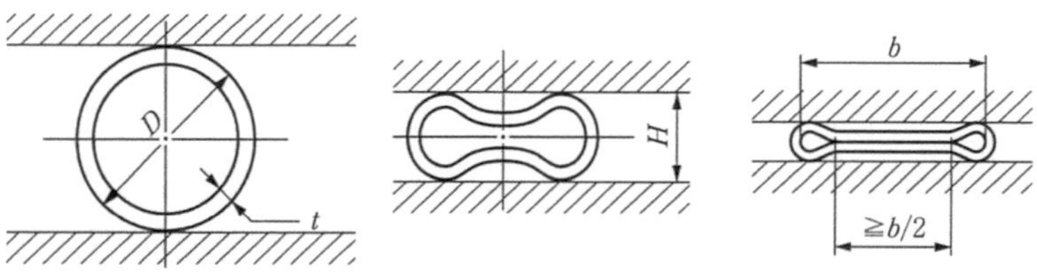

Fig. 2.13 Tube flattening test (ISO 8492)

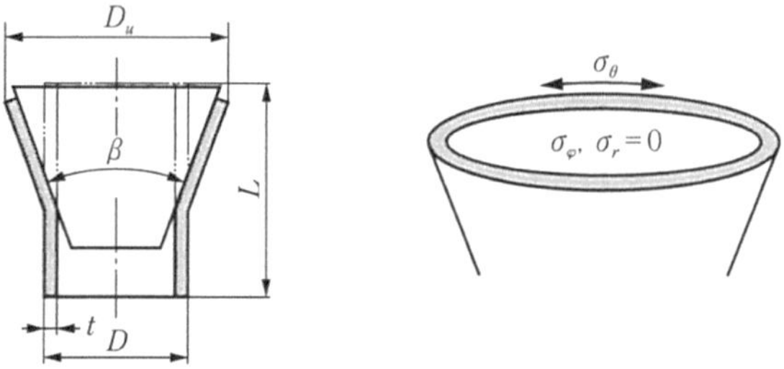

Fig. 2.14 Drift-expanding test for tube (ISO 8493) and stress state at tube end

Figure 2.14 shows the tube flaring test (ISO 8493). The flaring test involves pressing a prescribed cone tool with an angle β into the tube end to expand it to a specified outer diameter D_u, and evaluating the presence of cracks or other defects. Additionally, by expanding the tube end until cracks occur, the maximum expansion amount can be evaluated. It is important to ensure that the tube end cut is consistent with an actual cut when conducting these evaluations. During expansion, fracture typically occurs at the end face. However, if microcracks exist or significant work-hardening has occurred at the end face during cutting, the likelihood of fracture increases. Therefore, attention must be paid to the cutting method and finishing. This is similar to the hole expansion test for sheets, where a conical tool is inserted into a pre-opened hole to expand it.

During the flaring test, as shown in Fig. 2.14, the tube end experiences only circumferential stress σ_θ and no axial stress σ_φ or radial stress σ_r. This results in a uniaxial tension state in the circumferential direction. Therefore, the forming limit of the tube end in the expansion test can be considered as the formability under circumferential tension, similar to the evaluation of forming limits in bulge tests where circumferential uniaxial tension occurs. However, in the case of bulge tests, the deformed region does not suffer from damage caused by cutting. Therefore, when evaluating the formability of the tube using flaring tests, it is necessary to minimize damage to the tube end by machining it rather than cutting it.

Figure 2.15 shows the tube flanging test (ISO 8494). Firstly, the tube end is flared using a conical tool with an angle β (typically 90°). Then, a flat tool with a guided radius R is used to form the flange up to the prescribed diameter D_u, and the presence

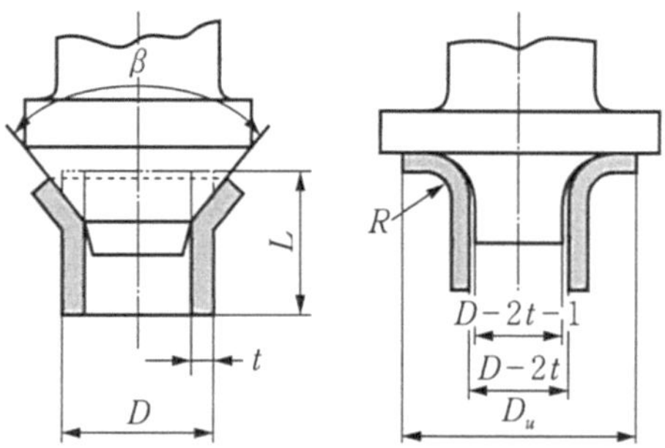

Fig. 2.15 Tube flanging test (ISO 8494)

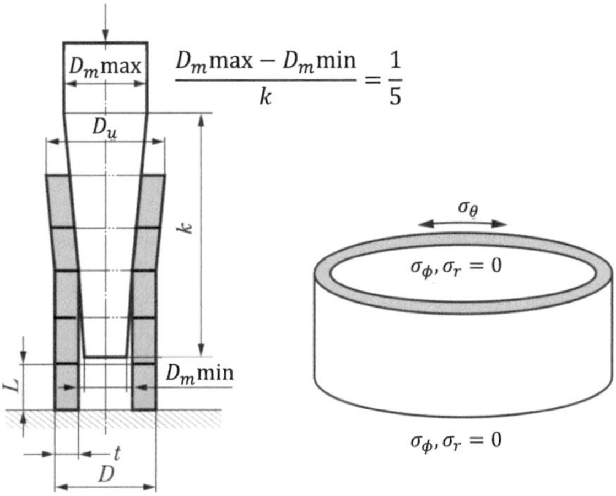

Fig. 2.16 Ring expanding test (ISO 8495) and ideal stress state of ring

of cracks or other defects is evaluated. The applicable range is specified as an outer diameter D of 150 mm or less and a thickness t of 10 mm or less. This test method closely resembles the actual flange processing of tubes, making it a convenient means of evaluating formability.

Figure 2.16 illustrates the ring expansion test for tubes (ISO 8495). This test involves cutting the tube into rings of height L and stacking them together, followed by expanding them using a taper tool. The specifications for the tube's outer diameter D, thickness t, and ring height L are defined as $18 \leq D \leq 150$ (mm), $2 \leq t \leq 16$ (mm), and $L = 10–16$ (mm). Additionally, the taper of the mandrel is defined as $(D_m\max - D_m\min)/k = 1/5$, considering the reduction of stress distribution in axial direction.

In contrast to the tube expansion test (ISO 8493), which exhibits uniaxial stress only at the tube end and stress distributes along the axis, the ring expansion test (ISO 8495) aims to achieve a nearly uniform uniaxial stress state throughout the entire length (Fig. 2.16). The finish of the ring end faces and lubrication of the tools are crucial, as they can influence the test results.

In hole expansion of sheets, the influence of the stress gradient on crack propagation has been studied. It has been shown that while crack initiation is the same, crack propagation is delayed when the stress gradient is high [38]. Ring expansion, compared to flaring, has a smaller stress gradient. As a result, crack propagation occurs more rapidly, leading to easier evaluation of the limit strain. Therefore, it is considered suitable for understanding the forming limit under circumferential uniaxial tension.

(4) **Bulge test**

The test conducted by hydraulic pressure instead of a tool for expanding the tube is called a bulge test. With the practical application of hydroforming, the importance of this test as an evaluation of its formability has increased. However, there are no standardized test methods for bulge testing of tubes. Table 2.6 shows various proposed bulge tests.

Table 2.6 Various bulge tests

Constraint	Loading	Remarks
Free bulge	Internal pressure	Free tube ends/fixed tube ends
	Internal pressure + axial load/feed	Free tube ends/fixed tube ends
Die bulge	Deformation of full cross-section	Formability for specified geometry
	Local deformation	Formability of specified part

Theoretical solutions regarding the rupture limit have been proposed for the free bulge test, which allows for fundamental investigations into the formability of tubes. The biaxial bulge test falls into this category, but it is a material characterization test where stress and strain are measured during the test, and internal pressure and axial force are controlled by feedback to achieve the desired load path. The die bulge test is used to evaluate formability using a representative shape in hydroforming and to assess the forming margin. The localized deformation die bulge test is a test method that limits deformation to specific areas to evaluate formability, as the free bulge test may lead to concentrated deformation in non-uniform sections with varying thickness and strength. For more details, please refer to Chap. 4.

2.4 Material Properties and Secondary Formability

The secondary formability of tubes varies significantly depending on the material properties, similar to the press formability of sheets or the forgeability of bulk materials. Secondary processes for tubes, such as end ironing, resemble deformation in forging. However, in general, these processes involve smaller thicknesses compared to representative dimensions, making them more similar to sheet forming. Additionally, the materials used for tube forming are often initially processed from sheets into tubular shapes. Therefore, in this section, we provide an overview of the secondary formability of tubes in comparison to sheet materials, while the specific details of each forming method will be described in their respective chapters.

2.4.1 n-*Value: Work-Hardening Characteristics*

The stress–strain curve of a metallic material obtained from tensile test is often expressed as

$$\sigma = C\varepsilon^n \tag{2.7}$$

where σ represents true stress and ε represents true strain. The coefficient of the strain exponent in this power-law relationship is referred to as the n-value. In the

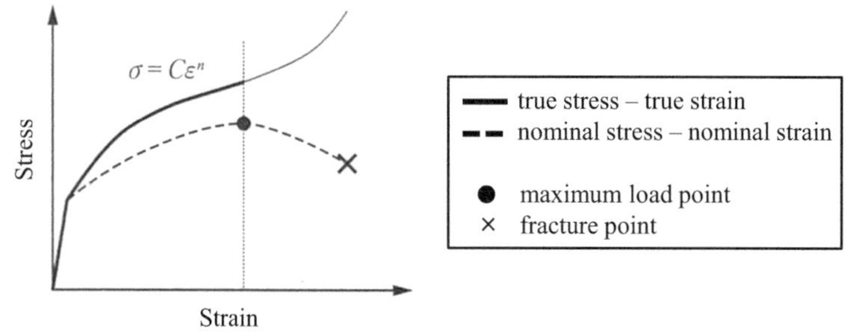

Fig. 2.17 Stress–strain curves of tensile test

range of large strains typically encountered in plastic deformation, the elastic strain is negligible compared to the plastic strain and can be disregarded. In such cases, ε can be considered as the plastic strain. For thin-walled cylindrical tubes or thin-walled spherical shells subjected to internal pressure, although stress state is not uniaxial, by introducing an equivalent stress $\overline{\sigma}$ and an equivalent strain $\overline{\varepsilon}$, the stress–strain relationship (constitutive equation) can be expressed as $\overline{\sigma} = C\overline{\varepsilon}^n$, similar to uniaxial state. Figure 2.17 illustrates a stress–strain diagram from a tensile test. Since the deformation becomes non-uniform beyond the point of maximum load, the n-value is determined using the data prior to the maximum load point.

The n-value corresponds to the degree of increase in stress (deformation resistance) when strain increases during forming (deformation), as represented by Eq. (2.7). It represents a work-hardening characteristic. When plastic deformation begins, significant strain occurs with a relatively small increase in stress compared to elastic deformation, causing strain (deformation) to concentrate in the areas where plastic deformation starts. However, with a higher n-value, there is deformation resistance in the areas. Therefore, the n-value represents the propagation characteristics of deformation. Figure 2.18 illustrates the differences in deformation due to the n-value in the bulge deformation by spherical punch of sheet metal. Although it is a schematic diagram, in actual forming, it can be observed that a higher n-value mitigates deformation concentration and allows deformation to propagate over a wider range.

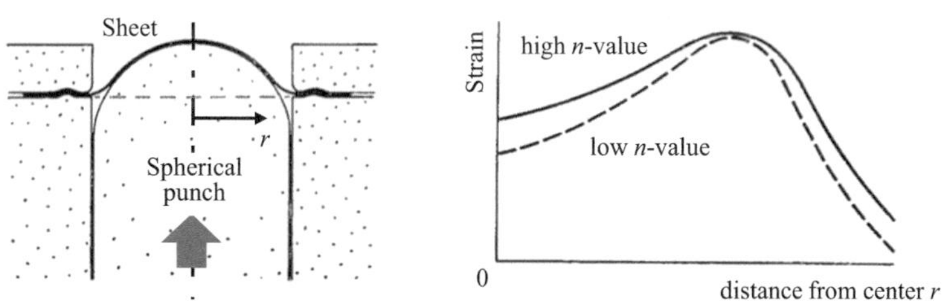

Fig. 2.18 Schematic strain distribution of bulge by spherical punch

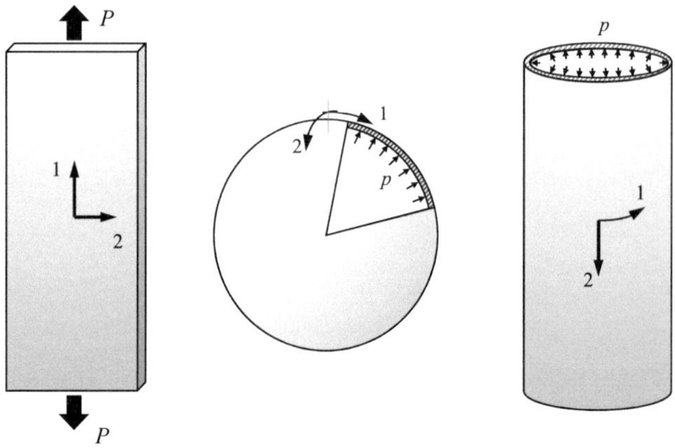

Fig. 2.19 Sheet under uniaxial tension, spherical shell under internal pressure, and tube under internal pressure

Figure 2.19 depicts uniaxial tension, thin-walled spherical shell under internal pressure, and thin-walled cylindrical tube under internal pressure. The strain at the points of maximum load in uniaxial tension (1), thin-walled spherical shell under internal pressure (2), and thin-walled cylindrical tube under internal pressure (3) is described by the n-value.

(1) **Maximum load point in uniaxial tension**

In uniaxial tension, the material undergoes work-hardening due to tensile deformation, causing an increase in load. However, as the width of the sheet decreases and the thickness decreases due to axial tension, it reaches a maximum load at a certain load, after which the load decreases. For a uniaxial tension test specimen with a cross-sectional area A, length l, and tensile load P,

the condition for maximum load, given by $dP = 0$, is $dP = \sigma dA + Ad\sigma = 0$;

the condition for constant volume, given by $d(Al) = 0$, is $d(Al) = ldA + Adl = 0$;

and using $d\varepsilon = dl/l$, the condition for maximum load is $d\sigma/d\varepsilon = \sigma$.

Here, let Z be defined as $(d\sigma/d\varepsilon)/\sigma = 1/Z$. In the case of uniaxial tension, $Z = 1$ is the condition for maximum load. If the stress–strain relationship is given by $\sigma = C\varepsilon^n$, then the strain ε_{cr} at the maximum load point is $\varepsilon_{cr} = n$. In this explanation, we used the condition for the maximum load point, but the same result can be obtained by considering the condition of plastic instability, where the decrease in cross-sectional area surpasses work-hardening.

Table 2.7 Strain in main deformation direction at maximum load for sheet, shell, and tube

	Stress	Strain	$Z = \overline{\varepsilon}_{cr}/n$	ε_{1cr}/n
Sheet under uni-axial tension	$\sigma_L = P/Wt$ $\sigma_W = \sigma_t = 0$	$\varepsilon_W = \varepsilon_t = -\frac{1}{2}\varepsilon_L$ $\overline{\varepsilon} = \varepsilon_L$	1	1
Spherical shell under internal pressure	$\sigma_\theta = \sigma_\phi = pr/2t$ $\sigma_t \approx 0$	$\varepsilon_t = -2\varepsilon_\theta = -2\varepsilon_\phi$ $\overline{\varepsilon} = \varepsilon_t$	2/3	1/3
Tube under internal pressure (closed ends)	$\sigma_\theta = pr/t$ $\sigma_\phi = \frac{1}{2}\sigma_\theta, \sigma_t \approx 0$	$\varepsilon_\phi = 0, \varepsilon_\theta = -\varepsilon_t$ $\overline{\varepsilon} = \left(2/\sqrt{3}\right)\varepsilon_\theta$	$1/\sqrt{3}$	1/2
Tube under internal pressure (open ends)	$\sigma_\theta = pr/t$ $\sigma_\phi = 0, \sigma_t \approx 0$	$\varepsilon_\phi = \varepsilon_t = -\frac{1}{2}\varepsilon_\theta$ $\overline{\varepsilon} = \varepsilon_\theta$	2/3	2/3

(2) **Maximum pressure point of thin-walled spherical shell under internal pressure**

When hydroforming is used to form branch pipes without a counterpunch, the dome top of the branch pipe undergoes deformation close to this condition. Let's consider a thin-walled spherical shell with a radius of *r*, thickness of *t*, and internal pressure of *p*. Since the thin-walled spherical shell expands uniformly, $\varepsilon_1 = \varepsilon_2 = -\frac{1}{2}\varepsilon_3$ and $\overline{\varepsilon} = -\varepsilon_3$. The stress is given by $\sigma_1 = \sigma_2 = pr/2t$. The equivalent strain at the maximum internal pressure is $\varepsilon_{cr} = (2/3)n$.

(3) **Maximum pressure point of thin-walled circular tube under internal pressure**

When the tube is uniformly bulge-formed, it reaches this condition. Let's consider a thin-walled circular tube with a radius of *r*, thickness of *t*, and internal pressure of *p*. In the case where the tube ends are closed, we have $\sigma_1 = pr/t, \sigma_2/\sigma_1 = 1/2$ and $\varepsilon_2 = 0$. From the condition of constant volume, we have $\varepsilon_1 = -\varepsilon_3$ and $\overline{\varepsilon} = \left(2/\sqrt{3}\right)\varepsilon_1$, resulting in an equivalent strain at the maximum internal pressure of $\overline{\varepsilon_{cr}} = \left(1/\sqrt{3}\right)n$. In the case where the tube ends are open, we have $\sigma_1 = pr/t$, and $\sigma_2 = 0$, and similarly, the equivalent strain at the maximum internal pressure is $\overline{\varepsilon_{cr}} = \left(2/\sqrt{3}\right)n$. Table 2.7 shows the stress, strain states for thin plates under uniaxial tension, thin-walled spherical shells under internal pressure, and thin-walled circular tubes under internal pressure, as well as the strains at the maximum load point. The strains at the maximum load point are normalized by *n*. In all cases, the strains at the maximum load point are expressed as multiples of *n*, which can be interpreted as the *n*-value that suppresses plastic instability [39].

2.4.2 *r*-Value: Anisotropy

Based on the tensile test of thin sheets (Fig. 2.20), where axial strain is ε_1, width strain is ε_w, and thickness strain is ε_t, the *r*-value is defined by the following equation.

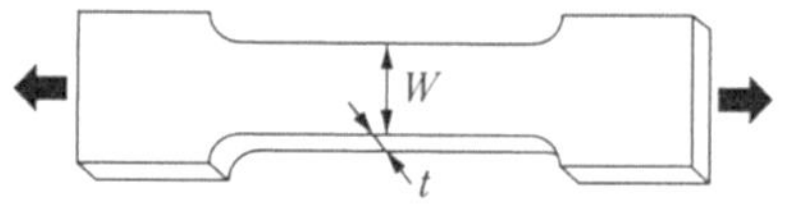

Fig. 2.20 Tensile test of sheet for definition of r-value

$$r = \varepsilon_w / \varepsilon_t \tag{2.8}$$

The r-value, also known as the Lankford value, is an indicator of the anisotropy of the plate. When a thin sheet is subjected to tension, it elongates in the tensile direction, contracts in the width direction, and the thickness decreases. Neglecting elastic strain, in plastic deformation, the volume remains constant, $\varepsilon_l + \varepsilon_w + \varepsilon_t = 0$. In the case of an isotropic material, $\varepsilon_w = \varepsilon_t$ and $\varepsilon_w = \varepsilon_t = -(1/2)\varepsilon_l$. In this case, the r-value is 1. In anisotropic materials, $r \neq 1$ and $\varepsilon_w \neq \varepsilon_t$, indicating different allocations of tensile strains in the width and thickness directions.

The r-value, defined by the equation, represents the magnitude of strain in the direction perpendicular to the main deformation direction (tensile direction) with respect to the reference strain in the thickness direction. It can be considered as an indicator of the ease of deformation in the direction orthogonal to the main deformation direction (width direction). In sheet metal forming, deformation characteristics are discussed by classifying deformation modes into four basic modes: deep drawing, bulging, stretch flange, and bending (Fig. 2.21) [40]. The r-value is regarded as an indicator of the deep drawing ability.

In deep drawing, the material flows from the flange into the vertical wall while undergoing circumferential contraction. This deformation at the flange is referred to as shrink flange deformation. In materials with a high r-value, the shrinkage deformation at the flange, which is orthogonal to the main deformation direction of

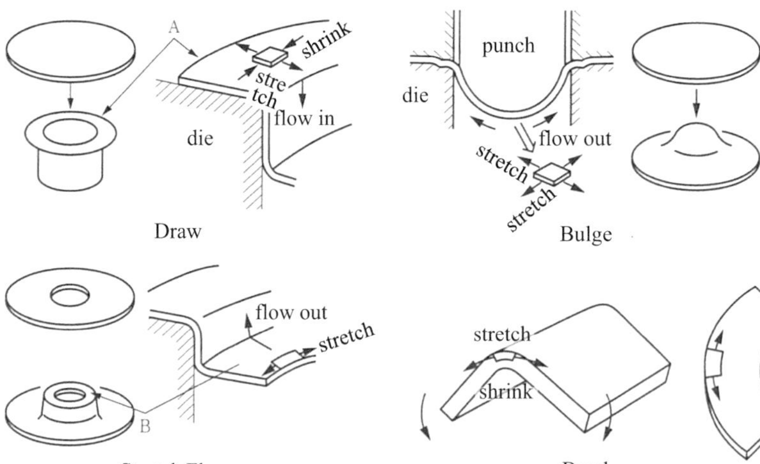

Fig. 2.21 Four basic deformation modes in sheet metal forming [40]

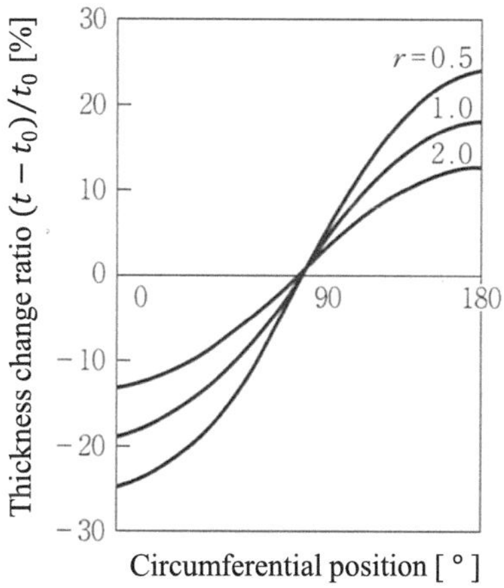

Fig. 2.22 Effect of r-value on thickness distribution in tube bending [41]

flow into the vertical wall, is easier, indicating better formability. More precisely, as the r-value increases, the shape of the yield locus becomes elongated in the equi-biaxial direction and narrow in the pure shear direction. This leads to a decrease in the strength of the flange and an increase in the strength of the punch shoulder, making the material easier to deform. However, to explain this phenomenon, it is necessary to consider both the deformation anisotropy and the strength anisotropy of the material. In this context, only the deformation anisotropy is discussed.

In tube forming, the thinning during bending of the tube varies depending on the r-value. Figure 2.22 shows the results of an elementary analysis considering anisotropy in bending [41]. The analysis assumes a circular cross-section and determines the bending strain based on the distance from the neutral axis. Similar to the case of sheet materials, the r-value is defined as the ratio of the circumferential strain ε_θ to the thickness strain ε_t ($r = \varepsilon_\theta/\varepsilon_t$) and it is found that a higher r-value suppresses the circumferential thickness distribution. This is because in bending deformation, the main deformation direction is axial, and in materials with a higher r-value, deformation in the circumferential direction orthogonal to it is easier, resulting in relatively smaller deformation in the thickness direction.

Furthermore, in the T-forming by hydroforming, which involves forming a branch tube from a mother tube, the forming limit height of the branch tube is shown to vary with the r-value according to FEM analysis results [42]. By varying the r-values in the axial direction (φ-axis) and circumferential direction (θ-axis) from 1.0 to 2.0, the influence of the r-value on the critical expansion ratio (R_e) of the branch tube is investigated, defining the forming limit as reaching the minimum thickness (t_{min}) equal to the specified thickness (in this case, 80% of the initial thickness, t_0) (see Fig. 2.23). In the figure, the triangles (△) represent the case where both axial and circumferential r-values are increased, the circles (○) represent the case where only the circumferential r-value is 1.0 and the axial r-value is increased to 1.5 and 2.0, and the squares (□) represent the case where the axial r-value is 1.0 and only the circumferential r-value is increased to 1.5 and 2.0.

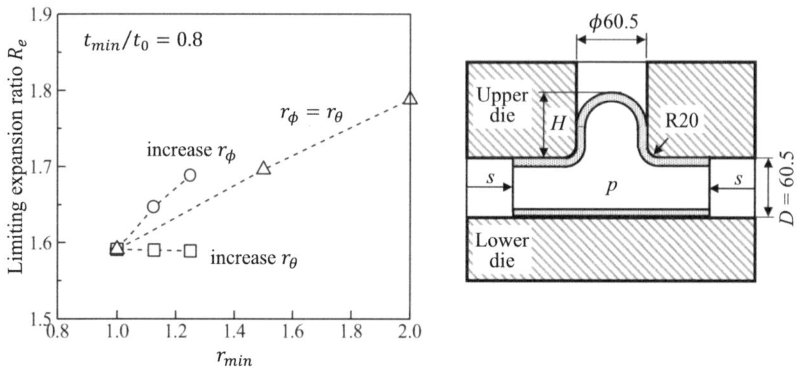

Fig. 2.23 Relation of limiting expansion ratio and r-value in T-forming of steel tubes [42]

When the axial r-value is increased, the forming limit rises; however, when only the circumferential r-value is increased, the forming limit does not increase. This can be explained by considering that in T-forming, the main deformation is due to axial compression and the ease of deformation into the branch tube orthogonal to it. Similar to deep drawing in press forming, in T-forming, the ease of deformation in the direction orthogonal to the axial flow deformation depends on the r-value, particularly the axial r-value. For a more detailed explanation based on the change of yield locus by the r-value, please refer to Chap. 4.

References

1. Iguchi, T.: J. JSTP **53–614**, 188–192 (2012)
2. Hasegwa, N., Hamatani, H., Fukami. T., Nakaji, T., Takeda, Y., Motoyoshi, T., Tanimoto, M., Ohsawa, T.: Nippon Steel Technical Report, vol. 397, pp. 118–124 (2013)
3. Inoue, T., Suzuki, M., Okabe, T., Matsui, Y.: JFE Tech. Rep. **18**, 18–22 (2013)
4. Tanabe, H., Anai, I., Yamazaki, K.: Tomokiyo. T.: Materia Japan **38–3**, 242–244 (1999)
5. Tanabe, H., Miyasaka, A., Yamazaki, K., Iwasaki, T., Akada, H.: Nippon Steel Technical Report, vol. 354, pp. 48–53 (1994)
6. Toyooka, T., Yorifuji, A., et al.: Materia Japan **41–1**, 51–53 (2002)
7. The Iron and Steel Institute of Japan (Ed.): Steel Handbook, vol. 2, 5th edn. The Iron and Steel Institute of Japan (2014)
8. The Japan Light Metal Association (Ed.): Aluminum Application Technologies for Automobiles: Materials, vol. 33 (1996)
9. Mizukoshi, H.: Proceedings of the 205th Symposium on the Technology of Plasticity, JSTP, pp. 65–70 (2001)
10. Mizukoshi, H.: Aluminum. Jpn. Alum. Assoc. **8–45**, 246–250 (2001)
11. Sugiyama, K., Sakaki, S.: J. JSTP **39–453**, 23–27 (1998)
12. Mizukoshi, H. et al.: Proceedings of the 2000 Autumn Meeting, the Japan Institute of Light Metals, vol. 99, pp. 221–222 (2000)
13. Uno, T.: Aluminum. Jpn. Alum. Assoc. **9–50**, 186–196 (2002)
14. Kuwabara, T., Yoshida, K., et al.: J. JSTP **44–506**, 281–286 (2003)

15. Fuchizawa, S., Tanaka, K., Shirayori, A.: Proceedings of the 53rd Joint Conference on the Technology of Plasticity, JSTP, (2002), 221–222.
16. Kuwabara, T.: J. JSTP **44–506**, 234–239 (2003)
17. The Japan Light Metal Association (Ed.): Production Technologies for Automotive Weight Reduction, vol. 33 (2003)
18. Kondo, K.: Press Working, vol. 41–7, pp. 34–38, Nikkan Kogyo Shimbun (2003)
19. Altan, T., Aue-U-Lan, Y., Palaniswamy, H., Kaya, S.: Proceedings of 8th International Conference on the Technology of Plasticity. Keynote Paper (2005)
20. Keihler, M., Bauer, H., Harrison, D., De Silva, A.K.M.: Mater. Process. Technol. **167**, 363–370 (2005)
21. Fukuchi, F., Hayashi, N., Ogawa, T., Yokoyama, O., Hori, I.: J. Jpn. Inst. Light. Metals **55–3**, 147–152 (2005)
22. Dick, P., Nagler, M., Zengen, K.H.: 塑性と加工. J. JSTP **39–453**, 1014–1018 (1998)
23. Watanabe, M., Hossogi, T., Shirai, T., Ishibashi, A.: R&D Kobe Steel Engineering Reports, vol. 58, 3rd edn, pp. 74–77 (2008)
24. Nakamura, M., Maruyama, K., et al.: Pipe Forming Techniques, 2nd edn, pp. 312–314. Nikkan Kogyo Shimbun (1998)
25. The Japan Society for technology of Plasticity (Ed.): Titanium Basics and Processing. Corona Publishing (2008)
26. The Japan Society for technology of Plasticity (Ed.): Magnesium Processing. Corona Publishing (2008)
27. Division of Processing for Quality Products Steering Committee, The Iron and Steel Institute of Japan: Final Report of the Research Committee on the Standardization of Steel Tube Formability Test Methods. The Iron and Steel Institute of Japan (2012)
28. Lode, W.: Z. Phys. **36**, 913–939 (1926)
29. Sautter, W., Kochendörfer, A., Dehlinger, U.: Z. Metallk. **44**, 442–553 (1953)
30. Taylor, G.I., Quinney, H.: Phil. Trans. R. Soc. A **230**, 323–362 (1932)
31. Marin, J., Hu, L.W.: Trans. ASME **75**, 1181–1190 (1953)
32. Kuwabara, T., Narihara, T., Yoshida, K., Takahasi, S.: J. JSTP **44–506**, 281–286 (2003)
33. Yoshida, K., Kuwabara, T., Narihara, T., Takahashi, S.: J. JSTP **45–517**, 123–128 (2004)
34. Yoshida, K., Kuwabara, T.: Tetsu-to-Hagané **92–1**, 36–45 (2006)
35. Yoshida, Y.: J. JSTP **57–669**, 940–944 (2016)
36. Yoshida, Y., Kume, S., et al.: CAMP-ISIJ **17**, 1054 (2004)
37. Kuwabara, T.: J. JSTP **54–624**, 18–24 (2013)
38. Ito, Y., Nakazawa, Y., Kuriyama, Y., Suzuki, K., Suzuki, N.: J. JSTP **57–660**, 53–59 (2016)
39. Backofen, W.A.: Deformation and Processing. Addison-Wesley Publishing (1972)
40. Japan Sheet Metal Forming Technology Research Group (ed.): Handbook of Sheet Metal Forming, 4th edn, 79 p. Nikkan Kogyo Shimbun (2017)
41. Sato, K., Takahashi, S.: J. JSTP **23–252**, 17–22 (1982)
42. Yoshida, T., Kuriyama, Y.: Proceedings of the 50th Joint Conference on the Technology of Plasticity, JSTP, pp. 447–448, (1999)

Chapter 3
Bending

Abstract In this chapter, we will discuss not only the bending of tubular materials such as circular tubes but also the bending of shapes and profiles. These sections can be classified into closed sections such as circular pipes and square pipes, open sections such as channel sections and H-sections, and asymmetrical sections that do not have an axis of symmetry. Generally, sections other than circular ones, such as square pipes, are classified as shapes.

3.1 Fundamentals

In this chapter, we will discuss not only the bending of tubular materials such as circular tubes but also the bending of shapes and profiles. These sections can be classified into closed sections such as circular pipes and square pipes, open sections such as channel sections and H-sections, and asymmetrical sections that do not have an axis of symmetry. Generally, sections other than circular ones, such as square pipes, are classified as shapes.

3.1.1 Introduction

The bending methods for circular tubes and shapes can be broadly divided into methods that apply bending moments to the materials and methods that induce a strain gradient along the axial direction of the material, resulting in bending. The majority of bending processes for circular tubes and shapes involve applying bending moments. In this bending method, deformations in the cross-sectional shape (flattening) and changes in wall thickness (eccentricity) occur from the beginning of the process. For example, circular tubes deform into an oval shape, reducing wall thickness in the tensile region on the outer side of the bend and increasing it in the compressive region on the inner side. These deformations have the potential to adversely affect the accuracy of the sectional shape after bending. Additionally, when bending shapes

Y. Kuriyama et al., *Tube Forming*, https://doi.org/10.1007/978-981-95-3421-0_3

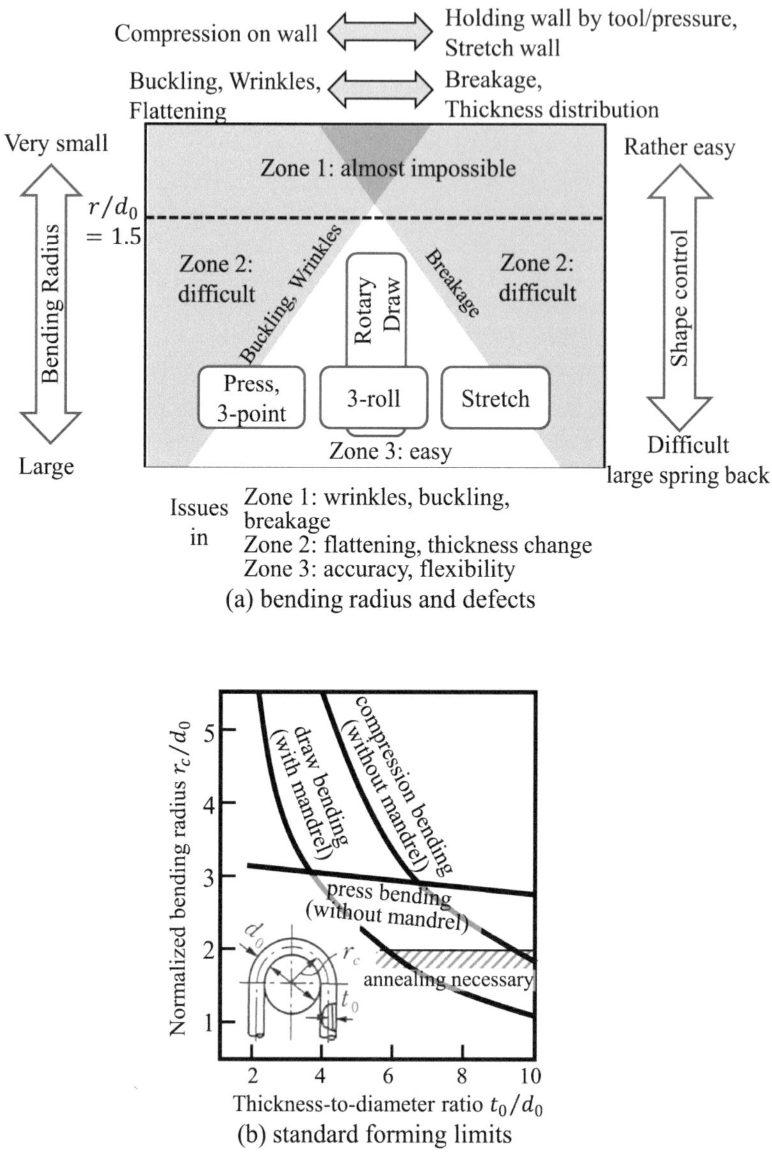

Fig. 3.1 Process window of bending [1, 2]

and profiles, it is important to consider the distortion of the section that occurs when bending in an asymmetrical direction, even if the section is symmetric. Furthermore, various defects can occur during the bending process, including local flattening that leads to yield on the cross-section, wrinkles caused by buckling on the inner side of the bend, as well as necking and cracking on the outer side of the bend, and springback after unloading the bending moment.

Such defects not only affect the dimensional accuracy of the product cross-section but also decrease the processing limits represented by functions such as the ratio of bending radius to outer diameter of the circular pipe (r/d_0) and the ratio of pipe thickness to outer diameter of the circular pipe (t_0/d_0). This, in turn, has a significant impact on productivity and product cost. There are numerous methods for bending circular tubes and shapes, each with its own strengths and weaknesses depending on the application. When performing bending for circular pipes and shapes, it is crucial to understand the characteristics of each bending method and select the appropriate one for the intended purpose. Various indicators, such as diagrams illustrating the characteristics of each bending method as shown in Fig. 3.1 [1, 2], and allowable ranges of bending conditions by different methods, can be used to assist in selecting the appropriate bending method.

3.1.2 Theory

In addition to the selection of processing methods, it is also important to select appropriate materials and to determine the geometry and dimensions of dies, jigs, and tools. Furthermore, various theoretical analyses have been attempted to reduce trial and error in prototyping and other processes. Elementary theories are primarily used for calculating the bending moment of circular tubes [3]. Numerical analyses have been employed to predict the thickness variation considering the r-value (anisotropy) of circular tubes, assuming no flattening deformation [4], as well as geometric evaluations of thickness variation in square tubes, such as waveguides [5]. Additionally, analyses of the curvature of wrinkles caused by bending in circular tubes [6] and plastic yield [7] using energy methods have also been attempted.

In recent years, there have been numerous cases where simulation using the finite element method has been applied not only to predict various defects caused by bending but also to predict change of cross-sectional shapes and thickness distributions of bent products, as well as to design the shapes and dimensions of tools and jigs. Software has also been developed that allows checking for interference in three-dimensional material handling or bending processes.[1]

[1] http://www.chiyoda-kogyo.co.jp/index.html (as of February 2019).

3.2 Processing Methods

3.2.1 Bending of Circular Tubes

The selection of bending methods depends on the dimensions of the original tube, such as diameter and thickness, as well as the specified bending radius. Various bending methods can be categorized into three main techniques, as shown in Table 3.1, and can also be classified based on factors such as processing temperature and applied axial force. Each method has its own principles and die configurations, as well as minimum bending radius (r/d) and allowable bending angle. Shear bending and elbow forming of pipe fittings are examples of techniques that can achieve small bending radii. Additionally, rotary draw bending has been improved in terms of method, machinery, and die design, enabling bending with r/d_0 ratios of 1. Some methods allow for a maximum bending angle of 360 degrees, allowing for spiral-shaped bends.

Table 3.1 Classification and comparison of bending methods

Classification	Summary	Typical bending method
Bending method	Apply bending moment	Press, rotary draw, compression, stretch
	Apply strain difference between outer and inner	Hamburg, eccentric plug
	Apply shear strain	Shear
Bending temperature	Cold	Rotary draw, stretch, flexible penetration
	Hot	Hamburg, high-frequency induction heating, quench
Axial load	Tension	Press, rotary draw, stretch
	Compression	Roll, flexible penetration, hamburg, quench

Tightness of bending		Typical bending method
Minimum bending radius (r/d_0)	Large ($r/d_0 > 6$)	Roll, quench
	Medium	Press, rotary draw, compression, stretch, penetration
	Small (possible $r/d_0 < 1$)	Shear, hot press, hot penetration
Limit bending angle	$\leq 90°$	Shear, high-frequency induction heating
	$\geq 360°$	Flexible penetration, hamburg, roll

In addition, there are representative methods that allow for continuous bending by incorporating CNC control, such as rotary draw bending and free penetration bending. Roll bending, free penetration bending, and high-frequency induction heating bending have the characteristic of not requiring specific bending dies, and they allow for variable bending radii during the bending process.

The selection of bending methods depends on the dimensions of the original tube (circular tube), such as diameter and thickness, as well as the specified bending radius, required quality, productivity, and yield rate. In terms of material yield, which is an important industrial consideration, the chuck length of the tube end may have an adverse effect on the yield. Rotary draw bending, push bending, and stretch bending methods require a certain chuck length. On the other hand, free intrusion bending, roll bending, and shear bending methods do not require chucking and allow for bending the entire length of the mother material. Furthermore, efforts are being made to develop new bending methods that enhance the precision and performance of bent products, while also pushing the boundaries of bending limits to accommodate compact piping configurations.

Below, we describe the characteristics of representative bending methods for circular tubes.

(1) **Press bending**

(a) **Overview**

Press bending is a bending method, one of the three-point bending method, in which a circular tube is fixed by two support rollers or support dies, and a bending die located at the center is pushed onto the tube using a press machine, as shown in Fig. 3.2. The bending moment is mainly applied to the tube, but depending on the support conditions of the tube, there may be tension applied in the axial direction.

(b) **Application guidelines**

This method is cost-effective, highly productive, and suitable for mass production. However, it is challenging to achieve high precision and perform continuous bending. In particular, for thin-walled tubes, there is significant flattening, resulting in poor dimensional accuracy. The maximum bending angle is around 160–165°, and the minimum bending radius is approximately $r_{\min} \fallingdotseq 2d_0$. The roller support method shown in Fig. 3.2a is commonly used for bending applications in fields such as gas pipes and electrical conduits. The support-type method shown in Fig. 3.2c can handle smaller bending radii due to the cushioning effect, making it widely adopted for bending automotive components.

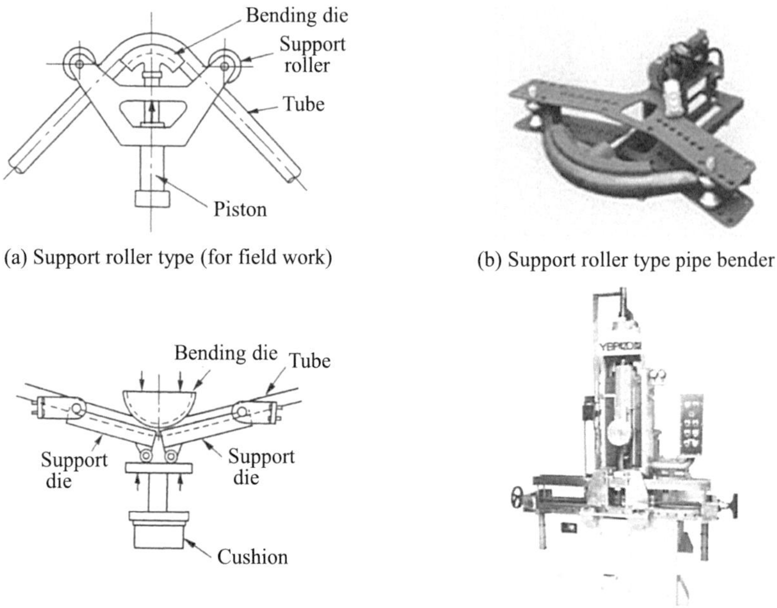

(a) Support roller type (for field work)

(b) Support roller type pipe bender

(c) Support die type

(d) Press type pipe bender [8]

Fig. 3.2 Press bending[2]

(c) **Processing machinery and processed product examples**

Figure 3.2b shows a roller support type bender. It is capable of bending gas pipes with a maximum bending angle of 90°. It is hydraulically operated and portable, allowing on-site processing using the available power source.[3] Figure 3.2d illustrates a press-type pipe bender with a support-type method [8]. It is a vertical hydraulic system composed of a cushion supporting the lower die (support die) and a hydraulic cylinder pressing the upper die (bending die). The equipment is simple and cost-effective. Multiple units, such as 2-bend and 3-bend configurations, are available depending on production volume and bending process requirements. Dedicated equipment tailored to the specifications of the product is commonly used. It offers high productivity and requires minimal space, making it suitable for integration into bending lines combined with end processing and assembly welding. For bending products, it is well-suited for bending short tubes with a single bend.

[2] http://taiyo-e-bender.com/ (as of February 2019).

[3] http://taiyo-e-bender.com/ (as of February 2019).

Fig. 3.3 Elbow forming by hot press bending

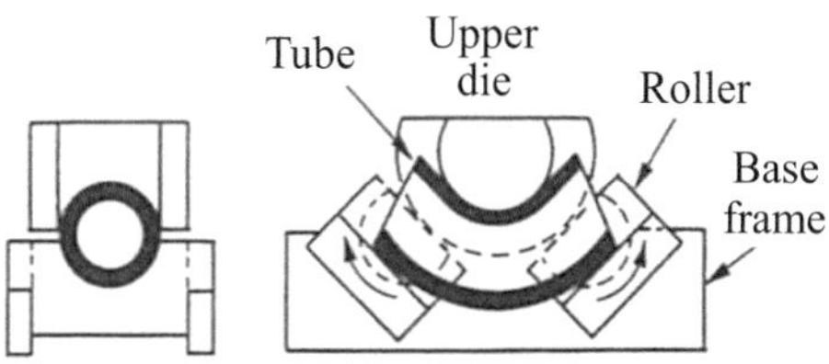

(d) **Manufacturing elbow by hot press bending**

This is a thick-walled elbow forming technique using hot press bending, which is widely used in the manufacturing of pipe fittings. The heated tube is placed on top of two support rollers and bent by pressing down the upper die, as shown in Fig. 3.3. It is also known as CD (concentric die or compression die) bending. The material used is thick-walled tube, with a ratio of thickness to outer diameter (t_0/d_0) of 8% or more, and thinner tubes with a ratio below this value are difficult. The material outer diameter is approximately 10% larger than the outer diameter of the finished product. Figure 3.4 shows the processing steps. The heated material is flattened (Fig. 3.4a). The flattened tube is then vertically pressed (Fig. 3.4b). Typically, press forming is performed twice, ensuring a large bending radius to prevent buckling on the compression side of the product, resulting in a side bulged oval form (Fig. 3.4c).

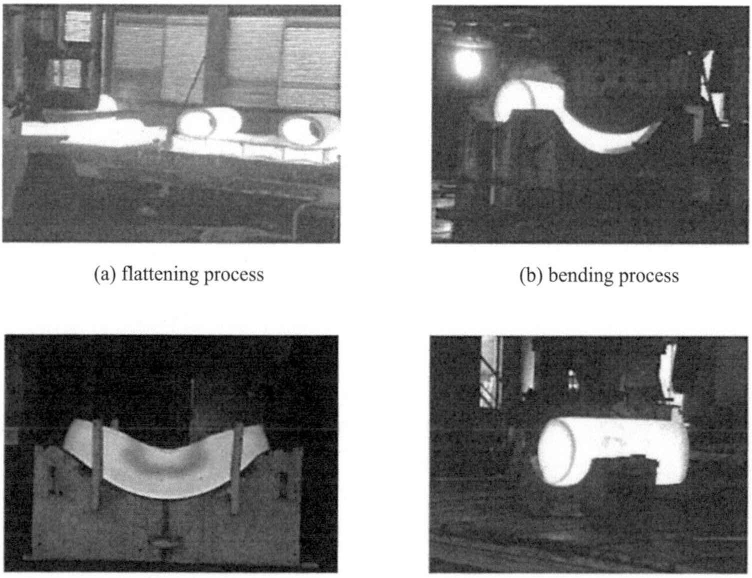

(a) flattening process

(b) bending process

(c) finish bending

(d) shape correction process

Fig. 3.4 Hot press (CD) bending process [9]

During the bending process, intentional bulge deformation on the side to induce buckling. Finally, the entire product is corrected and shaped (Fig. 3.4d) [9].

(2) **Rotary draw bending**

(a) **Overview**

Rotary draw bending is a bending process where one end of a tube is clamped to a die and the bending die is rotated while applying tension to the tube, preventing deformation of the cross-section between the bending die and the pressure die. As shown in Fig. 3.5, it is a highly precise bending method that allows for the suppression of flattening using (1) a mandrel, (2) a wiper die to suppress wrinkles, and (3) a pressure die or a booster die located at the rear to reduce thinning through axial force control.

(b) **Application guidelines**

The maximum bending angle is 180°, and the minimum bending radius with axial force control using a booster die is approximately $r_{\min} \fallingdotseq 1.0\,d_0$. With the CNC control of the equipment, continuous bending is possible, resulting in high work efficiency and superior productivity. It can be considered the most widely used bending method in various industries.

(c) **Processing machinery and processed product examples**

Machines using the rotary draw bending method have enabled three-dimensional continuous bending by employing servo motors and hydraulic servo valves to control the bending axis for controlling the bending angle, the feeding axis for controlling the pipe feeding quantity, and the tilting axis for controlling the twisting angle of the bend. With the addition of automatic feeding and extraction devices, fully automated unmanned operation becomes possible. Furthermore, 5-axis controlled benders [10], incorporating pressure control of chuck boosters and pressure boosters to achieve

Fig. 3.5 Rotary draw bending

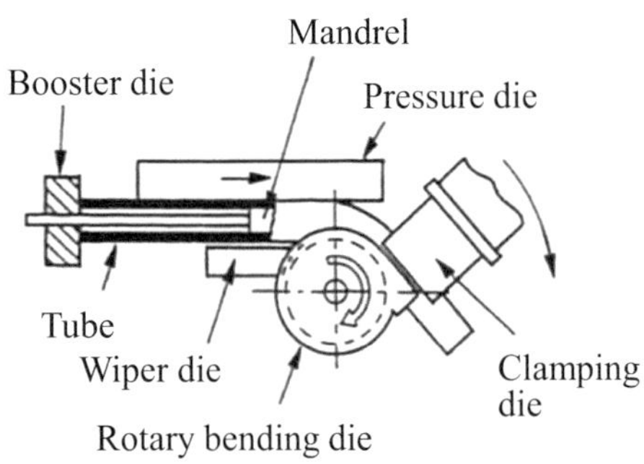

Fig. 3.6 CNC pipe bender [10]

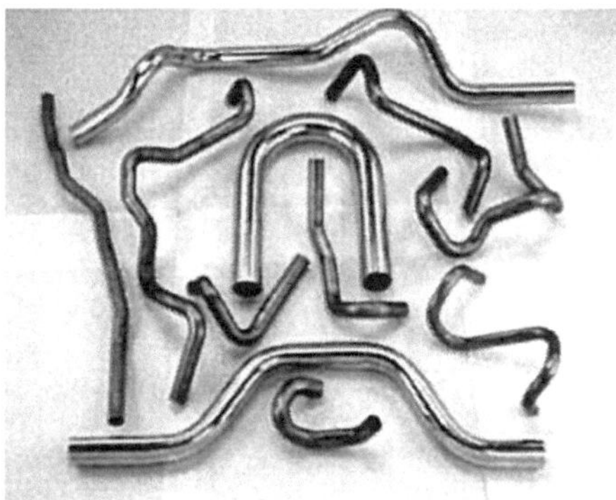

Fig. 3.7 Bent examples by CNC pipe bender [10]

small bending radii, have also been developed. Figure 3.6 illustrates such a device, and examples of processed products are shown in Fig. 3.7 [10].

(d) **Notable application example (bending of double tubes)**

In automotive exhaust pipes, double-walled tubes, as shown in Fig. 3.8, are used to achieve exhaust gas purification and improve fuel efficiency through heat utilization. An example of the bending process for double-walled tubes is illustrated in Fig. 3.9. The end of the outer tube is narrowed, the inner and outer tubes are fitted together, the ends are joined, and a filler material is inserted between the inner and outer tubes. Subsequently, the product is bent, and once the filler material is removed, the final product is completed. Common filler materials include ice, sand (shot beads), and low-melting-point alloys [11].

(e) **Improvement and development examples of rotary draw bending**

Rotary draw bending has been widely adopted, leading to numerous improvement and development examples. The Push Rotary Bending (PRB) method, as depicted in Fig. 3.10, is an enhanced form of rotary draw bending that reduces the outer diameter

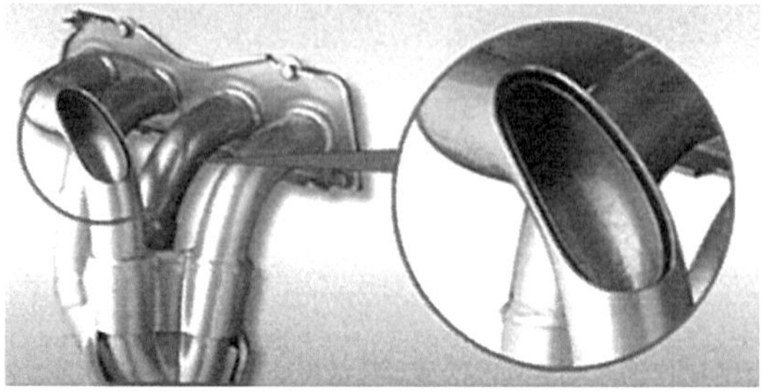

Fig. 3.8 Example of bent double-walled product[4]

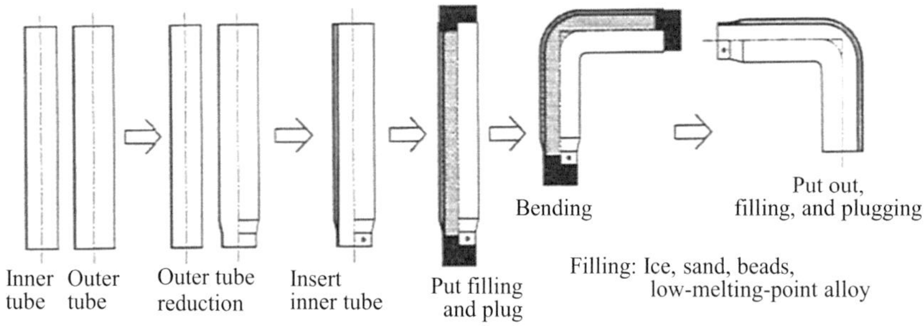

Fig. 3.9 Bending process example for double-walled tube [11]

using a pressure die. It has the advantage of preventing wrinkles even without a wiper die or mandrel. It is suitable for bending high-strength materials, and although there may be some flattening of the cross-section, it demonstrates minimal eccentricity as a characteristic feature. An application example of a 780 MPa grade high-strength thin-walled steel tube using this method and hydroforming is shown in Fig. 3.11 [12].

In addition, there are examples of flattening suppression by adding auxiliary dies (side compression) [13], and cases where ultrasonic vibration is applied to the mandrel to improve bending accuracy through friction reduction [14]. Furthermore, there are development examples such as the hydrobend method shown in Fig. 3.12,

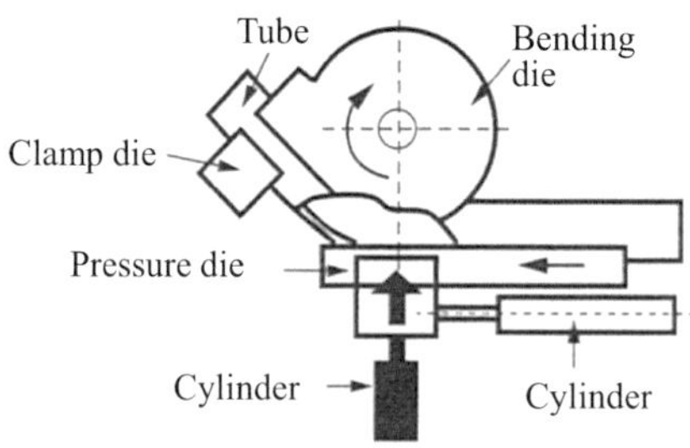

Fig. 3.10 Schematic of PRB method

[4] http://www.sango.jp/ (as of February 2019).

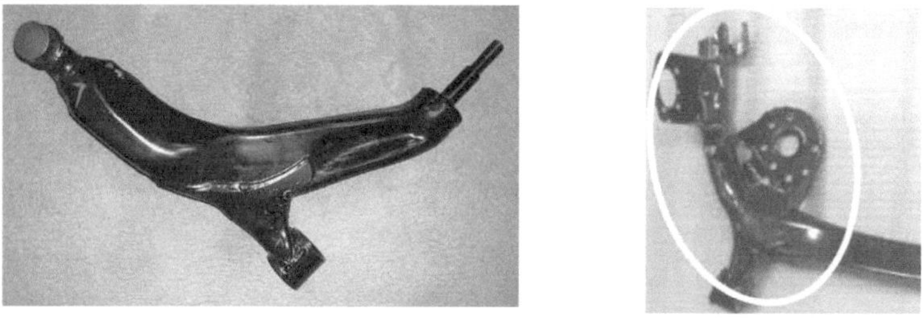

Lower arm made of 780 MPa steel tube

Trailing arm made of 590 MPa steel tube

Fig. 3.11 Examples of PRB method application

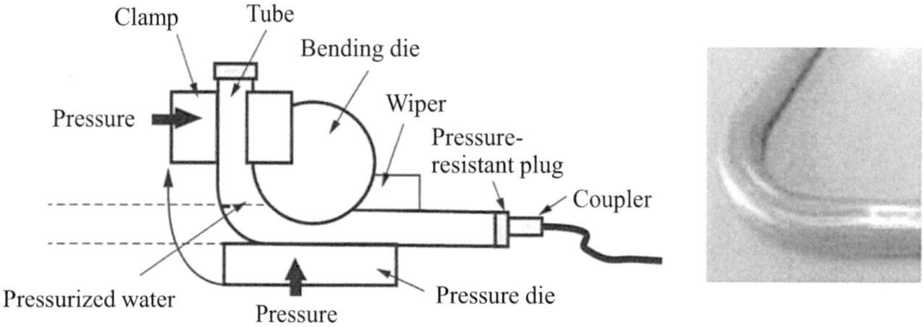

Fig. 3.12 Hydrobend method

where hydraulic pressure is applied to reduce frictional resistance and enable the bending of thin-walled SUS pipes [15].

(3) **Compression bending**

(a) **Overview**

It is a method of bending a tube by pushing it against a movable pressing die while fixing one end to a fixed die. Figure 3.13 shows a schematic diagram and a processing photograph [16]. When using a mandrel to suppress flattening, the mandrel itself must be rotated along with the pressing die.

(b) **Application guidelines**

The maximum bending angle is 180°, and the minimum bending radius is approximately $r_{\min} \fallingdotseq 2.0d_0$ when using a mandrel. There is a risk of wrinkling with a small bending radius. The work efficiency is relatively high, and it can be processed at a low cost, making it suitable for mass production.

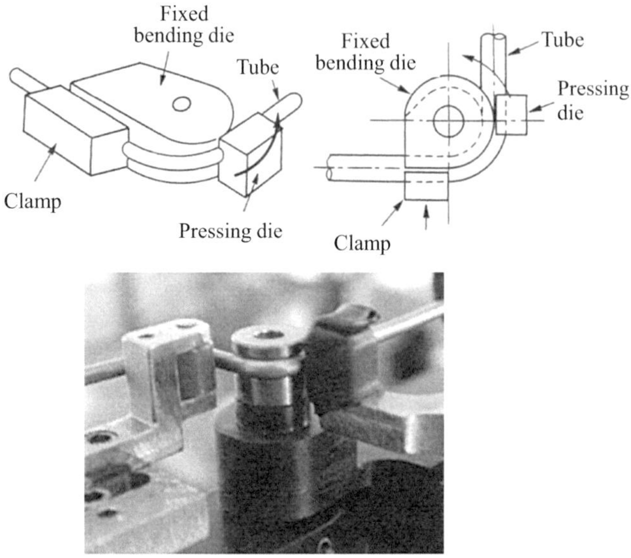

Fig. 3.13 Compression bending

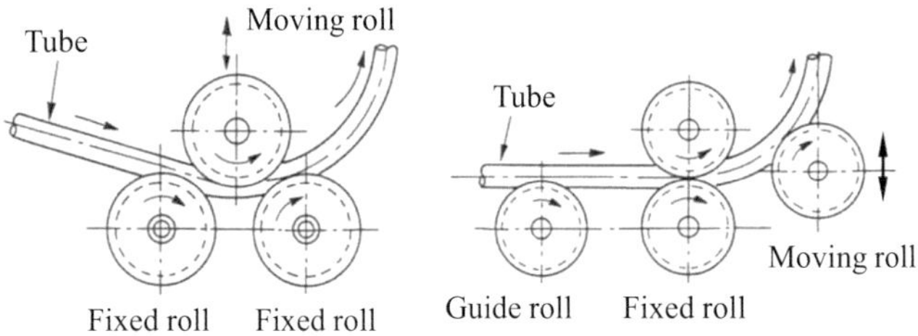

Fig. 3.14 Roll bending

(4) **Roll bending**

(a) **Overview**

It is a method of bending using three driven rolls, as shown schematically in Fig. 3.14. Depending on the arrangement of the rolls, it can be divided into (1) Pyramid style (Fig. 3.14a) and (2) Pinch style (Fig. 3.14b). By adjusting the spacing between the rolls, the curvature can be changed, and by installing another set of rolls on the

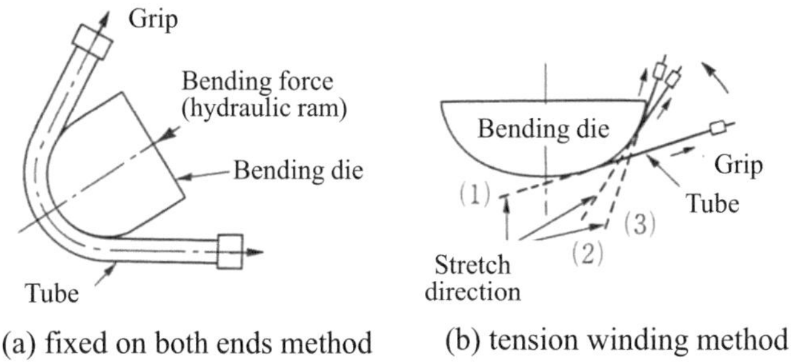

Fig. 3.15 Stretch bending

exit side and performing out-of-plane bending, continuous spiral bending can be achieved.

(b) **Application guidelines**

The maximum bending angle is 360° (spiral bending) and continuous bending is possible. However, small bending radii are difficult, and the minimum bending radius is twice the roll diameter. By dynamically adjusting the spacing between the rolls through CNC control during the bending process, gradual changes in the bending radius can be achieved.

(5) **Stretch bending**

(a) **Overview**

It is a method of bending a tube along the die while stretching it in the axial direction.

There are several variations of this method depending on the combination of chuck and die fixation and rotation, as shown in Fig. 3.15. Essentially, it follows the same principle as stretch forming of sheet materials and is suitable for larger bending radii. Additionally, it has the effect of suppressing springback, making it useful for purposes such as curvature correction. Figure 3.15a depicts the method where the chuck position is fixed and the die is pushed out, while Fig. 3.15b shows the method where the die is fixed and the chuck moves during the tension winding process.

(b) **Application guidelines**

The maximum bending angle is 180°, and the bending radius cannot be made very small. The tube is subjected to bending moments and axial tension. It has the advantage of being able to bend into changing curvatures such as elliptic and parabolic curves. However, due to the need for fixed parts, the yield is lower, and it is difficult to achieve precise cross-sectional shape accuracy due to deformation caused by axial forces.

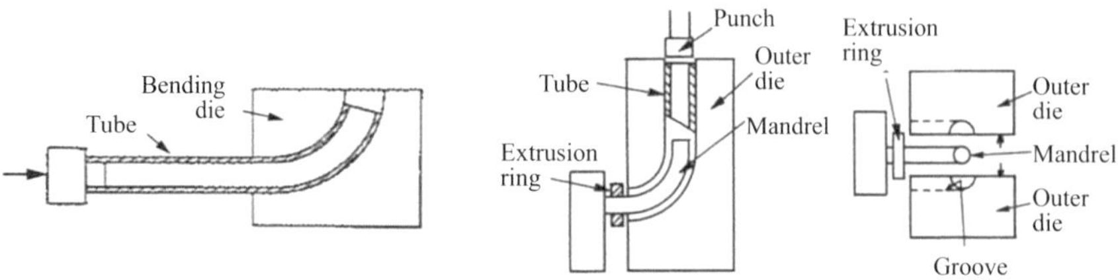

(a) Basic tool (b) Improved tool

Fig. 3.16 Penetration bending

(6) **Penetration bending**

(a) **Overview**

It is a method of bending a tube by pushing it into a mold with a cavity corresponding to the desired outer diameter of the bent tube. The process is illustrated schematically in Fig. 3.16. Figure 3.16a represents the basic form, while Fig. 3.16b shows the use of a split die as the outer shape (bending die), along with the use of a mandrel to suppress flattening and achieve smaller bending radii. The tube is subjected to bending moments and axial compression.

(b) **Application guidelines**

The maximum bending angle is approximately 90°, and the bending shape is determined by the shape of the mold. It is commonly used for manufacturing short-length elbows and may not be suitable for bending a portion of long tubes.

(c) **Processing machinery and processed product examples**

Specialized presses tailored to the specific application of the products are often used in this method. It is applied to automotive exhaust pipe components, and simultaneous drawing and bending processes are also performed.

(d) **Elbow forming**

In the fitting industry, this technique is applied to the fabrication of thin-walled stainless steel elbows. As shown in Fig. 3.17, a tube with its ends cut at an angle is pressed and bent while being constrained by a die and rollers on the outer side and a mandrel on the inner side. One distinctive feature of this technique is the angled cut at the end of the tube, which helps suppress thinning on the outer side of the bend. Careful considerations are given to the shape of the die, the mandrel and outer bend profiles, and the clearance between the mandrel and outer die at the bending start and end points, the central section of the bend. These measures help prevent wrinkling

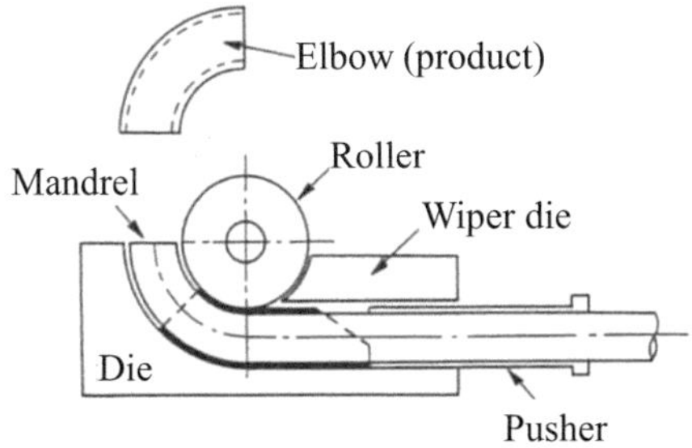

Fig. 3.17 Elbow forming by penetration bending

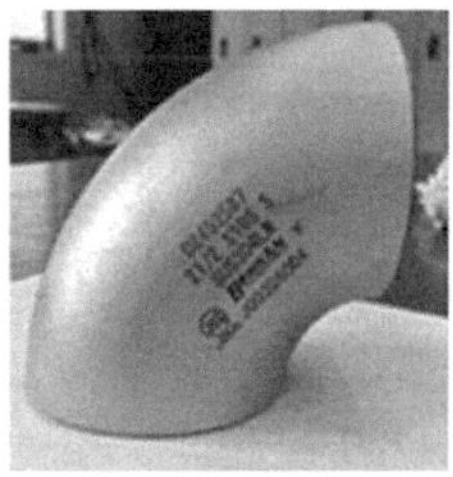

Fig. 3.18 Example of produced elbow

and ensure the desired bending shape. Lubrication also plays a significant role in the process feasibility. Figure 3.18 depicts an example of an elbow product [17] where the bending radius r is equal to the tube diameter d_0, specifically $r = 1.0\, d_0$.

(7) **Flexible penetration bending**

(a) **Overview**

The process of bending a circular tube freely into a desired shape involves continuous bending while simultaneously controlling the "bending radius", "bending angle", and "bending direction" to specified values. Flexible penetration bending involves pushing the circular tube through a die and moving the die to positions corresponding to the desired bending radius, enabling three-dimensional bending of the tube. The

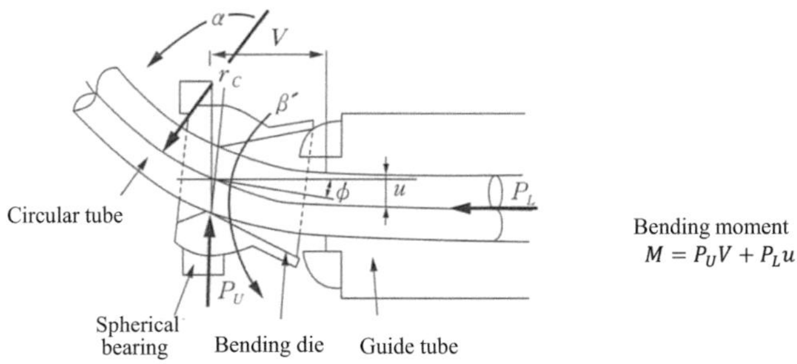

Fig. 3.19 MOS bending (Forming principle) [18]

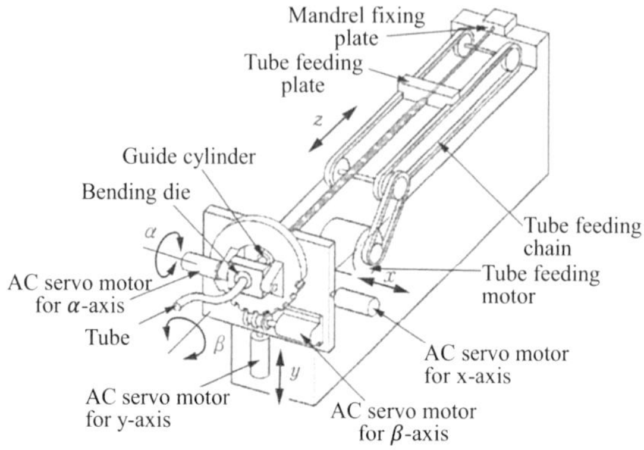

Fig. 3.20 MOS bending machine

processing principle is shown in Fig. 3.19 [18]. This technique is also known as MOS bending, named after the three developers. Since the outer circumference of the tube is constrained by the die in this method, flattening of the tube is prevented. Additionally, applying axial compressive force results in minimal reduction in thickness on the outer side of the bend but a tendency to increase thickness on the inner side [19].

(b) **Application guidelines**

The bending angle can be continuously adjusted up to 360° (spiral bending) and allows for variation in bending radius, enabling three-dimensional shapes with flexibility. Moreover, the use of dies ensures a smooth surface finish on the workpiece. The minimum bending radius is approximately $r_{\min} \fallingdotseq 3.0d_0$.

(c) **Processing machinery and processed product examples**

Figure 3.20 provides an overview of the processing machine used to realize this technique [20]. By controlling the z-axis for tube feeding, the x-axis, y-axis, and the angular axis (a-axis, b-axis) for die positioning through CNC control, free-form bent tubes can be achieved. Figure 3.21 shows examples of processed products.

(8) **Hamburg bending** (named after the place of development)

(a) **Overview**

Hamburg bending, as depicted in Fig. 3.22, is a technique for bending a thick-walled tube with a small inner diameter by expanding the tube using a conically shaped curved mandrel pushing into it. Typically, this process is carried out at elevated temperatures. It is often referred to as hot mandrel bending [21]. The mandrel is divided into the tube guiding section, expansion section, and product guiding section, as shown in the figure. The expansion section has a curvature radius larger than the desired bending radius, and its cross-section perpendicular to the longitudinal axis

Fig. 3.21 Example by MOS bending [20]

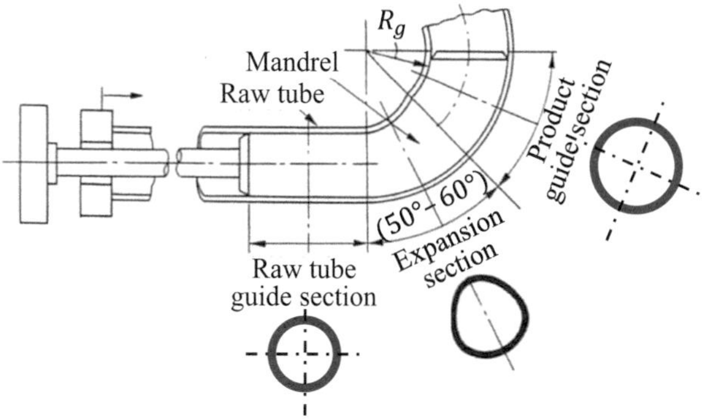

Fig. 3.22 Hamburg bending

has an oblate shape (egg shape). This cross-sectional shape is formed by smoothly blending a triangle-based profile on the inner side of the bend with a semicircular profile on the outer side, resulting in a rounded and continuous contour. Additionally, it gradually transitions from an oblate shape to an oval shape towards the latter half of the expansion section. This is because deformation mainly occurs on the inner side during expansion, causing the inner side of the bend to undergo axial shrink and circumferential stretch. This process allows for the production of curved tubes with reduced thickness variation.

(b) **Application guidelines**

The maximum bending angle for this technique is 180°, and the minimum bending radius is approximately $r_{\min} \fallingdotseq 1.0d_0$. The notable features of this process are as follows: (1) Minimal thickness variation for any cross-section, (2) Absence of flattening of cross-section, and (3) Capability to handle small bending radii. In the

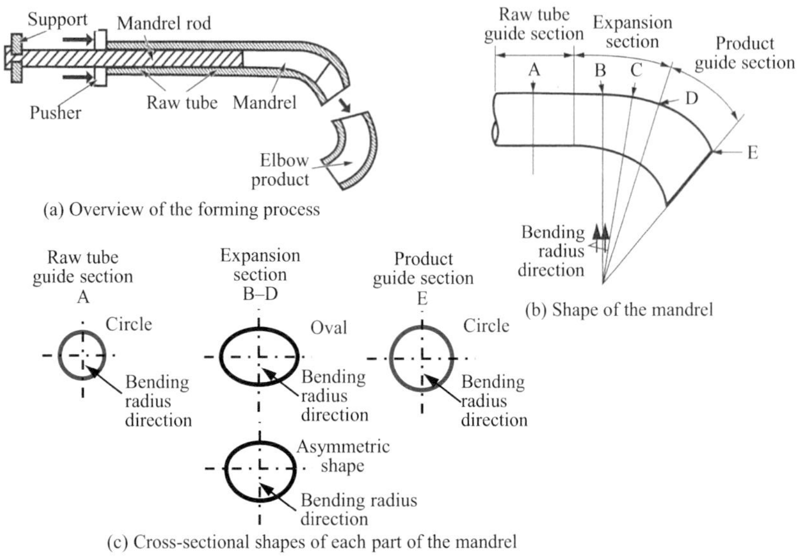

Fig. 3.23 Elbow forming by cold Hamburg bending

case of hot processing, there is available information for product sizes (outer diameter) ranging from small diameters (1/2 inch) to large diameters (48 inches).[5] The maximum expansion ratio for hot processing is approximately 50%, and the ratio t_0/d_0 is 3–15%. For the hot bending of large-diameter products, the bending shape is determined solely by the mandrel, and the pushing load on the material tube is relatively low due to the hot processing. This makes it a suitable method for bending large-diameter products compared to other techniques [9]. Large-diameter elbow products are commonly used as piping components in various applications such as chemical plants, power facilities, oil refining facilities, ships such as tankers, and buildings.

(c) Cold Hamburg bending

To address the cost and complexity associated with heating, temperature control, and heating location in hot Hamburg bending, a forming method for elbows using cold Hamburg bending has been devised. Figure 3.23 illustrates cold Hamburg bending [22]. Figure 3.23a provides an overview of the processing method, where the material tubes are continuously put on the mandrel and mandrel support rod, and are pushed by a pusher. As a result, the tube is formed and pushed out individually for each elbow while being expanded by the mandrel. Figure 3.23b shows the mandrel shape,

[5] http://www.tkbend.co.kr/ (as of February 2019).

which is divided into tube guiding section (circular shape), expansion section (oval shape), and product guiding section (circular shape). The respective cross-sectional shapes at each position are depicted in Fig. 3.23c [22]. Due to the lower processing limits and higher processing force compared to hot processing, excellent lubrication is necessary. Therefore, it is not suitable for large diameters, and the maximum expansion ratio is around 20% and the bending radius exceeds $1.5d_0$. But it significantly improves forming efficiency and processing yield.

(9) **Eccentric plug bending**

(a) **Overview**

Eccentric plug bending is a bending process that involves inserting an eccentric spherical plug into a cylindrical tube, and pushing it out while locally expanding the tube as shown in Fig. 3.24. The processing curvature is determined by the eccentricity of the spherical plug tip, and the tube diameter is limited by the plug's diameter. A method called floating expansion plug bending, where the plug is allowed to float and bending force is applied immediately after expansion, has also been devised [23].

(b) **Application guidelines**

It is similar to Hamburg bending. When cold processing, the deformation is smaller compared to Hamburg bending. However, it results in better cross-sectional shape with reduced flatness and thickness variation. Furthermore, by using eccentric plug shapes such as oval or non-uniform shapes, it becomes possible to process tubes with non-uniform cross-sectional shapes. One notable feature is that the processing curvature can be adjusted by the eccentricity of the sphere, allowing for a reduction in the number of required dies [24–26].

(c) **Processing machinery**

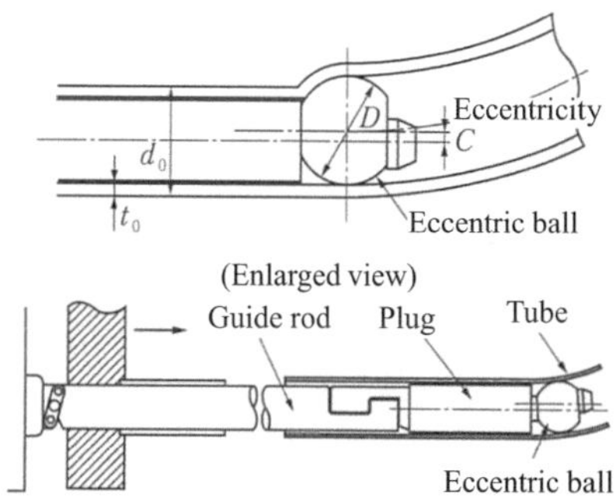

Fig. 3.24 Eccentric plug bending

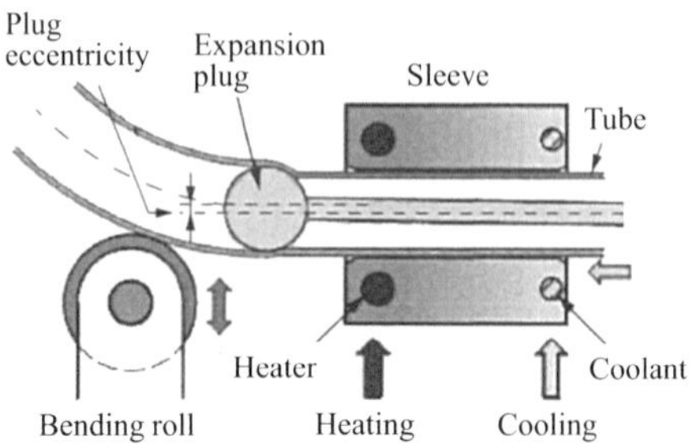

Fig. 3.25 Warm floating expansion plug bending

Figure 3.25 depicts a floating expansion plug bending device incorporating a heating and cooling mechanism to enable warm processing for magnesium alloy tubes [27]. In this device, a heater is added to the entrance side of the sleeve, while a water cooling mechanism is added to the exit side. The minimum bending radius for cylindrical tubes is $1.7d_0$.

(10) **Shear bending**

(a) **Overview**

It is difficult to achieve a bending radius of zero by bending processes due to their deformation mechanism. Shear bending is a processing method that applies shear deformation to the tube, resulting in a small bending radius. The principle of this method is illustrated in Fig. 3.26, where a split die is used to restrain the outer circumference of the tube while continuously subjecting it to shear deformation as it is fed through. In the past, high hydraulic pressure was used inside the tube to prevent buckling [28]. However, this method results in high equipment costs and long processing times. To address these issues, a new method using insert mandrels instead of hydraulic pressure was developed [29], leading to downsizing of equipment and dies as well as reduced processing time. An overview of this method is presented in Fig. 3.27 [30]. By positioning the inserted mandrels from both ends of the tube in close proximity at the boundary of the die, their relative positions are maintained. The right die is fixed, while the left die is pushed down as the tube is fed, enabling shear bending. With the use of these split mandrels, deformation of the cross-section of the pipe during bending can be prevented.

(b) **Application guidelines**

The bending outer radius, r_0, is determined by the mandrel tip radius, and $r_0 = (0.3\text{–}0.5)d_0$, while the bending inner radius, r_i, is determined by the die radius, and $r_i \leq 0.1d_0$. The bending angle ranges 45–90°, and bending angles exceeding 90° are theoretically difficult. As for the material used in the process, there are successful applications with steels and non-ferrous materials. This method is less susceptible to the influence of material properties compared to other techniques. The achievable sizes are not constrained as long as the durability of the mandrel is ensured, even for small-diameter thick-walled tubes. In addition to round tubes, there are also

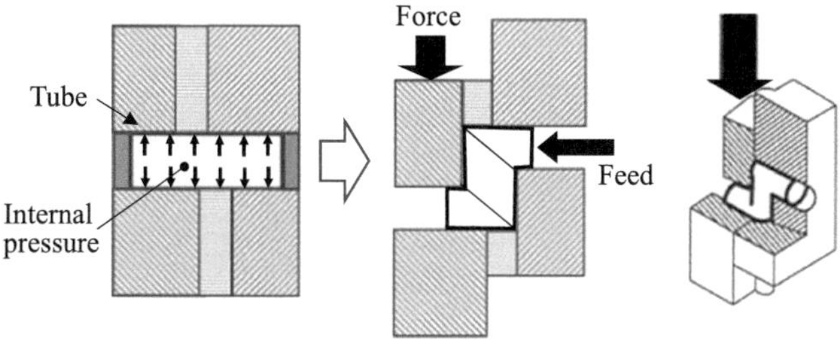

Fig. 3.26 Principle of shear bending

Fig. 3.27 Shear bending using mandrel [30]

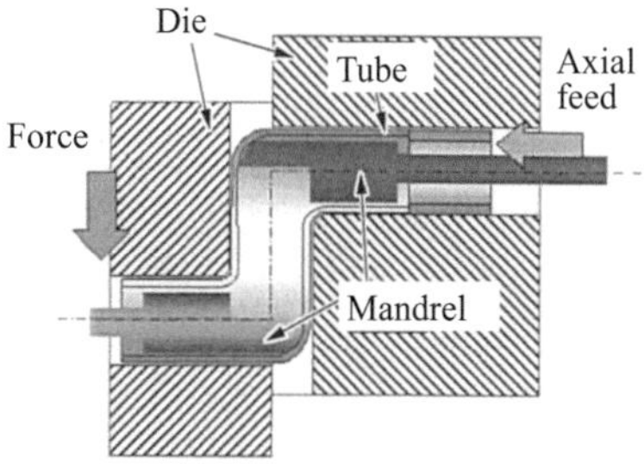

records of application on shaped tubes (such as oval tubes and square tubes). In terms of precision, the deformation of the cross-section is small, and the springback is minimal [30].

(c) **Processing machinery and processed product examples**

In this method, the relationship between the pressing force of the shear press and the material feed on the material insertion side is crucial. Therefore, control of their respective positional relationships is performed. Both general-purpose presses and dedicated presses have been used for this method. In the case of general-purpose presses, mechanical control using cams embedded in the die is employed for pressurization and material feed position control. In the case of dedicated presses, pressurization and material feed position are controlled by CNC, resulting in a simple die structure. Figure 3.28 shows examples of processed products [30].

(11) **High-frequency induction heating bending**

(a) **Overview**

Figure 3.29 shows a schematic illustration of the forming process. One end of a raw tube is clamped using a bending arm, and while high-frequency induction heating is applied, the bending arm is rotated from point A to point B around a pivot. The raw

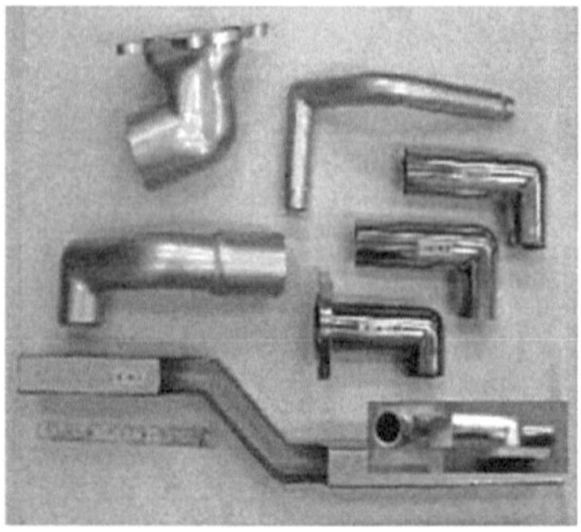

Fig. 3.28 Examples of shear bending [30]

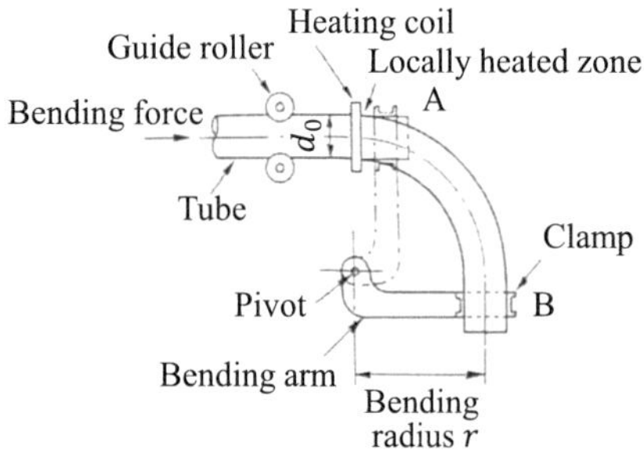

Fig. 3.29 High-frequency induction heating bending

tube, supported by guide rollers, is pushed forwards through the heating coil, but since the tube end is fixed to the bending arm, its trajectory forms an arc centered around the pivot. The bending deformation occurs in the narrow section of the heating area due to the resulting bending moment. The raw tube undergoes progressive deformation, and cumulative bending forms a large bend shape. By narrowing the coil width and cooling the vicinity of the heating area, the heating width is minimized, reducing flattening and preventing buckling by the rigidity of the unheated (cooling) region of the raw tube. Additionally, due to the axial compressive force during bending, the reduction in wall thickness on the outer side of the bend is also minimized. Figure 3.30 shows an example of processing a thin-walled tube. The material used is ferritic stainless steel. Figure 3.30a demonstrates an example where a tube with an outer diameter of 38.1 mm and a thickness of 2 mm is bent to a radius of 35 mm. The characteristic feature is the increased wall thickness on the inner side of the bend, measuring 4 mm after the bending. Figure 3.30b depicts an example where a tube with an outer diameter of 54 mm and a thickness of 1.2 mm is bent to a radius of 60 mm. Since the material is thin-walled, this example showcases bending with wrinkles using this method, and the wrinkle shape is stably formed [31].

(b) **Application guidelines**

The maximum bending angle is 180°, and the minimum bending radius is approximately $r_{\min} \fallingdotseq 1.0d_0$. It is possible to bend continuously and vary the bending radius by using CNC control for the material feed axis, pivot position, and the twist axis

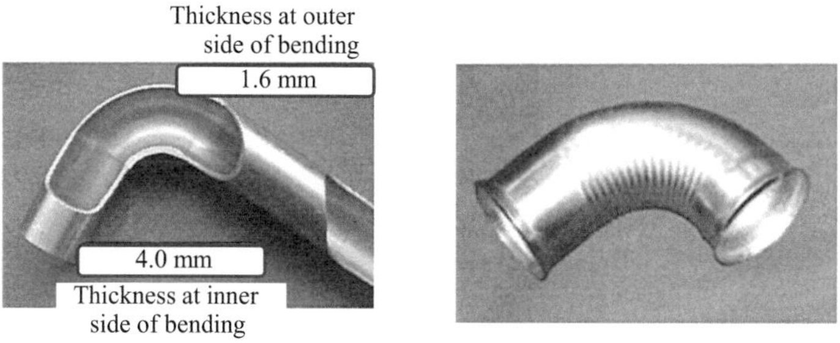

(a) (b)

Fig. 3.30 Examples of high-frequency induction heating bent tubes

at material chuck without using a bending die. Additionally, it is capable of bending shapes and profiles. This method is a dieless forming technique that does not require a dedicated die, resulting in reduced tooling costs. It finds many applications in the bending of large components for volume production such as shipbuilding piping, oil refining, gas processing, and environmental plants. However, it is important to minimize the heating area, and there are challenges when applying this method to aluminum or copper tubes with high thermal conductivity.

(c) **Processing machinery and processed product examples**

The processing machinery consists of high-frequency induction heating equipment, cooling equipment, material feed systems, and bending axes (pivot position axes). By utilizing CNC control for the feed axis, bending axis, and twisting axis, it is possible to achieve continuous bending and vary the bending radius. The processed products are primarily used in power generation plants, and boiler and turbine facilities, and involve large-diameter thick-walled tubes.

(12) **Hot bending and direct quenching (3DQ) technique**

(a) **Overview**

The three dimensional hot bending and direct quenching (3DQ) technique is a sequential forming process that involves the bending and quenching of steel tubes

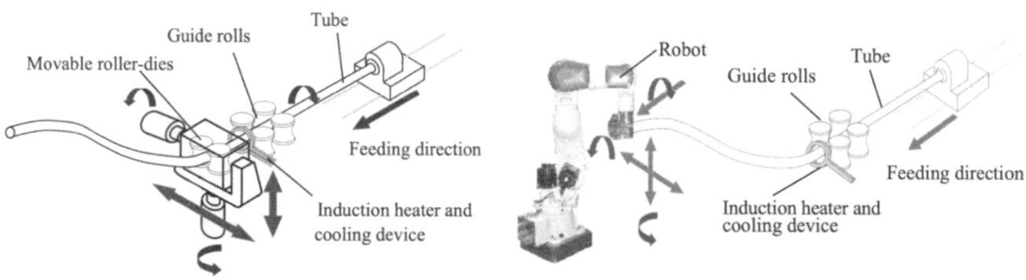

(a) Movable roller-dies type (b) Robot type

Fig. 3.31 Schematic illustration of 3DQ forming machine

by applying bending moments while locally heating them, followed by water cooling after heating. Figure 3.31 illustrates an overview of the processing machine. Figure 3.31a shows the movable roller die method, and Fig. 3.31b shows the robot method, both of which restrain the end of the steel tube and apply the required bending moment for the bending process. Both methods can process tub into any three-dimensional bending shape through CNC control. The robot method adopts a multi-joint general purpose robot, enabling practical application with reduced cost [32].

(b) **Application guidelines**

The 3DQ technique possesses several distinctive features: (1) quenching enables the production of three-dimensional bent components with 1470 MPa or higher strength, (2) it exhibits superior geometrical accuracy compared to cold working due to smaller springback achieved by a hot working process, and (3) it is a dieless forming method, making the equipment compact. Various cross-sectional shapes of steel tubes and profiles can be used as work materials [32]. Additionally, partial quenching is also possible through controlled high-frequency heating. The bendable range of 3DQ is determined by the limit of wrinkling on the inner side of the bend, and for round tubes, the limit strain ε_{cr} of wrinkling can be expressed as $(t_0/d_0)^2$, as shown in Fig. 3.32. The forming range can be specified within the range of $(t_0/d_0)^2 \leq 0.006$. In the range where $(t_0/d_0)^2 > 0.006$, the wrinkling occurrence mode differs, and plastic yielding of the cross-section occurs. In this method, the influence of the temperature distribution on the limit of wrinkling is significant, and a narrower width of the high-temperature region exhibits a wrinkle suppression effect [33].

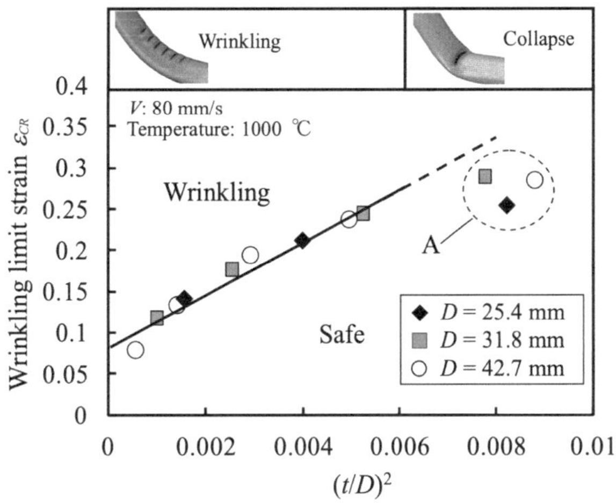

Fig. 3.32 Wrinkling limit (FEM)

Fig. 3.33 Appearance of 3DQ machine (robot type)

(c) **Processing machinery and processed product examples**

Figure 3.33 shows the appearance of a 3DQ processing machine utilizing a 6-axis general-purpose robot [34].

3.2.2 *Bending of Shapes and Profiles*

Shapes and profiles refer to shaped long products with a profile of specific cross-section. There are a variety of profile sections, such as channel, angle, and H-shape, which consisting of bent sections and straight sections [35]. These profiles are made of steel and also light metals. In particular, for steel materials, the term "structural steel" is used, which can be broadly divided into two categories: heavy structural steel produced through hot rolling and lightweight structural steel produced by cold bending thin steel sheets. On the other hand, aluminum alloys and copper alloys have shaped materials formed through extrusion, and these are referred to as extruded structural shapes. In this section, we will discuss representative structural shapes and their bending processes.

(1) **Bending of shapes and its characteristics**

(a) **Defects and challenges in bending of shapes**

The bending of shapes presents different phenomena compared to the bending of flat sheets due to their complex cross-sectional shapes. Moreover, the issues associated with the bending of shapes become more complex compared to flat sheets. Figure 3.34 illustrates the defects that occur during the bending process of shapes. Springback is a concern for shapes as for flat sheets, in the case of shapes. The bending stiffness is higher due to the cross-sectional shape, resulting in a smaller springback amount than that of sheet, but it also makes it challenging to rectify in subsequent processes, emphasizing the need for higher precision in bending angles. Buckling and changes in the cross-sectional shape are issues specific to shapes and not encountered in flat sheets. These factors often determine the bending limits and require careful attention. Generally, buckling occurs on the inner side of the bend; however, as shown in Fig. 3.35, it can also occur on the outer side. Inner buckling occurs during the bending process, while outer buckling occurs during unloading [36]. Fracture, similar to flat sheets, becomes a concern when there is stretching strain on the outer side of the bend.

(b) **Characteristics of cross-sectional shapes**

Shapes and profiles have more complex cross-sectional shapes compared to cylindrical tubes. Based on their cross-sectional shape, they can be classified into open sections (such as channels) and closed sections (such as square tubes). However, in actual applications such as construction materials and automotive components, there

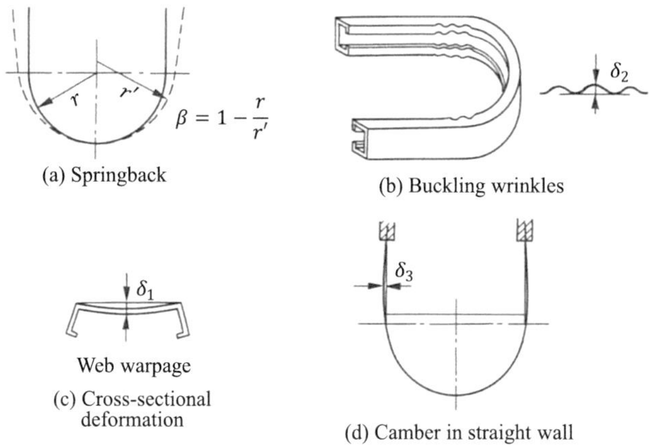

Fig. 3.34 Forming defects in bending of profiles

Fig. 3.35 Buckling example in bending of profiles

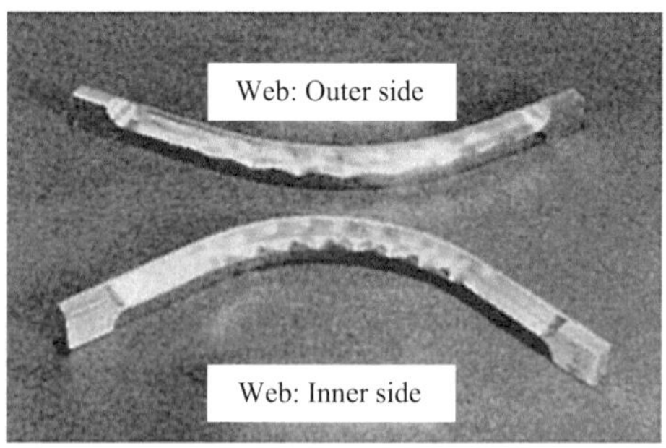

are many composite sections that combine elements of both open and closed sections. Additionally, based on the relationship between the bending center and the cross-sectional shape, bending can be classified into symmetric bending and asymmetric bending. Asymmetric bending occurs when bending a shape that does not have a symmetric axis or when applying bending moments to a plane that does not include the symmetric axis, even for shapes that have a symmetric axis [37]. For example, in the case of the H-shaped beam shown in Fig. 3.36, asymmetric bending occurs when the bending direction is shifted away from the axis of symmetry ($\varphi = 45°$), resulting in deformations of the web that would not occur in symmetric bending [38–40].

(2) **Square tubes** (square and rectangular tubes, and ribbed rectangular tubes)

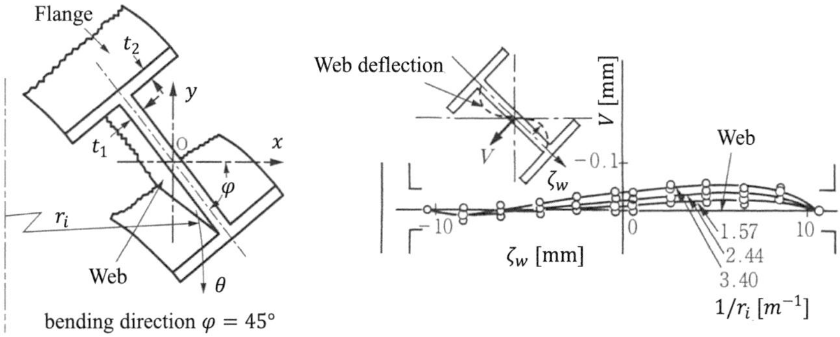

Fig. 3.36 Deformation of web in asymmetric bending of H-shaped beam [40]

(a) Compression bending machine [16]

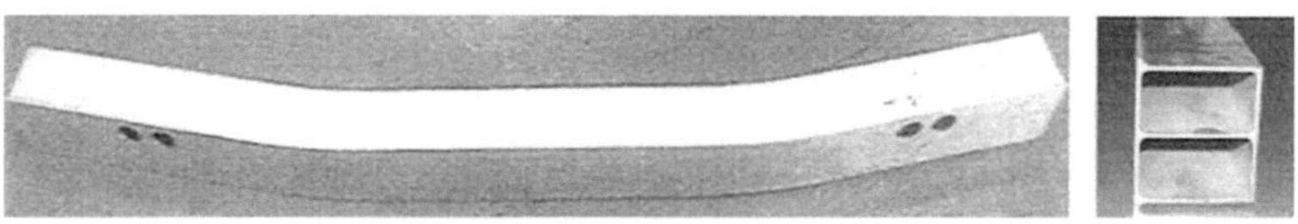

(b) Example of compression bending [41]

Fig. 3.37 Compression bending of rectangular tube [16, 41]

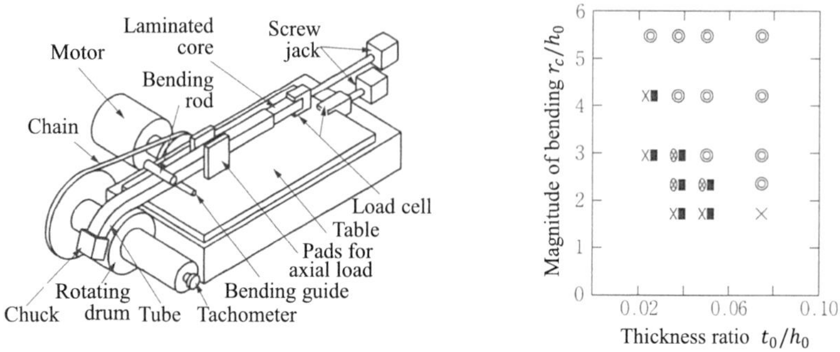

Fig. 3.38 Rotary bending apparatus and deformation mode diagram of rectangular tube bending (A 6061S-O, laminated elastic core, axial tension) [42]

Square tubes are one of the fundamental shapes. They can be classified into square tubes and rectangular tubes based on the width and height of the cross-sectional shape. There are also tubes with ribs in the rectangular hollow section for reinforcement.

(a) **Compression bending**

Push bending is commonly applied to square tubes. Figure 3.37 shows an example of the machinery and the processing of an extruded aluminum alloy for an automotive bumper reinforcement component [41].

(b) **Rotary draw bending**

The rotary draw bending process is widely applied to various shapes such as square tubes and channel sections, as it allows for easy application of mandrels and axial tension. A deformation mode diagram of an extruded aluminum alloy square tube obtained using the apparatus shown in Fig. 3.38 has been proposed. In the deformation mode diagram, "×" represents fracture, "⊗" represents necking, "■" represents wrinkling at the web, and "◎" indicates successful bending without defects. In this bending process, the application of a laminated mandrel and the application of axial tension greatly expand the range of feasible bending conditions. Additionally, it has been observed that the effect of axial tension increases as the thickness t_0 (or h_0 for square tubes) decreases [42]. Furthermore, there have been cases where a wiper die was applied to the rotary draw bending of square tubes [43]. The wiper die helps reduce the pitch of wrinkles caused by buckling, thereby increasing the processing limits. Figure 3.39 illustrates the mechanism of wrinkle suppression by the wiper die. Due to the space where wrinkles grow between the compression flange and the rotating die in the unprocessed region, compressive stress in Region I induces

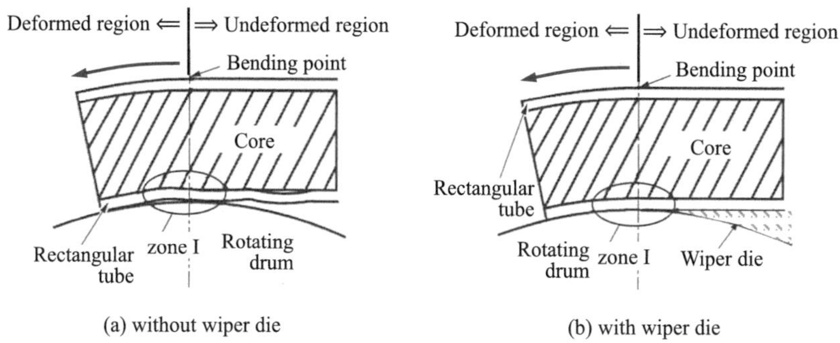

Fig. 3.39 Effect of wiper die

wrinkling in the unprocessed region. However, the wrinkle growing space from the bending point to the unprocessed region is filled by wiper die, preventing growth of wrinkles and particularly suppressing wrinkles on the flange.

(3) Channel section

Channel section, also known as a groove section or U-section, has a cross-sectional shape that resembles the letter "U". It has higher deformation flexibility for each side of section compared to regular tubular sections, making it prone to irregular deformations. The position of the bending center determines whether it undergoes symmetrical or asymmetrical bending.

(a) Rotary draw bending

The deformation characteristics of a channel section vary significantly depending on the position of the bending center, more than with square tubes. To improve the bending limits, the application of dedicated mandrels and jigs, as shown in Fig. 3.40, is being considered [44]. When bending without dedicated jigs such as the anti-twist plates shown in Fig. 3.41, deformation of the section and twisting occur. Furthermore, the application of mandrels and wiper dies does not improve the deformation of the section. In contrast, axial tension with dedicated mandrels and jigs suppresses various types of deformation [44, 45].

(b) Stretch bending

Stretch bending, as shown in Fig. 3.42a, is a process where the workpiece is wrapped around a die while applying tension T to both ends of the workpiece in the longitudinal direction. An example of the die and the cross-section of a channel for bending is shown in Fig. 3.42b [46]. This bending method, similar to bending flat sheets, only requires a male die, eliminating concerns of surface damage on the workpiece

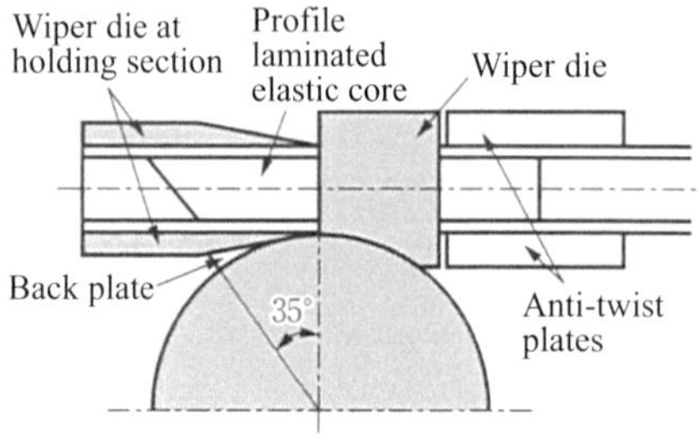

Fig. 3.40 Bending jigs for asymmetrical channels [44]

(A 6061S-O, $t_0/h_0 = 0.038$, $t_0 = 1.5$ mm)

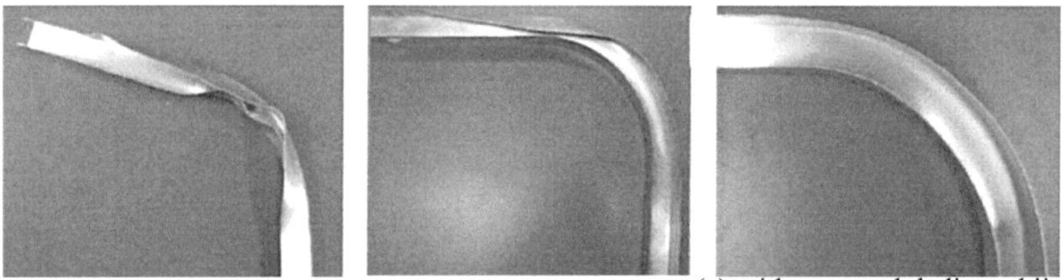

(a) without core and jigs (b) with core and wiper die (c) with core and dedicated jigs, applying axial load

Fig. 3.41 Bent examples of asymmetric channel profiles [44, 45]

and resulting in minimal springback. Additionally, the application of tension in the bending of sections is effective in restraining buckling as it causes the neutral axis

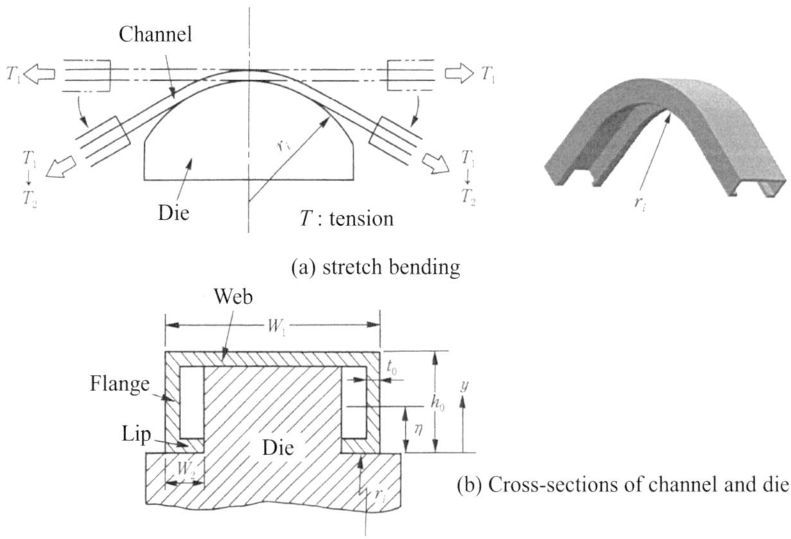

(a) stretch bending

(b) Cross-sections of channel and die

Fig. 3.42 Stretch bending of channel

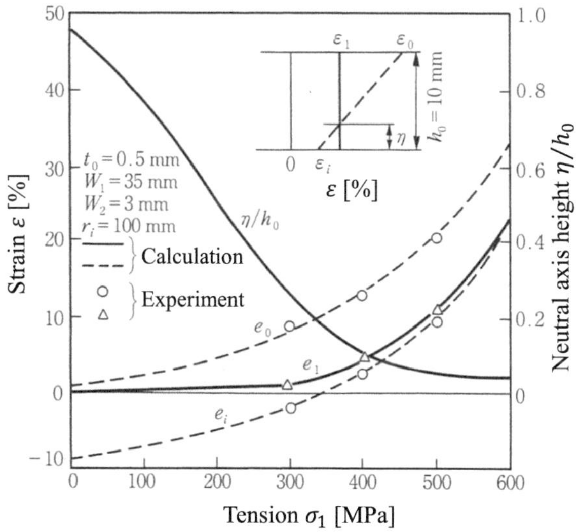

Fig. 3.43 Relation of tension, neutral axis height, and strain [46]

of bending to shift inwards, and it also reduces secondary deformations of the cross-section [47]. The mechanics of stretch bending of section is fundamentally similar to that of flat sheet. However, in the case of sections, it becomes slightly more complex due to variations in width along the height direction.

The (longitudinal) strain and neutral axis position for each section when bending a channel section with thickness $t_0 = 0.5$ mm, height $h_0 = 10$ mm, web width $W_1 = 35$ mm, and lip width $W_2 = 3$ mm with a radius of curvature $r_i = 100$ mm are shown in Fig. 3.43. As can be seen from this figure, the neutral axis position h/h_0 moves inwards rapidly with an increase in tensile stress σ_1. With this movement, the longitudinal strains in each section also increase. In particular, the longitudinal strain is significant at the outer periphery, resulting in a decrease in the width and thickness of the channel section (approximately half of the longitudinal strain in isotropic materials). This effect needs to be taken into account. The relationship between springback and tensile stresses σ_1, σ_2 in stretch bending is shown in Fig. 3.44. The T-M processing shown in the figure refers to the conventional stretch bending applying tension while bending, and the T-M-T processing means bending with applied tension followed by additional larger tension T_2. Similar to the case of the stretch bending of flat sheets, the springback decreases as the initial tensile stress σ_1 increases, and the additional tensile stress σ_2 has significant effect compared to σ_1. It should be noted that the experimental value of the initial tension T_1 in a small tension range is smaller than the calculated value due to the influence of buckling.

Due to variations in material mechanical properties, sheet thickness, and other dimensions, springback fluctuates. To minimize this variation, the method of maintaining a constant initial elongation has been proposed [46].

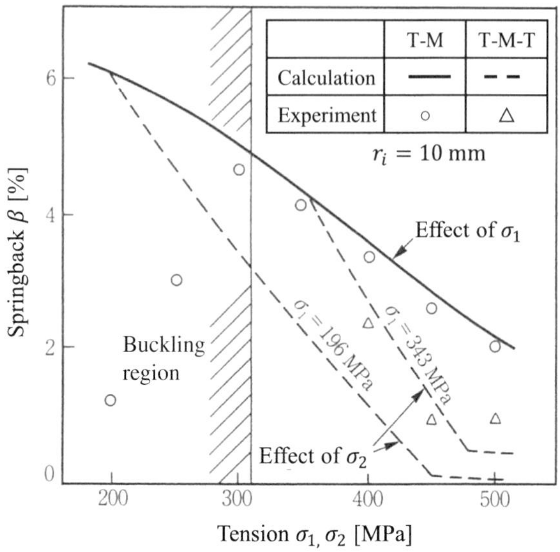

Fig. 3.44 Relation of tension and springback [46]

(4) Various other sections

Other sections are classified as open sections and are often bent using dedicated bending dies because applying mandrels is difficult.

(a) Rolling bending

Rolling bending of shapes is performed by rolling a portion of the section. This method is broadly divided into wedge rolling bending [48] and rolling bending using

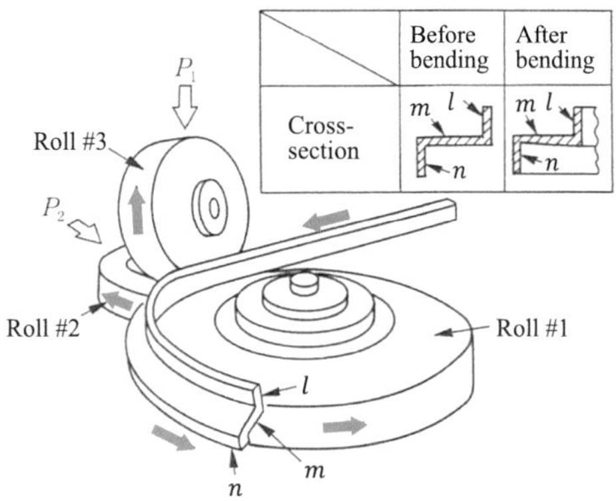

Fig. 3.45 Rolling bending by wedge roll [48]

parallel rolls [49]. Figure 3.45 shows an outline of wedge rolling bending, which achieves a wedge-shaped section by varying the reduction ratio in the width direction during rolling. In this method, all of the ***l***, ***m***, and ***n*** parts have a uniform thickness before bending, the ***m*** part of the section is wedge-rolled, and the ***n*** part is rolled in parallel. Figure 3.46 shows the roll arrangement for bending L, Z, and U-shaped materials.

The basic idea is as follows: out-of-plane bending area and the inner circumferences are not rolled whereas the outer circumferences are rolled in parallel, and all in-plane-bending area is wedge-rolled. Figure 3.47 shows the processing limits of wedge rolling bending. The horizontal axis represents the ratio of the material width b to the sheet thickness t_0, and the vertical axis represents the ratio of the material width to the bending radius r_0. As we move towards the upper right direction in the figure, the processing becomes more difficult. From this, it can be understood that this processing method is suitable for bending thin-walled materials.

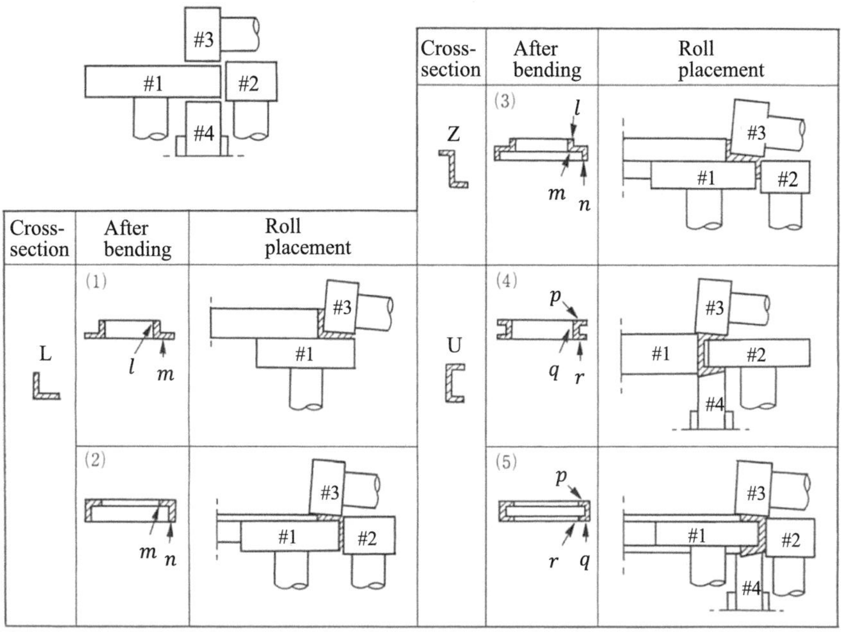

Fig. 3.46 Roll placements in wedge rolling [48]

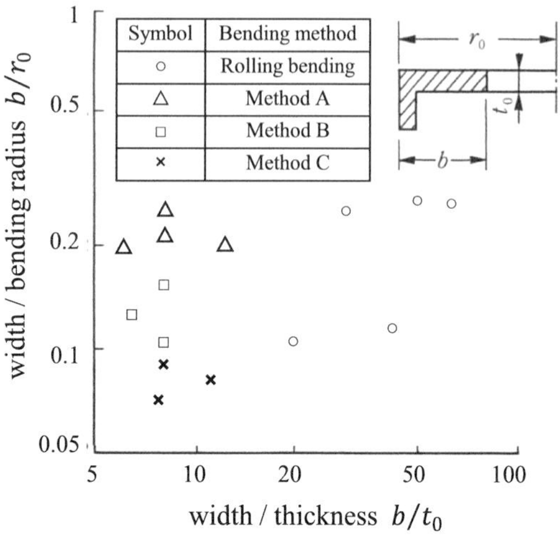

Fig. 3.47 Forming limit of rolling bending and other methods [48]

(b) **Stretch bending**

The processing limit of shapes in stretch bending is largely determined by fracture at the outer periphery and buckling at the inner periphery during bending. Tables 3.2, 3.3 and 3.4 show the bending limits of various shapes [50]. (Please refer to Figs. 3.48, 3.49 and 3.50 for the cross-sectional shape and the direction of bending.)

Table 3.2 Forming limit of stretch (inward) bending of angle and channel [50]

Material	Fracture limit h/r_0										
	Temp. [°C]		h/t_0								
			3	5	7	10	15	25	30	40	50
2024 T-3	R.T.	h/r_0	.23	.22	.21	.19	.17	.14	.14		
		r_0/t_0	13	22.8	34	52.5	88	172	217		
Ti-8-1-1	R.T.	h/r_0	.21	.20	.18	.17	.16	.13			
		r_0/t_0	14.3	25.4	38.1	58.8	96.1	188			
TZM Moly	R.T.	h/r_0	.23	.22	.21	.19	.17	.15	.14	.13	
		r_0/t_0	13	23.2	33.9	52.5	86.5	168	212	316	
Cb-752	R.T.	h/r_0	.32	.30	.29	.26	.24	.21	.20		
		r_0/t_0	9.4	16.6	24.5	37.8	62	117.6	150		
PH 15-7 Mo	260	h/r_0	.38	.35	.33	.31	.29	.25	.24	.22	.20
		r_0/t_0	8.2	14.3	21	31.8	51.6	98.5	124	183	250
L-605	R.T.	h/r_0	.51	.49	.47	.46	.43	.39	.38	.35	.33
		r_0/t_0	5.9	10.2	14.8	21.9	35	63.5	80	114	150
Ti-8-1-1	R.T.	h/r_0	.22	.20	.19	.17	.15	.13	.12		
		r_0/t_0	13.7	25	37.4	57.5	98.4	192	244		
Ti-13-11-3	260	h/r_0	.31	.29	.27	.25	.23	.20			
		r_0/t_0	9.7	17.5	25.9	39.4	65	125			
TZM Moly	R.T.	h/r_0	.27	.26	.25	.23	.21	.18	.17	.15	
		r_0/t_0	10.9	19.3	28.2	42.3	72.7	139	180	269	
Cb-752	R.T.	h/r_0	.32	.30	.29	.26	.24	.21	.20	.18	
		r_0/t_0	9.5	16.4	24.5	37.8	62	117.6	150	222	

Material	Buckling limit h/t_0										
	Temp. [°C]		h/r_i								
			.02	.04	.06	.08	.10	.15	.20	.30	.40
2024 T-3	R.T.	h/t_0	67	48	39	36	36	36			
		r_0/t_0	3417	1248	689	386	396	276			
Ti-8-1-1	R.T.	h/t_0	52	36	30	29	29				
		r_0/t_0	2652	936	530	392	319				
TZM Moly	R.T.	h/t_0	85	60	49	44	44				
		r_0/t_0	4335	1560	866	594	484				
Cb-752	R.T.	h/t_0	72	51	41	38	33	38	38		
		r_0/t_0	3672	1326	725	513	418	291	228		
PH 15-7 Mo	R.T.	h/t_0	104	75	61	53	53	52	52		
		r_0/t_0	5304	1950	1076	716	583	399	312		
L-605	427	h/t_0	121	85	70	60	58	58	58	58	58
		r_0/t_0	6171	2210	1235	810	640	444	348	251	203
Ti-8-1-1	260	h/t_0	60	42	34	32	32				
		r_0/t_0	3060	1092	601	432	352				
Ti-13-11-3	260	h/t_0	52	37	30	29	29	29	29		
		r_0/t_0	2652	963	530	392	319	222	174		
TZM Moly	260	h/t_0	92	65	53	46	46	46			
		r_0/t_0	4692	1690	937	621	506	353			
Cb-752	427	h/t_0	85	60	49	44	44	44	44		
		r_0/t_0	4335	1560	866	594	484	337	264		

Table 3.3 Forming limit of stretch (outward) bending of angle and channel [50]

Material	Temp. [°C]	Fracture limit h/r_0									
		h/t_0									
			2	5	7	10	15	20	25	30	40
2024 T-3	R.T.	h/r_0 r_0/t_0	.48 4.17	.42 11.9	.39 17.9	.36 27.8	.31 48.4	.28 71.4	.15 172		
Ti-8-1-1	R.T.	h/r_0 r_0/t_0	.22 9.10	.19 26.3	.18 38.9	.16 62.5	.14 107	.13 154			
TZM Moly	R.T.	h/r_0 r_0/t_0	.26 7.70	.23 21.7	.21 33.4	.19 52.6	.17 88.3	.16 125	.12 203	.11 270	
Cb-752	R.T.	h/r_0 r_0/t_0	.48 4.17	.42 11.9	.39 17.9	.35 28.6	.31 48.4	.28 71.4	.17 150		
PH 15-7 Mo	260	h/r_0 r_0/t_0	.63 3.17	.55 9.1	.51 13.7	.47 21.3	.42 35.7	.38 52.6	.20 125	.19 160	
L-605	R.T.	h/r_0 r_0/t_0	1.2 1.67	1.04 4.8	.98 7.14	.90 11.1	.79 19	.70 28.6	.32 77	.31 96.6	
Ti-8-1-1	R.T.	h/r_0 r_0/t_0	.22 9.10	.19 26.3	.18 38.9	.16 62.5	.14 107	.13 154	.11 233		
Ti-13-11-3	260	h/r_0 r_0/t_0	.44 4.54	.38 13.1	.36 20	.32 31.2	.28 53.6	.25 80			
TZM Moly	R.T.	h/r_0 r_0/t_0	.27 7.40	.23 21.7	.22 31.8	.19 52.6	.17 88.3	.16 125	.12 203	.12 260	.10 403
Cb-752	R.T.	h/r_0 r_0/t_0	.48 4.17	.42 11.9	.39 17.9	.35 28.6	.31 48.4	.28 71.4	.17 150	.15 202	

Material	Temp. [°C]	Buckling limit h/t_0									
		h/r_i									
			.02	.04	.06	.08	.10	.15	.20	.30	.40
2024 T-3	R.T.	h/t_0 r_0/t_0	102 6800	73 2435	57 1140	45 563	33 220	29 145			
Ti-8-1-1	R.T.	h/t_0 r_0/t_0	83 5530	60 2000	46 920	37 463					
TZM Moly	R.T.	h/t_0 r_0/t_0	130 8670	94 3130	73 1460	58 725					
Cb-752	R.T.	h/t_0 r_0/t_0	115 7660	82 2733	64 1280	51 638	37 246	32 160	26 156		
PH 15-7 Mo	R.T.	h/t_0 r_0/t_0	161 10710	115 3835	90 1800	70 875	52 347	44 220	37 217		
L-605	427	h/t_0 r_0/t_0	185 12320	131 4370	102 2040	80 1000	60 400	52 260	42 252	35 152	31 108
Ti-8-1-1	260	h/t_0 r_0/t_0	95 6330	67 2232	52 1104	41 512					
Ti-13-11-3	260	h/t_0 r_0/t_0	83 5540	59 1965	46 920	36 450	30 300	26 173	23 115		
TZM Moly	260	h/t_0 r_0/t_0	138 9200	98 3265	76 1520	60 750					
Cb-752	260	h/t_0 r_0/t_0	138 9200	98 3625	76 1520	60 750	44 293	33 190			

Table 3.4 Forming limit of stretch (outward) bending of hat section [50]

Material	Temp. [°C]	Fracture limit h/r_0						
		h/t_0:	3	5	8	10	15	20
2024 T-3	R.T.	h/r_0	.23	.21	.19	.19	.17	.16
		r_0/t_0	13.3	24	41.3	53.4	86.4	128
Ti-8-1-1	R.T.	h/r_0	.17	.16	.15	.14	.13	.12
		r_0/t_0	17.3	31.3	55	72.5	115	174
TZM Moly	R.T.	h/r_0	.19	.18	.17	.16	.14	.13
		r_0/t_0	15.5	27.7	48	62.6	103	153
Cb-752	R.T.	h/r_0	.25	.24	.22	.21	.19	.18
		r_0/t_0	11.8	20.6	35.6	47	77.5	111
PH 15-7 Mo	260	h/r_0	.30	.28	.26	.25	.23	.22
		r_0/t_0	10.1	17.8	30.2	39.4	65	92.7
L-605	R.T.	h/r_0	.44	.42	.40	.39	.36	.34
		r_0/t_0	6.8	11.9	20.1	25.8	41.3	58.4
Ti-8-1-1	R.T.	h/r_0	.17	.16	.15	.14	.13	.12
		r_0/t_0	17.3	31.3	53.7	70.6	115	174
Ti-13-11-3	260	h/r_0	.25	.24	.22	.21	.19	
		r_0/t_0	11.8	21.2	36.5	47.8	80.3	
TZM Moly	R.T.	h/r_0	.19	.18	.17	.15	.14	.13
		r_0/t_0	15.5	27.7	48	70	203	149
Cb-752	R.T.	h/r_0	.25	.24	.22	.21	.19	.17
		r_0/t_0	11.8	21.2	36.5	47.8	80.3	115

Material	Temp. [°C]	Buckling limit h/t_0						
		h/r_i:	.02	.04	.06	.08	.10	.15
2024 T-3	R.T.	h/t_0	66	48	40	35	32	26
		r_0/t_0	3366	1248	667	472	352	199
Ti-8-1-1	R.T.	h/t_0	52	38	32	28	25	
		r_0/t_0	2652	988	565	378	275	
TZM Moly	R.T.	h/t_0	83	61	51	44	40	
		r_0/t_0	4233	1584	901	594	440	
Cb-752	R.T.	h/t_0	72	52	44	38	34	29
		r_0/t_0	3672	1352	777	514	374	222
PH 15-7 Mo	R.T.	h/t_0	103	76	64	55	50	42
		r_0/t_0	5253	1976	1130	743	550	322
L-605	427	h/t_0	120	86	72	63	57	48
		r_0/t_0	6120	2231	1272	851	627	368
Ti-8-1-1	260	h/t_0	62	46	38	33	30	
		r_0/t_0	3162	1196	672	446	330	
Ti-13-11-3	260	h/t_0	54	40	34	29	27	22
		r_0/t_0	2754	1040	601	392	297	169
TZM Moly	260	h/t_0	90	65	55	48	44	
		r_0/t_0	4590	1688	971	648	484	
Cb-752	260	h/t_0	90	65	55	48	44	36
		r_0/t_0	4590	1688	971	648	484	276

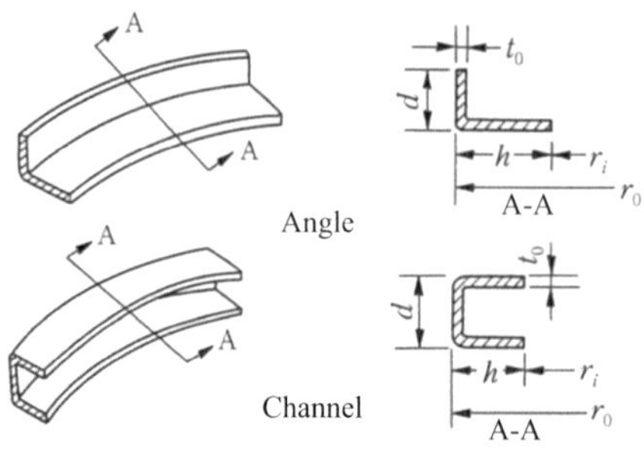

Fig. 3.48 Inward bending [50]

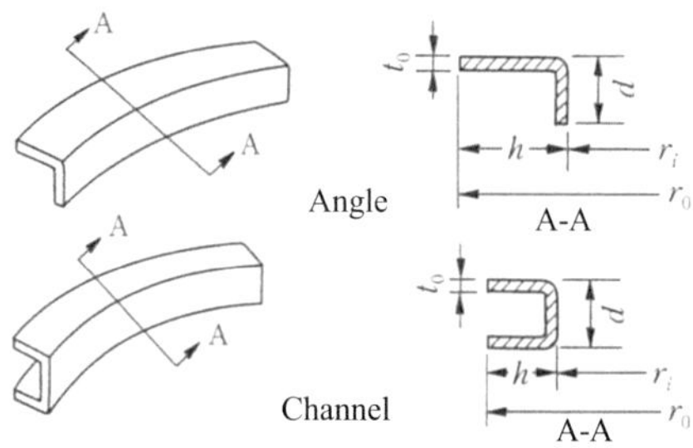

Fig. 3.49 Outward bending [50]

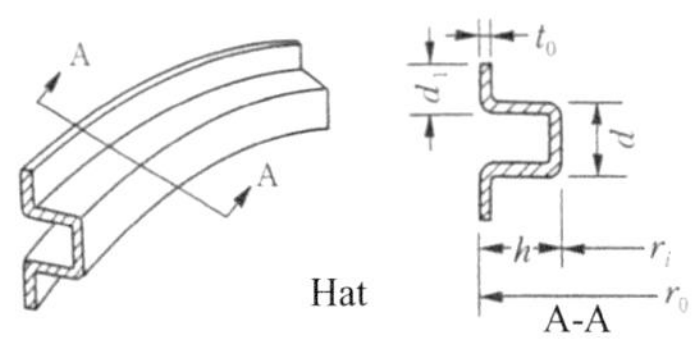

Fig. 3.50 Outward bending [50]

3.3 Processing Force

The required bending force is determined by the required plastic bending moment and the length of the arm providing that moment. Although the plastic bending moment may remain constant, the length of the arm can vary during bending. Therefore, the maximum value of the processing force can be estimated by dividing the plastic bending moment by the minimum length of the arm. To improve the processing limit, secondary external forces are often added in addition to the bending moment. However, the fundamental bending moment, M_0, is the one without secondary external forces and is generally expressed by the following equation.

$$M_0 = \int_A \sigma y dA \tag{3.1}$$

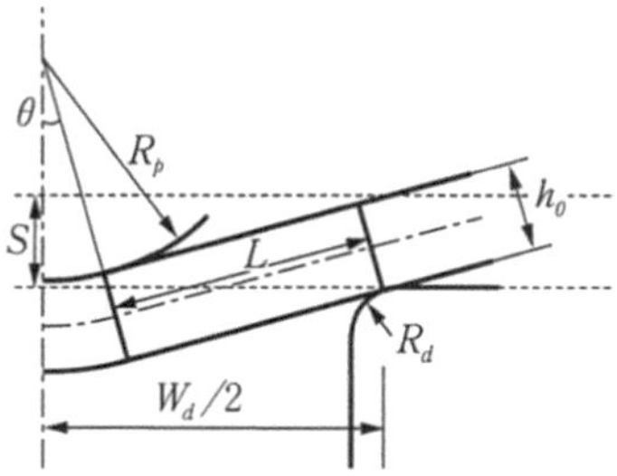

Fig. 3.51 Geometrical relation of stroke and bending angle in press bending [51]

Here, y represents the distance from the neutral axis, σ is the bending stress, and A is the cross-sectional area. On the other hand, the length of the arm providing the moment is determined by the relative motion between the tool that provides the bending radius and the tool that applies the bending force. In widely used configuration, the former tool is a punch or drum-shaped die, and the latter tool is a support die. In this case, considering the press bending (push bending) shown in Fig. 3.51 as an example, Eq. (3.2) is derived from the relationship between the punch stroke S and the bending angle 2θ, and the length of the arm L is given by Eq. (3.3) [51].

$$\theta = \sin^{-1}\frac{R_p + R_d + h_0}{\sqrt{(W_d/2)^2 + \left(R_p + R_d + h_0 - S\right)^2}} - \tan^{-1}\frac{R_p + R_d + h_0 - S}{W_d/2} \tag{3.2}$$

$$L = \frac{W_d/2 - \left(R_p + R_d + h_0\right)\sin\theta}{\cos\theta} \tag{3.3}$$

Here, R_p is the punch radius, R_d is the die shoulder radius, h_0 is the cross-sectional height of the workpiece, and W_d is the die shoulder width.

3.3.1 Rigid Perfectly Plastic Materials [52]

The fully plastic bending moment, M_0, neglecting cross-sectional flattening and uneven thickness in the bending of a circular tube, is given by the following equation.

$$M_0 = \frac{1}{6}\left(d_o^3 - d_i^3\right)\sigma_s \tag{3.4}$$

Here, σ_s represents the yield stress of the material, d_0 is the outer diameter of the tube, and d_i is the inner diameter. In the case of thin-walled tubes, considering the average diameter d_m and wall thickness t_0, the fully plastic bending moment, M_0, is given by Eq. (3.5).

$$M_0 = d_m^2 t_0 \sigma_s \tag{3.5}$$

Similarly, for thin-walled square tubes with side length a and wall thickness t_0, Eq. (3.6) applies.

$$M_0 = \frac{3}{2}a^2 t_0 \sigma_s \tag{3.6}$$

3.3.2 *Work-Hardening Materials* (Neglecting Flattening) *[3]*

Assuming the material follows the power hardening rule $\sigma = C\varepsilon^n$, and neglecting the cross-sectional flattening, the bending moment M of a circular tube can be determined using the following equation.

$$M = \frac{1}{4}C\left(\frac{d_o}{2r_c}\right)^n \cdot d_0^3 \left\{1 - \left(\frac{d_i}{d_o}\right)^{n+3}\right\} \cdot \mathrm{B}\left(\frac{n+2}{2}, \frac{3}{2}\right) \tag{3.7}$$

Here, d_0 is the outer diameter of the tube, d_i is the inner diameter, and r_c is the bending radius of the tube's central axis. The beta function B{(n + 2)/2, 3/2} is shown in Fig. 3.52 [53].

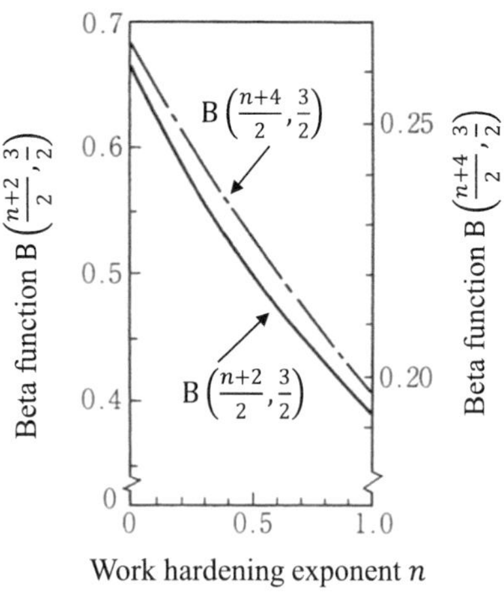

Fig. 3.52 Beta function diagram [54]

3.3.3 *Work-Hardening Materials* (Considering Flattening) *[54]*

Assuming the material follows the power hardening rule $\sigma = C\varepsilon^{n}$ and considering the cross-sectional flattening, the bending moment can be expressed by the following equation, which corrects the bending moment M without considering flattening. Here, D_f represents the flattening ratio. D_f can be determined using the experimental formula (3.11) mentioned later. The beta function B{(n + 4)/2, 3/2} is shown in Fig. 3.52 [53].

$$
\begin{aligned}
M_r &= M\left(1 - \xi D_f\right) \\
\xi &= \frac{1}{2}\left\{(n+7) - 2(n+5)\mathrm{B}\left(\frac{n+4}{2}, \frac{3}{2}\right) \Big/ \mathrm{B}\left(\frac{n+2}{2}, \frac{3}{2}\right)\right\}
\end{aligned}
\tag{3.8}
$$

Figure 3.53 shows the experimental results of bending pure aluminum tubes compared with the calculated results using Eqs. (3.7) and (3.8). The calculated results match the experimental data in the range of small curvature, but as the curvature increases, the consideration of flattening deviates slightly from the experimental values, although the difference is small [54].

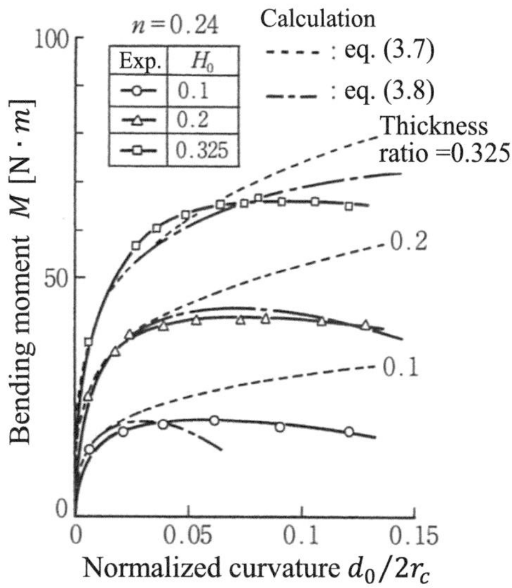

Fig. 3.53 Bending moment and curvature ($H_0 = 2t_0/d_0$) [54]

3.3.4 Empirical Formula for Bending Moment in Rotary Bending [53]

The following empirical formula for bending moment in rotary bending has been proposed.

$$M = \mu Z \sigma_B \sqrt[3]{d_o / r_c} \tag{3.9}$$

Here, M represents the bending moment [N·mm], d_0 is the outer diameter of the tube [mm], d_i is the inner diameter of the tube [mm], r_c is the bending radius of the tube's central axis [mm], and σ_B is the tensile strength [MPa]. Z takes the following values based on the wall thickness t_0:

For $t_0 > 0.06\, d_o$: $Z = 0.1({d_0}^4 - {d_i}^4)/d_0$ [mm^3]
For $t_0 < 0.06\, d_0$ (thin-walled tube): $Z = 0.8\, t_0(d_0 - t_0)^2$ [mm^3]
μ takes the following values based on the mandrel:
μ = 1.0: No mandrel
μ = 2.0: Hemispherical mandrel, good lubrication
μ = 3.0: Floating mandrel, good lubrication
μ = 5–8: Mandrel used, poor lubrication.

3.4 Defects in Bending Process

In bending processes of tubes, unless there is deliberate cross-sectional deformation, such as crushing and flattening in the direction of bending radius, the following results are generally desirable:

1. Minimal deformation of the cross-section.
2. Good conformance to the desired bending radius (good formability).
3. Minimal springback, achieving the desired bending angle (good shape fixability).

Therefore, if the above results are not achieved, it is considered a processing defect.

3.4.1 Deformation of Cross-Section

In bending of tubes, due to the application of shear forces and the resultant axial force directed towards the neutral axis within the cross-section, circular tubes tend to flatten [55]. Square tubes, on the other hand, deform by causing both compression and tension flanges to sink towards the neutral plane of bending, resulting in the web curving outwards [56]. These deformations are collectively referred to as flattening deformation.

3.4.2 Specific Examples of Processing Defects [57, 58]

(1) **Folding** (Fig. 3.54a)

Folding refers to deformation where a tube fails to conform to the shape of a punch and instead localizes the deformation. It is more likely to occur in materials with a low n-value in their mechanical properties, indicating low uniform deformation capability and poor punch conformity. In the case of square tubes, it involves inward deformation and wrinkling of the compression flange and web. In terms of bending, it is a state where localized bending occurs before conforming to the punch. The use of mandrels to suppress localized deformation or the application of counter pressure to help the material conform to the tool can sometimes mitigate folding.

(2) **Fracture** (Fig. 3.54b)

Fracture generally occurs when a material reaches its fracture limit on the tensile side, and low ductility is considered the main cause. After folding occurs, further bending can progress and lead to fracture. In materials exhibiting special deformation resistance, such as magnesium alloys, cracking may occur on the compression side,

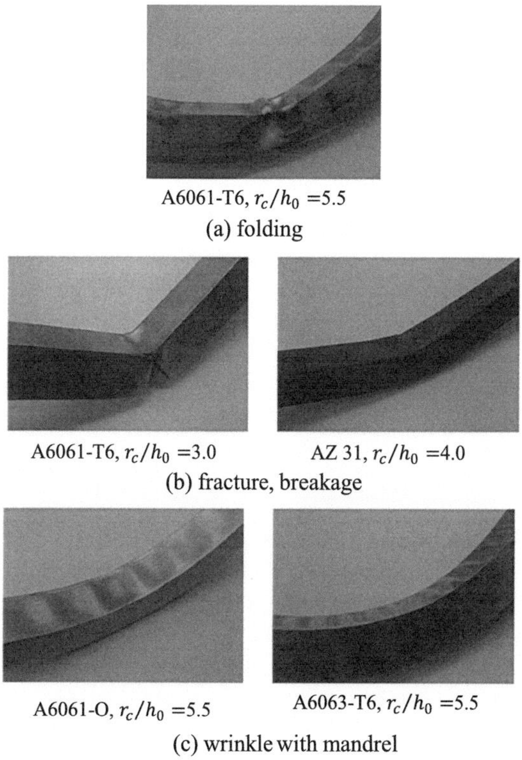

Fig. 3.54 Forming defects in rectangular tube bending

reaching the fracture limit [59]. The use of mandrels to suppress localized deformation or the application of counter pressure to help the material conform to the tool can sometimes mitigate fracture.

(3) **Wrinkling** (Fig. 3.54c)

Wrinkling is a plastic buckling phenomenon observed on the compression side of circular tubes, the compression flange of square tubes, or the compression side of the web. Even if the material conforms to the punch radius globally, periodic undulations occur in the compressed region of the tubes during bending. This phenomenon falls under this category. It represents a state where the material has reached the buckling limit, despite having a high n-value and good uniform deformation capability, due to its low yield stress. On the other hand, even when mandrels are used to suppress localized deformation and avoid fracture for materials with high flow stress, there may be instances where gaps form between the material, the tool, and the punch. For materials with low flow stress, restraining the inward deformation or wrinkling of the cross-section using mandrels is effective in suppressing wrinkle formation.

In Fig. 3.54, examples of aluminum alloy square tubes (with square cross-sections of 40 mm side length and 2 mm thickness) and magnesium alloy square tubes (with square cross-sections of 30 mm side length and 2 mm thickness) processed under various conditions by press bending are shown. r_c represents the bending radius, and h_0 represents the cross-sectional height of the square tube.

3.4.3 Springback (Shape Fixability)

In plastic processing, elastic deformation occurs during unloading after applying plastic deformation to the material, which is commonly known as springback, and it is unavoidable. This becomes a significant issue, particularly in bending of tubes and shapes. Generally, springback increases with higher required bending moments and lower bending stiffness. If a tube undergoes bending with a curvature radius of $1/r_1$ under applied forces and then undergoes elastic recovery during unloading, resulting in a residual curvature radius of $1/r_1$, it can be described by Eq. (3.10).

$$^{1}/_{r_1} - {}^{1}/_{r_2} = {}^{M}/_{EI} \tag{3.10}$$

Here, r_1 represents the curvature radius during loading, r_2 represents the curvature radius after unloading, M is the bending moment, and EI is the bending stiffness.

Variations in the amount of springback due to differences in processing methods, the position of mandrel or pressure die, and other processing conditions have been reported for a long time. Figure 3.55 presents measurement results of springback in rotary draw bending and compression bending [60]. Figure 3.56 illustrates the variation of springback when changing the insertion position of mandrels, while Fig. 3.57 shows the impact of pressure die position on springback, both using rotary draw bending [61]. In addition, studies have been conducted on the influence of axial

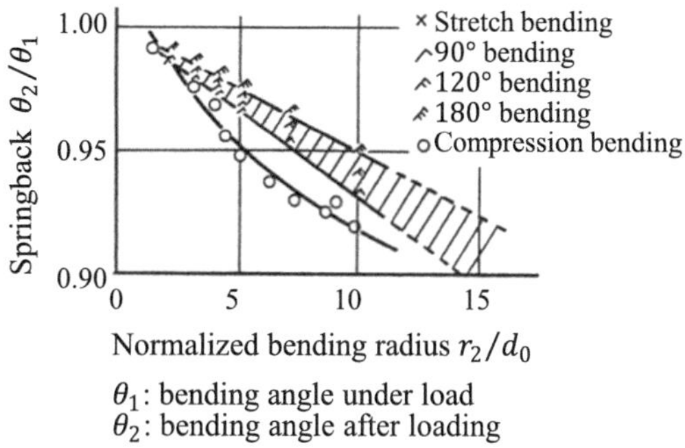

Fig. 3.55 Springback in various bending methods [60]

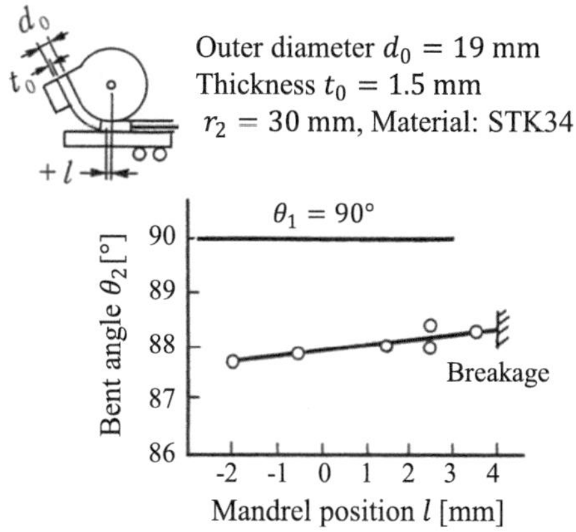

Fig. 3.56 Springback and mandrel position [61]

thrust force in electric resistance welded steel tubes during rotary draw bending [62] and the influence of mandrel in A6000 series square tubes during uniform bending [63], but it has been reported that these factors have minimal effect on springback. Furthermore, it is said that springback tends to increase in large radius bending. This increase is primarily geometric and not due to bending effects; it is generally determined by the bending moment and bending stiffness as described in Eq. (3.10). Larger bending radii result in smaller bending moments and smaller springback per unit length. However, under constant bending angles, larger bending radii lead to longer arcs, resulting in greater overall observed springback. Additionally, research has reported the calculation results of springback in the plastic range using elementary theory in press bending [64], as well as experimental results of springback compiled based on the yield strength of A6000 series alloys [65]. Figure 3.58 demonstrates springback in compression bending organized by proof stress, showing that higher proof stress correlate with increased springback. In recent years, simulation results considering springback in roll bending of shapes have also been presented [66].

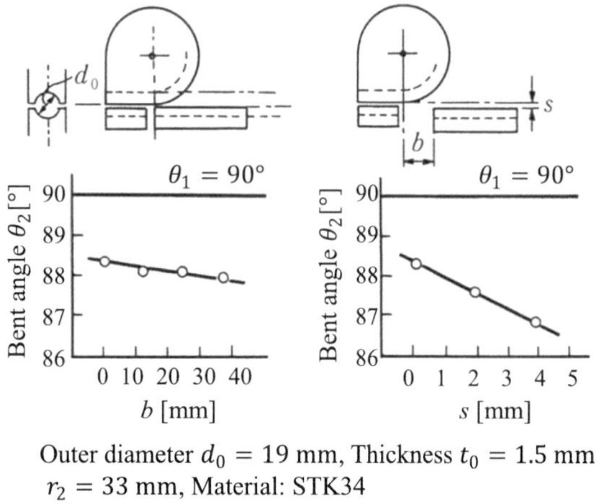

Fig. 3.57 Springback control by pressure die [61]

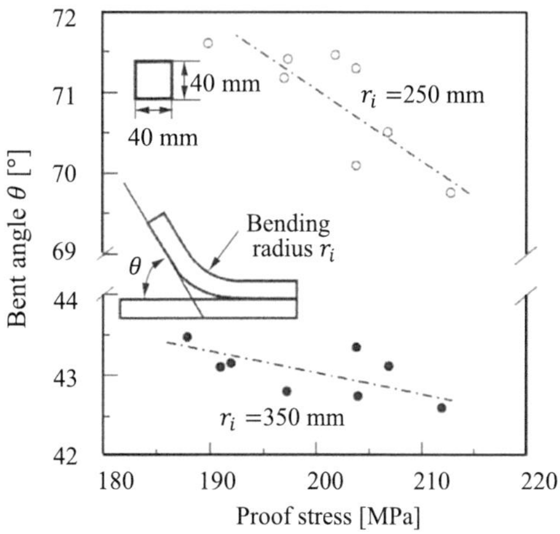

Fig. 3.58 Proof stress and bent angle [65]

3.5 Bending Die, Mandrel, Jigs, and Tools

3.5.1 Bending Dies and Dieless Bending

To achieve curvature in circular tubes or shapes, bending dies are generally required. Examples of such dies include compression bending dies and rotary draw bending dies. The accuracy of the curvature is determined by the bending die, and the required number of bending dies is the number of desired curvatures. In the case of rotary draw bending, as shown in Fig. 3.59 [5], in addition to the bending die, clamps and pressure dies are the basic tools, and mandrels, wiper dies, and booster dies (as shown in Fig. 3.5) are applied as supports to suppress cross-sectional deformation of the workpiece. On the other hand, in dieless bending, which does not require

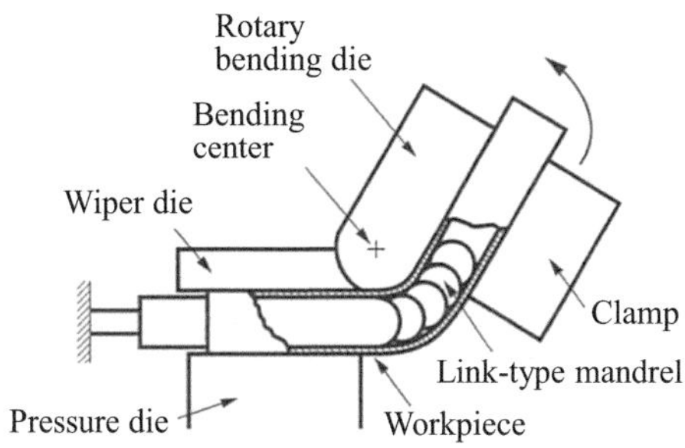

Fig. 3.59 Arrangement of tools in draw bending

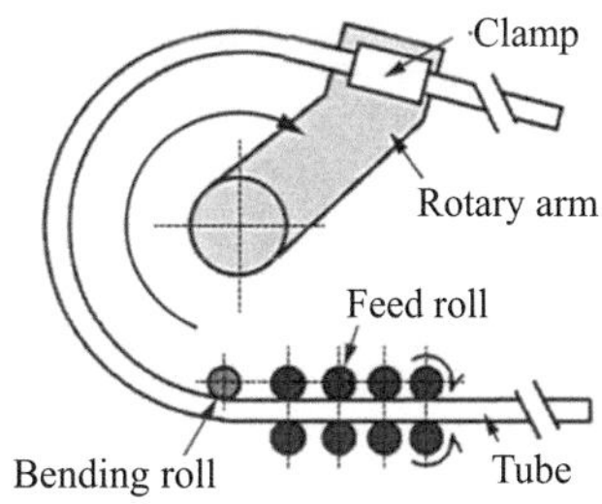

Fig. 3.60 Schematic of dieless U-bending [67]

bending dies, high-frequency induction heating bending was traditionally used as a representative method. In recent years, dieless U-bending and 3DQ, extended from free penetration bending, have been developed. In the former method, the ratio of after springback radius to diameter (r/d_0) ranges from 14 to 130 [67]. As shown in Fig. 3.60, dieless bending applies bending moments using a rotating arm, gripping devices, bending rollers, and supporting rolls in the rear to obtain the desired curvature. When the gripping devices are numerically controlled, such as with robotic systems, the curvature can be freely adjusted. Figure 3.61 demonstrates the accuracy of the achieved bending radii in dieless U-bending.

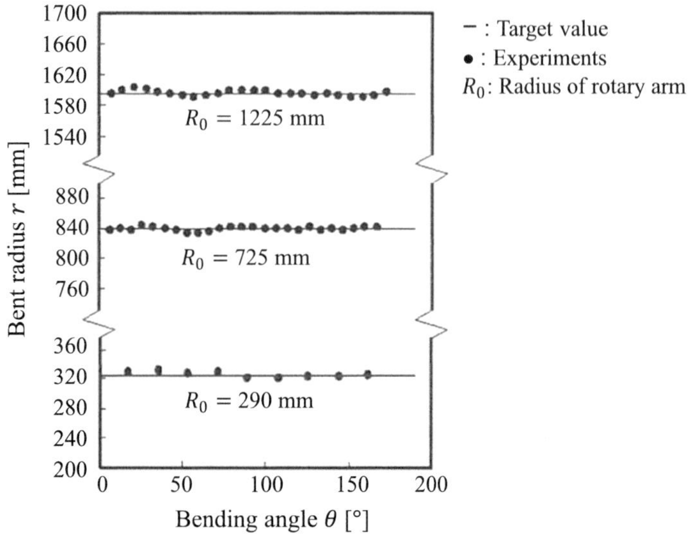

Fig. 3.61 Bent radius of dieless U-bending [67]

3.5.2 *Mandrels and Fixtures*

There are several methods to suppress cross-sectional deformation caused by bending of a workpiece. Essentially, these methods can be divided into restraining the inner side of the workpiece and restraining the outer side. The restraining methods can involve applying external forces or constraining displacement.

(1) **Method of restraining inner side of tubes**

In this method, mandrels [5] are often used, as shown in Fig. 3.62. Figure 3.63 illustrates the processing limits achieved by applying a mandrel. Table 3.5 presents the selection criteria for suitable mandrels in rotary bending [50], while Table 3.6

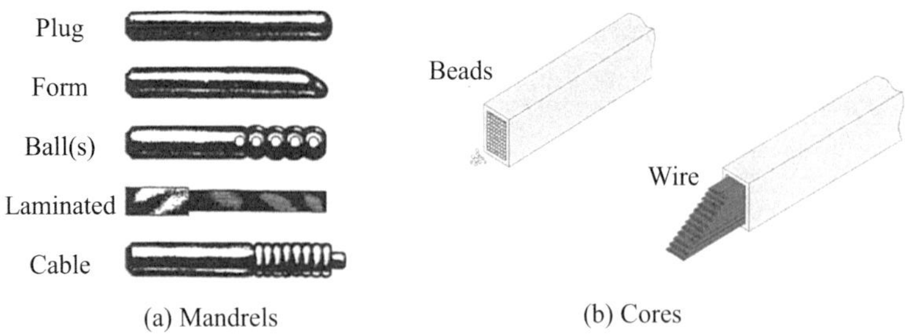

Fig. 3.62 Various mandrels and core materials [5]

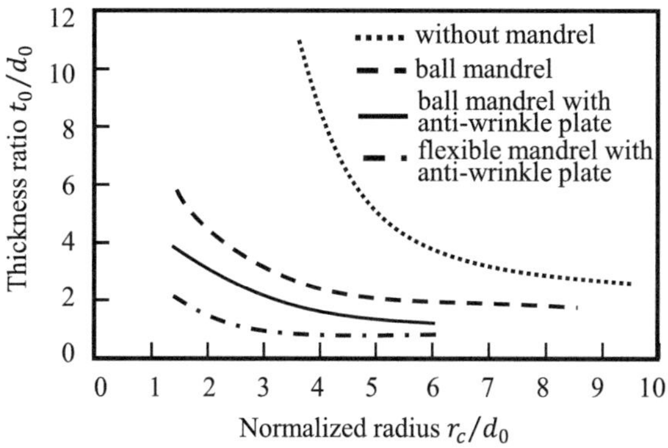

Fig. 3.63 Forming limit of rotary draw bending

shows the relationship between the mandrel protrusion and bending characteristics [68]. In addition to these mandrels, granular materials, wire materials, low-melting-point alloys, hydraulics, and other methods are also employed. They are all effective for circular cross-sections, and for square tubes, mandrels such as square forms, link forms, and laminated forms are used. As a special mandrel, there is a triangular plug used in floating expansion plug bending [69], which allows bending while expanding from a circular cross-section to a triangular cross-section. However, when applying these mandrels, attention must be paid to the friction between the tube and the mandrel, as well as the flexibility of the mandrel.

Table 3.5 Selection criteria for mandrel of rotary draw bending

Outer diameter [mm]	Thickness [mm]														
	0.406	0.508	0.635	0.711	0.889	1.067	1.245	1.473	1.651	1.829	1.981	2.108	2.413	2.769	3.048
15.88	1	1	1	1	1	F	F	F	F	–	–	–	–	–	–
19.05	1	1	1	1	1	1	1	F	F	–	–	–	–	–	–
22.23	1	1	1	1	1	1	1	F	F	F	–	–	–	–	–
25.40	2	2	2	2	1	1	1	F	F	F	F	–	–	–	–
31.75	3	3	2	2	2	2	2	2	2	1	1	F	F	F	F
38.10	3	3	2	2	2	2	2	2	2	2	2	1	1	F	F
44.45	4	4	3	3	3	3	3	3	3	2	2	2	1	1	F
50.80	4	4	3	3	3	3	3	3	3	2	2	2	2	1	1
57.15	4	4	4	4	3	3	3	3	3	2	2	2	2	2	1
63.5	4	4	4	4	3	3	3	3	3	2	2	2	2	2	2
76.2	4	4	4	4	3	3	3	3	3	2	2	2	2	2	2
88.9	5	5	5	5	4	4	4	3	3	2	2	2	2	2	2
101.6	5	5	5	5	4	4	4	3	3	2	2	2	2	2	2
114.3	6	6	5	5	4	4	4	3	3	3	3	3	2	2	2
127.0	6	6	6	6	5	5	5	4	4	4	4	4	3	3	3
139.7	6	6	6	6	5	5	5	4	4	4	4	4	3	3	3
152.4	6	6	6	6	6	6	5	5	5	4	4	4	4	3	3

Form mandrel: F, Ball mandrel: (number) = number of balls
Cable mandrel: 2×(number)+1 = number of cables, in lower left box 2×(number)+2 = number of cables

Table 3.6 Protrusion amount of mandrel and bendability [68]

Protrusion amount of mandrel	Without mandrel	~0.2 mm	0.5 mm	1.5 mm
Maximum thickness reduction rate		21%	22%	32%
Flattening rate	51%	21%	18%	13%
Outlook	Outer: dent Inner: wrinkle	Good	Good	Deformation by mandrel at end of bend

(2) **Method of restraining outer side of tubes**

With this method, there have been cases where improvements in the shape accuracy of the cross-section were achieved by carefully designing the pressure die load and shape in rotary bending processes [12, 70]. By applying loads in the circumferential direction of the tube, it is possible to suppress thinning on the tensile side. Additionally, in press bending of square tubes, the use of a mandrel can suppress flattening but results in a trapezoidal cross-section. Therefore, by applying a load to a portion of the compression side of the square tube, it is possible to reduce the deformation of the cross-section that could not be fully suppressed by using a mandrel [71]. In summary, the application of a mandrel is effective in suppressing flattening, while external restraint helps stabilize the shape accuracy of the cross-section.

(3) **Other methods**

Applying the aforementioned mandrels can reduce the occurrence of wrinkles to some extent, but it is not a fundamental suppression method. Wrinkles occur when compressive stress exceeds the buckling limit, so it is necessary to reduce this compressive stress. Therefore, rotary bending or stretch bending, which can bend the workpiece while applying tension in the longitudinal direction, is effective in suppressing wrinkles. However, caution is required as excessive tension can result in significant thinning and potential fracture of the flange portion on the tensile side of the workpiece. Conversely, theoretical analysis has shown that loading axial compression from the rear of the tube can suppress the wall thickness reduction on the tensile side [4]. Figure 3.64 shows an example of controlling axial force by applying a booster die in rotary bending and changing the wall thickness of the tube [72]. In addition, the application of a wiper die explained in Fig. 3.39 is also effective in suppressing wrinkles. By restraining the compression flange between the pressure die and wiper die through the mandrel, wrinkles can be suppressed.

However, adjusting the contact point between the bending point and the wiper die is not easy. Additionally, the tip of the wiper die is thin and susceptible to wear and damage. Therefore, a hinge-style tool integrating the bending die and the wiper die, as shown in Fig. 3.65, has been developed to address these issues [73].

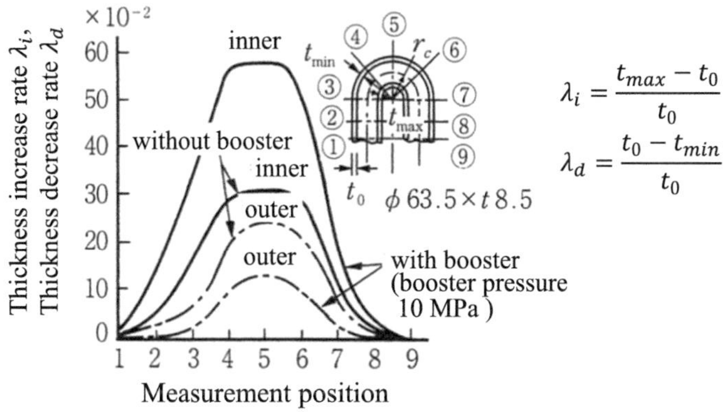

Fig. 3.64 Effect of booster die on thickness deviation [72]

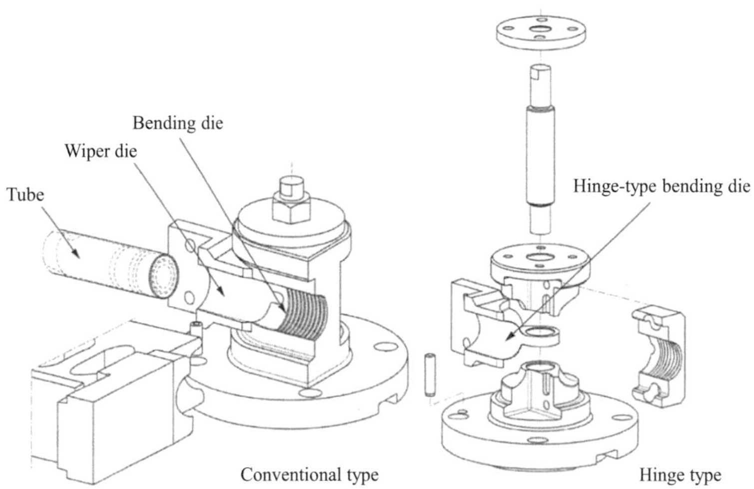

Fig. 3.65 Conventional and hinge-type bending dies [73]

3.6 Processing Limits

The bending processing limits of circular tubes and shapes are sometimes defined by shape accuracy, such as cross-sectional deformation and thickness variation, but generally they are defined by the occurrence of defects such as necking, fracture, yield, and wrinkles.

3.6.1 Cross-Sectional Deformation

(1) **Flattening**

The flatness ratio $D_f = (d_{\max} - d_{\min})/d_0$ is used as a measure of flattening caused by bending. Here, $d_{\max}$ is the maximum outer diameter of the flattened cross-section, $d_{\min}$ is the minimum outer diameter, and d_0 is the initial outer diameter of the circular tube. In some cases, the change ratio of the major axis is expressed as $D_l = (d_{\max} - d_0)/d_0$, and the change ratio of the minor axis is expressed as $D_s = (d_0 - d_{\min})/d_0$. The flatness ratio in the uniform bending of circular tubes is given by the following empirical equation [3].

$$D_f = A \cdot \left(\frac{d_o}{2r_c}\right)^B \tag{3.11}$$

Here, r_c is the bending radius, and A and B are constants determined by the thickness ratio $H_0 = 2t_0/d_0$ and various materials as shown in Table 3.7. From Eq. (3.11), it can be observed that the flatness ratio increases as the thickness decreases and the bending radius becomes smaller.

$$P_{tn} = P_{cn} = t_0 W_0 C \left(\frac{h_0}{2r_c}\right)^n d\theta \tag{3.12}$$

In the flattening of square tubes, as shown in Fig. 3.66, the tension and compression flanges recess within the cross-section while the webs bulge outward. The following equation provides a mechanical analysis of this flat deformation [56].

In the equation, P_{tn} and P_{cn} represent the tension and compression components of flattening forces for a small angle $d\theta$, r_c is the curvature radius, h_0 is the height of the square tube, W_0 is the width, t_0 is the thickness, n is the work-hardening exponent, and C is the plastic coefficient. Equation (3.12) indicates that as the bending progresses and the curvature radius r_c becomes smaller, the flattening force increase, leading to greater flattening.

Table 3.7 Constants A and B in Eq. (3.11) for flattening [3]

Material	A	B
A 1100	$\dfrac{0.07}{(H_0 - 0.05)^{2.67}}$	$\dfrac{1.4}{(H_0 - 0.05)^{0.167}}$
A 5056	$\dfrac{0.03}{(H_0 - 0.017)^{3.73}}$	$\dfrac{0.07}{(H_0 - 0.012)^{0.267}}$
Pure Copper	$\dfrac{0.08}{(H_0 + 0.4)^{9.51}}$	$\dfrac{0.07}{(H_0 - 0.08)^{0.042}}$

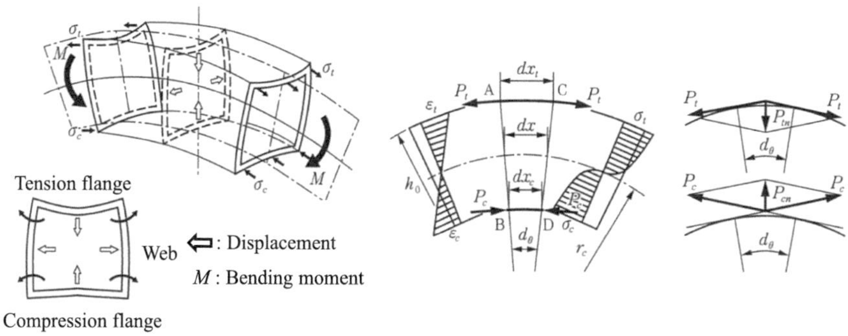

Fig. 3.66 Flattening deformation in bending square tube [56]

(2) **Deformation of web**

The deformation of the web, as shown in Fig. 3.67, refers to the bending of the web of a channel section towards the outer side or the inner side. There are cases where the web tilts outwards for conditions with relatively a short web or thicker sections, while thinner sections may cause the web to tilt inwards, accompanied by wrinkles [74].

(3) **Torsional deformation**

The distortion of the cross-section shown in Fig. 3.41a is an irregular deformation that occurs in sections with asymmetric profiles. When a bending moment is applied to a section with an asymmetric profile, torsion occurs due to the misalignment of the centroid and the shear center (twist center) [44].

(4) **Thickness variation**

When circular tubes or shapes are bent, the tension side generally experiences thinning, while the compression side undergoes thickening. This uneven thickness distribution in the processed material is referred to as thickness variation or thickness eccentricity. The thickness eccentricity ratio $\lambda_t = (t_{max} - t_{min})/t_0$ is used as a measure of thickness variation. Here, t_{max} is the maximum thickness, t_{min} is the minimum

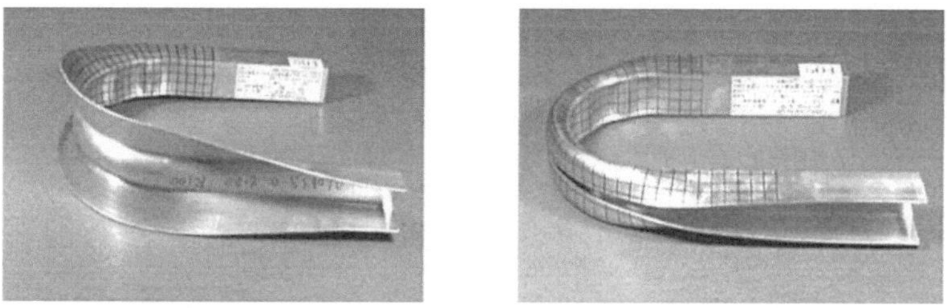

(a) outward (b) inward

Fig. 3.67 Web deformation in bending of channel section [74]

Table 3.8 Constants C and D in Eq. (3.13) for thickness deviation [3]

Material	C	D
A 1100	$0.092H_0^{-0.324}$	$0.158H_0^{-0.324}$
A 5056	$0.26H_0^{0.35}$	$0.30H_0^{0.11}$
Pure copper	$0.31H_0^{0.43}$	$0.37H_0^{0.26}$

thickness, and t_0 is the original thickness of the material. Sometimes, the thickening ratio $\lambda_i = (t_{max} - t_0)/t_0$ and the thinning ratio $\lambda_d = (t_0 - t_{min})/t_0$ are used. For uniform bending of circular tubes, the relationship between the thickness eccentricity ratio λ_t and the dimensionless curvature can be described by the following empirical equation.

$$\lambda_t = C \log_{10}\left(\frac{d_o}{2r_c}\right) + D \tag{3.13}$$

Here, r_c is the bending radius, and C and D are constants determined by the thickness ratio $H_0 = 2t_0/d_0$ and materials as shown in Table 3.8. The equation reveals that thinner sections and smaller bending radii result in larger thickness eccentricity ratios.

Furthermore, the calculation of the thickness strain in circular tubes, which matches well with experimental results, has been organized through finite element analysis for rotary draw bending [75].

3.6.2 *Fracture*

Fracture generally occurs in the tensile region on the outer side of the bend, except for special cases, such as square tubes accompanied by yield at relatively small bending radii or magnesium alloy tubes on the compression side [50]. In general, the fracture limit is determined by the uniform elongation limit based on tensile tests. In bending processes, the fracture limit due to bending is higher than the uniform elongation limit because there is a strain gradient between the compression and tension sides from the neutral axis. The outermost tensile strain ε_o in bending of circular tubes can be expressed as follows, assuming no flattening:

$$\varepsilon_o = \frac{0.5\, d_o + e}{r_c - e} \tag{3.14}$$

where d_0 represents outer diameter of circular tube, r_c represents bending radius, and e represents the distance between center axis and neutral axis. For the fracture limit of square tubes, we can consider d_0 as the height h_0 of the section in Eq. (3.14). Additionally, for rough estimations, it can be assumed that the neutral axis and center axis coincide ($e = 0$), and Eq. (3.14) becomes $\varepsilon_0 = d_0/2r_c$. Furthermore, in stretch bending where longitudinal tension is applied to circular tubes or shapes, the neutral axis shifts to the inner side (compression side), causing e to have a value. As a result, the strain on the tension side increases, leading to a decrease in the fracture limit. To prevent excessive strain, a processing method called crush bending is used, which involves flattening the tube to reduce the strain on the tension and compression sides before bending [76].

3.6.3 Folding

Yield of cross-section occurs in bending when flattening locally increases, taking place in circular tubes, square tubes, and other sections. It can also occur due to the formation of wrinkles on the inner side of the bend, resulting in a decrease in the cross-sectional second moment of the tube. Figure 3.68 shows typical examples of each case [3, 77]. The theoretical analysis for determining the limit of yield, where localized increase in flattening occurs, provides an approximate expression for the critical curvature radius r_b in uniform bending:

$$r_m/r_b = 4.8\,H_m^{2.0} n^{-0.3H_m^{-0.21}} \tag{3.15}$$

where the average radius of the circular tube is $r_m = (r_o + r_i)/2$, the thickness ratio is $H_\mathrm{m} = t_0/r_m$, the thickness is t_0 and n is the strain-hardening exponent. According to Eq. (3.15), yield is more likely to occur in thin-walled tubes and materials with a high strain-hardening exponent. Additionally, in cases of rotary bending for square tubes, it is known that yield is more likely to occur under severe bending conditions with smaller radii. Furthermore, it is common to apply a mandrel to suppress yield caused by flattening.

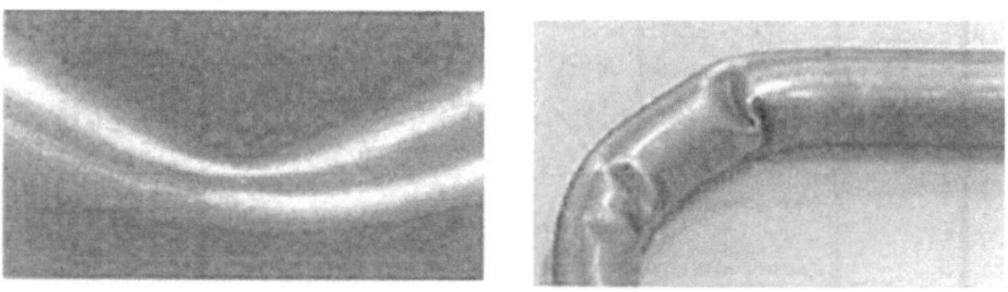

(a) folding caused by flattening [3] (b) folding caused by wrinkles [77]

Fig. 3.68 Examples of folding

3.6.4 Wrinkling

Wrinkling is a buckling phenomenon that occurs on the compression side of the bend. A theoretical analysis for uniform bending of circular tubes [6] provides an equation which expresses the bending radius r_w at which wrinkling initiates:

$$r_m/r_w = K \cdot H_m^{2.0} n^{-0.46} \tag{3.16}$$

where the average radius of the circular tube is $r_m = (r_o + r_i)/2$, the thickness ratio is $H_{\mathrm{m}} = (t_0/r_m)$, the thickness is t_{o}, and n is the strain-hardening exponent. Similar to yield, Eq. (3.16) indicates that thinner-walled tubes and materials with a higher strain-hardening exponent are more prone to wrinkling.

Since wrinkling is a buckling phenomenon, the buckling limit is generally determined by the boundary conditions, material length, cross-sectional second moment, and material properties, as can be understood from the Euler buckling load in the elastic range.

The minimum bending radii for rotary bending of circular and square tubes are shown in Tables 3.9, 3.10 and 3.11 [78]. There are also cases comparing finite element analysis and theoretical analysis for the wrinkling limit based on rotary bending [79]. Furthermore, there are instances where the bending limits using the push rotary bender shown in Fig. 3.10 are predicted by the following equation [41], considering the diameter of the circular tube d_0, thickness t_0, and bending radius r_c as processing conditions:

$$\kappa = {}^{r_c}/_{d_o} \cdot {}^{t_0}/_{d_o} \tag{3.17}$$

By varying the value of κ in the range of 0.005–0.800, predictions can be made regarding the application and type of mandrel, the application of a wiper die, and suitable material properties for bending. Smaller κ values correspond to thinner thicknesses and smaller bending radii, indicating more stringent requirements for the application of fixtures and material properties.

Table 3.9 Minimum bending radius of rotary draw bend for steel tube [78]

Outer diameter [mm]	Thickness [mm]	Minimum bending radius r_0/d_0			
		Without mandrel	Cylindrical mandrel	Spheric mandrel	Shoe and spheric mandrel
12.7–22.225	0.89	$6^1/_2$	$2^1/_2$	3	$1^1/_2$
	1.25	$5^1/_2$	2	$2^1/_2$	$1^1/_4$
	1.65	4	$1^1/_2$	$1^3/_4$	1
25.4–38.1	0.89	9	3	$4^1/_2$	2
	1.25	$7^1/_2$	$2^1/_2$	3	$1^3/_4$
	1.65	6	2	$2^1/_2$	$1^1/_2$
41.275–53.975	1.25	$8^1/_2$	$3^1/_2$	$4^1/_2$	$2^1/_4$
	1.65	7	3	$3^1/_2$	$1^3/_4$
	2.11	6	$2^1/_2$	3	$1^1/_2$
57.15–76.2	1.65	9	$3^1/_2$	4	$2^1/_2$
	2.11	8	3	$3^1/_2$	$2^1/_4$
	2.77	7	$2^1/_2$	3	2
88.9–101.6	2.11	9	$3^1/_2$	$4^1/_2$	3
	2.77	8	3	4	$2^1/_2$

Table 3.10 Minimum bending radius of rotary bending for square welded steel tube [50]

Side length	Thickness			
	2.11	1.65	1.24	0.89
12.7	41.28	44.45	47.63	50.0
19.05	50.8	50.8	63.5	76.2
25.4	76.2	76.2	88.9	101.6
28.58	76.2	76.2	88.9	101.6
31.75	88.9	88.9	101.6	–
38.10	114.3	114.3	127.0	–
44.45	152.4	165.1	177.8	–
50.80	177.8	215.9	228.6	–
63.5	228.6	266.7	–	–
76.2	304.8	381.0	–	–

[mm]

Table 3.11 Minimum bending radius of rotary bending for steel tube [78]

Outer diameter [mm]	Thickness [mm]	Minimum bending radius r_0/d_0			
		With mandrel		w/o mandrel	
		Bending angle		Bending angle	
		90°	180°	90°	180°
9.53	0.71	2	2.3	3.3	6.7
	0.81	2	2.3	3.3	4.7
	0.89	1.7	2.0	3.0	4.7
	1.25	1.5	2.0	2.7	4.0
12.70	0.81	2	2.25	3.0	4.0
	0.89	1.8	2.0	3.0	4.0
	1.25	1.5	1.8	2.8	3.5
	1.65	1.5	1.8	2.5	3.5
15.88	0.89	1.6	2.0	3.2	4.0
	1.25	1.4	2.0	2.8	4.0
	1.65	1.2	1.8	2.4	3.6
19.05	0.89	1.8	2.0	3.0	4.0
	1.25	1.7	1.8	3.0	4.0
	1.65	1.5	1.7	2.7	3.3
	2.12	1.3	1.5	2.7	3.3
22.23	0.89	1.9	2.0	2.9	4.0
	1.25	1.6	1.9	2.9	3.7
	1.65	1.4	1.8	2.6	3.7
	2.12	1.3	1.6	2.6	3.4
25.40	1.25	1.6	1.9	3.0	4.5
	1.65	1.5	1.8	3.0	4.0
	2.12	1.25	1.8	2.75	4.0
	2.77	1.25	1.65	2.75	3.75
28.58	1.25	1.7	2.1	3.1	4.0
	1.65	1.6	2.0	2.8	4.0
	2.12	1.4	1.0	2.7	3.8
	2.77	1.4	1.8	2.7	3.8
	3.05	1.5	1.8	2.7	3.8
31.75	1.25	1.8	2.1	3.2	4.0
	1.65	1.7	2.0	3.0	3.8
	2.12	1.5	1.8	3.0	3.6
	2.77	1.4	1.7	2.8	3.6
	3.05	1.4	1.7	2.8	3.6

(continued)

Table 3.11 (continued)

Outer diameter [mm]	Thickness [mm]	Minimum bending radius r_0/d_0			
		With mandrel		w/o mandrel	
		Bending angle		Bending angle	
		90°	180°	90°	180°
34.93	1.25	1.8	2.1	3.3	4.4
	1.65	1.7	2.0	3.1	4.0
	2.12	1.6	1.9	3.0	4.0
	2.77	1.5	1.7	2.9	3.6
	3.05	1.5	1.6	2.7	3.6
38.10	1.25	1.8	2.1	3.2	4.7
	1.65	1.75	1.9	3.0	4.3
	2.12	1.5	1.8	2.8	4.3
	2.77	1.4	1.7	2.8	4.0
	3.05	1.4	1.6	2.7	4.0
44.45	1.25	2.0	2.1	3.1	4.6
	1.65	1.7	1.9	3.0	4.3
	2.12	1.5	1.6	3.0	4.3
	2.77	1.3	1.6	2.9	4.0
	3.05	1.2	1.5	2.9	4.0
50.80	1.25	2.0	2.1	3.25	4.5
	1.65	1.75	1.9	3.0	4.25
	2.12	1.75	1.9	3.0	4.25
	2.77	1.7	1.8	3.0	4.0
	3.05	1.6	1.75	2.9	4.0
57.15	1.25	1.9	2.1	4.4	5.3
	1.65	1.8	1.9	3.1	4.2
	2.12	1.8	1.9	3.0	4.2
	2.77	1.7	1.8	3.0	4.0
	3.05	1.7	1.8	2.9	4.0
63.5	1.65	1.8	1.9	3.2	4.0
	2.12	1.75	1.9	3.0	4.0
	3.05	1.7	1.8	2.8	3.8
76.2	1.65	1.8	2.0	–	–
	2.12	1.75	1.8	3.0	–
	3.05	1.7	1.75	3.0	–

(continued)

Table 3.11 (continued)

Outer diameter [mm]	Thickness [mm]	Minimum bending radius r_0/d_0			
		With mandrel		w/o mandrel	
		Bending angle		Bending angle	
		90°	180°	90°	180°
88.9	1.65	2.1	2.3	–	–
	2.12	2.0	2.2	–	–
	3.05	2.0	2.1	–	–
101.6	1.65	2.25	2.4	–	–
	2.12	2.1	2.25	–	–
	3.05	2.0	2.1	–	–
127	1.65	2.5	2.5	–	–
	2.12	2.4	2.45	–	–
	3.05	2.3	2.4	–	–

3.7 Processing Examples and Others

In the medical and measurement/analysis industries, there is a demand for products with small bending radii in three-dimensional bending and sub-mm scale diameters, as shown in Fig. 3.69.[6] Furthermore, in industries such as aerospace, aviation, and shipbuilding, there is a growing demand for bending products aimed at reducing weight, including waveguides used as radar components in addition to structural components.

In recent years, materials that have been applied to bending have included aluminum alloy extrusions, ERW high-strength steel tubes [12], and AZ31 magnesium alloy extrusions [59], primarily in the automotive industry, aiming for

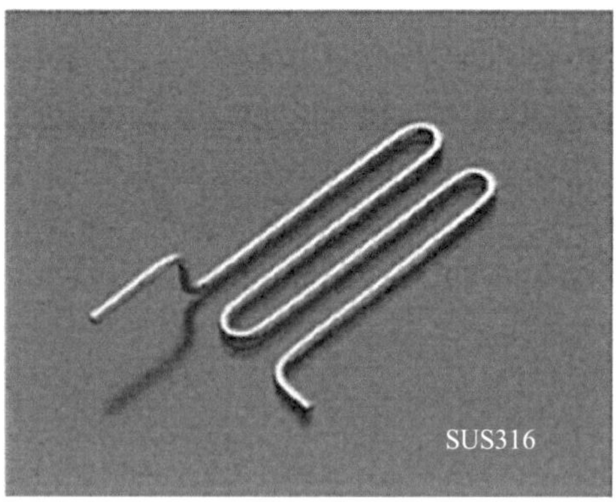

Fig. 3.69 Three-dimensional bending of small-diameter tube (see Footnote 6)

[6] http://www.tsudass.jp/ (as of February 2019).

Fig. 3.70 Three-dimensional bending of small-rectangular tube (see Footnote 7)

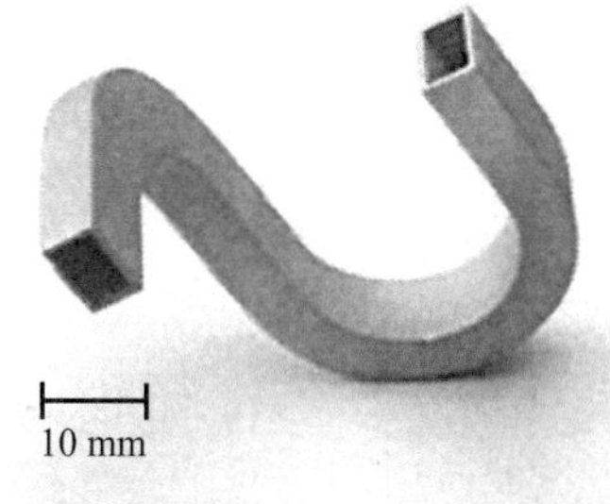

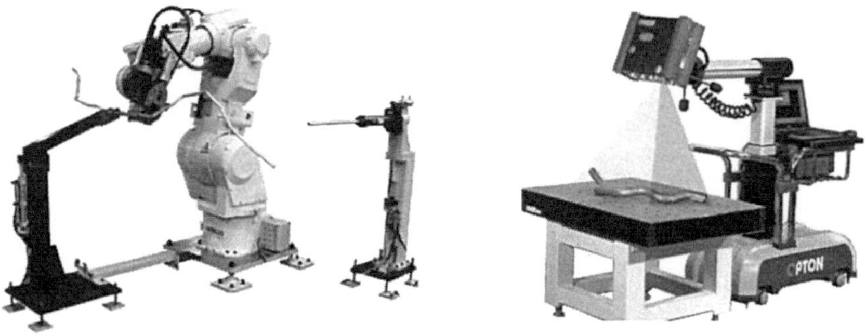

Fig. 3.71 Robot bender and bent geometry measurement system [8]

lightweight and high rigidity. Figure 3.70 illustrates an example of three-dimensional bending of aluminum alloy extruded square tubes.[7] Additionally, tubular sections with ribs or fins have been applied in heat exchangers to enhance rigidity and cooling efficiency [80].

Bending facilities are equipped with arm-type robots for material handling and applying bending moments in addition to CNC control. Moreover, with the remarkable advances in measurement technology using cameras, sensors, and other devices, as shown in Fig. 3.71, precise 3D dimensions, including bending angles, can be obtained, allowing for real-time adjustments during bending.[8]

---

[7] http://www.yuhara.co.jp/ (as of February 2019).

[8] https://opton.co.jp/index.html/ (as of February 2019).

References

1. Alwood, J.M., Duncan, S.R., Cao, J., Groche, P., Hirt, G., Kinsey, B., Kuboki, T., Liewald, M., Sterzing, A., Tekkaya, A.B.: CIRP Ann. Manuf. Technol. **65**, 573–596 (2016)
2. Nakamura, M., Sasaki, S.: Press Working, vol. 12-8, pp. 99–104. Nikkan Kogyo Shimbun (1974)
3. Murota, T., Endo, J.: J. JSTP **23–255**, 343–350 (1982)
4. Sato, K., Takahashi, M.: J. JSTP **23–252**, 17–22 (1982)
5. Fuchs, F.J., Jr.: Bel. Syst. Tech. J. **38**, 1457–1484 (1959)
6. Endo, J., Murota, T.: J. JSTP **18–202**, 930–937 (1977)
7. Endo, J., Murota, T.: J. JSTP **23–258**, 708–713 (1982)
8. Yatsubihsi Co.Ltd., Catalog (2002)
9. Yoshimoto, S.: Technical Report (2017), Benkan Kikoh Corporation
10. Chiyoda Industries Co. Ltd.: Catalog (1995)
11. Kato, K.: Preprints of 139th Tube Forming Committee Research Meeting, pp. 9–15. JSTP (2015)
12. Suzuki, T., Toyoda, S., Sato, A., Ueno, Y., Okada, M.: JFE Tech. Report **17**, 52–58 (2007)
13. Takahashi, K., Kuboki, T., Murata, M., Yano, K.: Proceedings of the 60th Joint Conference on the Technology of Plasticity, pp. 377–378. JSTP (2009)
14. Yuhara, M., Hoshino, M., Wada, O., Iso, Y., Jin, M.: J. JSTP **53–618**, 646–650 (2012)
15. Chuma, K., Nishio, K., Kurobe, A.: Proceedings of the 2011 Spring Meeting, pp. 5–56. JSTP (2011)
16. Okada, M.: Seminar materials on technology of plasticity, practical techniques and key points of bending. JSTP (2009)
17. Benkan Kikoh Corporation: “Butt Welding Fittings” catalog (2017)
18. Murata, M.: J. JSTP **54–635**, 39–40 (2013)
19. Murata, M., Ohashi, S., Suzuki, H.: Transactions of the Japan society of mechanical engineers. C 55–517, 2488–2492 (1989)
20. Murata, M., Yamamoto, O., Suzuki, H.: J. JSTP **35–398**, 262–267 (1994)
21. Kano, Y.: Proceedings of the 106th Symposium on the Technology of Plasticity, pp. 35–39. JSTP (1986)
22. Nippon Steel Sumikin Kikoh: Jpn. Pat. Appl. Publ. No. 206–247664
23. Nakamura, M., Maki, S., Nakajima, M., Nakamura, Y.: Proceedings of the 1989 Spring Meeting, pp. 287–290. JSTP (1989)
24. Nakamura, M., Maki, S., Harada, Y., Hayashi, K., Nakajima, M.: J. JSTP **37–424**, 540–545 (1996)
25. Nakamura, M., Maki, S., Koyama, H.: J. JSTP **46–538**, 1044–1049 (2005)
26. Nakamura, M., Asai, M., Maki, S., Harda, Y., Takeuchi, K.: Proceedings of the 1989 Spring Meeting, pp. 269–270. JSTP (1989)
27. Aida, S., Shirakawa, M., Yamazaki, E.: Report of the Industrial Research Institute of Niigata Prefecture, vol. 38, pp. 10–17 (2009)
28. Tanaka, M., Koike, M., Watanabe, M., Michino, M., Sano, K., Narita, M.: Proceedings of the 42nd Joint Conference on the Technology of Plasticity II, pp. 719–722. JSTP (1991)
29. Sekido, Y.: Jpn. Pat. No. 3000017
30. Kato, K.: Preprints of 99th Tube Forming Committee Research Meeting, pp. 5–10. JSTP (2002)
31. Kato, K.: Seminar Materials on Technology of Plasticity, Practical Techniques and Key Points of Bending, pp. 11–12. JSTP (1998)
32. Tomizawa, A., Matsuda, H., Mori, H., Hara, S., Kuwayama, S., Suyama. T., Kinoshita, Y.: Proceedings of the 63rd Joint Conference on the Technology of Plasticity, pp. 171–172. JSTP (2012)
33. Kubota, H., Tomizawa, A., Okada, N., Yamamoto, K., Hama, T., Takuda, H.: J. JSTP **57–668**, 879–885 (2016)
34. Tomizawa, A., Shimada, N., Kubota, H., et al.: Nippon Steel Technical Report, vol. 397, pp. 83–89 (2013)

35. The Japan Society for Technology of Plasticity (Ed.): Technology of Plasticity terminology dictionary, vol. 143, Corona Publishing (2011)
36. Ueno, K., Ueda, M., Kobayashi, M.: J. JSTP **30–336**, 26–30 (1989)
37. The Japan Society for Technology of Plasticity (Ed.): Technology of Plasticity Handbook, vol. 583. Corona Publishing (2006)
38. Yamashita., I., Murota, T., Endo, J.: J. JSTP **28–313**, 172–179 (1987)
39. Endo, J., Murota, T., Yamashita, I.: J. JSTP **31–352**, 664–670 (1990)
40. Endo, J., Murota, T.: J. Jpn. Inst. Light Metals **5–42**, 257–262 (1992)
41. Taiyo Corporation, Pipe Bender catalog (2006)
42. Sakaki, S., Utsumi, N.: J. JSTP **52–604**, 574–578 (2011)
43. Okude, Y., Sakaki, S., Yoshihara, S.: J. JSTP **53–614**, 241–245 (2012)
44. Okude, Y.: Doctoral Thesis (Yamanashi University), (2013)
45. Okude, Y., Sakaki, S., Yoshihara, S.: J. Jpn. Inst. Light Metals **61–9**, 435–439 (2011)
46. Ueda, M., Ueno, K., Kagamida, M., Kobayashi, M.: J. JSTP **22–248**, 904–911 (1981)
47. Ueda, M., Ueno, K., Kobayashi, M.: J. JSTP **25–284**, 793–798 (1984)
48. Yamada, O., Murakami, H., Takasaki, M.: J. JSTP **30–336**, 37–42 (1989)
49. Ueno, Y.: Proceedings of the 25th Joint Conference on the Technology of Plasticity, pp. 237–238. JSTP (1974)
50. Kervich, R.J., Springborn, R.K.: Cold bending and tube forming and other sections. ASTME (1966)
51. Hasegawa, O., Manabe, K., Nishimura, H.: J. Jpn. Inst. Light Metals **57–6**, 245–249 (2007)
52. Endo, J., Nishimura, H., Manabe, K.: J. JSTP **19–212**, 742–749 (1978)
53. Oehler, G.: Tech. Int. **56**, 51 (1967)
54. Endo, J., Murota, T.: J. JSTP **27–301**, 314–316 (1986)
55. Tanabe, H., Miyasaka, A., Yamazaki, K., Tokunaga, Y.: J. JSTP **36–413**, 651 (1995)
56. Sakaki, S., Utsumi, N.: J. JSTP **46–529**, 160–164 (2005)
57. Hasegawa, O., Nishimura, H.: J. Jpn. Inst. Light Metals **46–5**, 243–248 (1996)
58. Working Group on Forming Methods of Extruded Profiles. Research Committee Report. The Japan Institute of Light Metals, vol. 32 (1996)
59. Hasegawa, O., Manabe, K., Inoue, N., Nishimura, H.: J. JSTP **48–556**, 422–426 (2007)
60. Franz, W.D.: Das Kalt-Biegen von Rohren, (1961), Springer-Verlag.
61. Aizu, S: Press Working, vol. 6–4, pp. 33–36, Nikkan Kogyo Shimbun (1968)
62. Takahashi, K., Watanabe, S., Kuboki, T., Murata, M., Ono, K., Yano, K. 塑性と加工, vol. 49–572, pp. 896–900 (2008)
63. Utsumi, N., Yoshida, M., Sakaki, S.: Bulletin of the faculty of education. Saitama Univ. **60–1**, 119–124 (2011)
64. Hasegawa, O., Soya, H., Nishimura, H.: Proceedings of the 9th Spring Meeting, the Japan Institute of Light Metals, pp. 261–262 (1999)
65. Kaida, K., Hirano, M., Fujii, T., Yoshida, M.: Kobe Steel Engineering Reports, vol. 47-2, pp. 17–20 (1997)
66. Murasato, Y., Teramae, T., Makiyama, T.: J. JSTP **57–668**, 886–891 (2016)
67. Kuboki, M., Furugen, M.: J. JSTP **42–491**, 1243–1247 (2001)
68. Ochiai, I.: Press Working, vol. 17-7, pp. 60–63, Nikkan Kogyo Shimbun (1979)
69. Koyama, H., Sakano, Y., Kobayashi, K.: Proceedings of the 66th Joint Conference on the Technology of Plasticity, pp. 327–328. JSTP (2015)
70. Takahashi, K., Watanabe, T., Kuboki, T., Murata, M., Ono, K., Yano, K.: J. JSTP **49–572**, 896–900 (2008)
71. Nakajima, K., Utsumi, N., Yoshida, M.: Int. J. Precis. Eng. Manuf. **14–6**, 965–970 (2013)
72. Aizu, S: Press Working, vol. 4–8, pp. 75–80, Nikkan Kogyo Shimbun (1966)
73. Nozu, K.: Proceedings of the 68th Joint Conference on the Technology of Plasticity, pp. 401–402. JSTP (2017)
74. Sakaki, S., Utsumi, N., Taguchi, K., Hasegawa, O.: J. Jpn. Inst. Light Metals **49–9**, 426–431 (1999)

75. Mizumura, M., Kuriyama, Y.: Proceedings of the 2002 Spring Meeting, pp. 247–248. JSTP (2007)
76. Nakamura, M., Maruyama, K., et al.: Pipe forming techniques, 2nd edn, pp. 118–119. Nikkan Kogyo Shimbun (1998)
77. Sakaki, S.: J. JSTP **50–581**, 466–470 (2009)
78. Schubert, P. B.: Pipe and Tube Bending. The Industrial Press (1953)
79. Yoshida, M., Fujiwara, A.: J. JSTP **41–468**, 74–78 (2000)
80. Kudo, D., Hoshino, T., Yamadate, T.: Proceedings of the 64th Joint Conference on the Technology of Plasticity, pp. 127–128. JSTP (2013)

Chapter 4
Hydroforming

4.1 Basics

4.1.1 Introduction

This chapter corresponds to the forming method previously referred to as bulge forming in the previous edition. However, in this book, we deliberately use the term "hydroforming". Bulge forming is a forming method that involves applying pressure inside a tube to expand it. Its history dates back to as early as 1940 when a patent for manufacturing copper tube T-joints was already registered in the United States [1]. In Japan, experimental research and development on various methods for manufacturing pipe joints were conducted by the Industrial Institute of Nagoya of the Agency of Industrial Science and Technology and the Bicycle Technical Center of the Japan Bicycle Promotion Institute in the late 1950s [2–7]. Subsequently, in the 1960s, industrialization of various pipe fittings began, and they have been widely used primarily for piping joints [8] and bicycle joints [9]. The term "hydroforming" began to be used to refer to bulge forming in the 1990s, particularly in the field of automotive parts [10, 11]. At that time, two global challenges surrounding automobiles emerged: improving crash safety and weight reduction of automobiles to prevent global warming [12]. Hollow closed-section structures attracted attention as one means of solving these two conflicting challenges due to their advantages in terms of strength and rigidity compared to open-section structures [13]. However, in order to accommodate them within the confined structure of automobiles, it was necessary to process the tubes into complex shapes. Therefore, hydroforming has evolved to enable the processing of considerably more complex shapes than the previous era of bulge forming by incorporating advanced processing machinery and control technologies using recent computer technology.

Hydroforming, which began to be applied primarily in Europe and North America, gained worldwide popularity following the Ultra Light Steel Auto Body (ULSAB) project initiated by the International Iron and Steel Institute in 1994 [14]. In Japan,

Y. Kuriyama et al., *Tube Forming*, https://doi.org/10.1007/978-981-95-3421-0_4

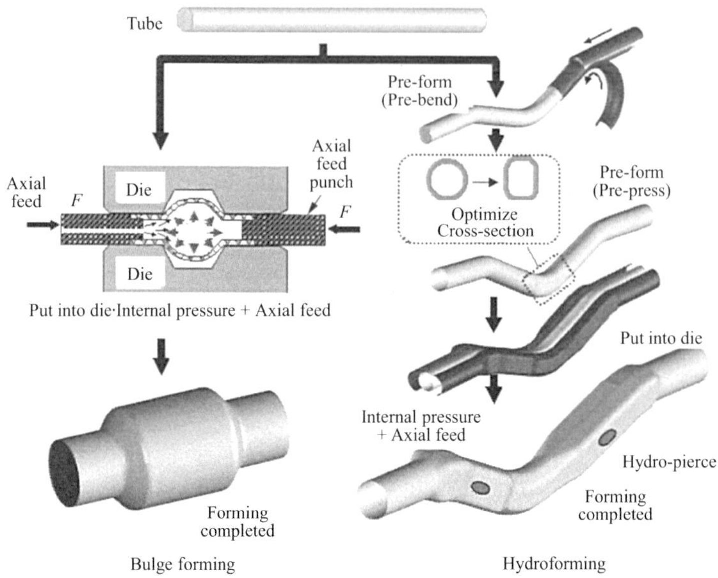

Fig. 4.1 Comparison of bulge forming and hydroforming [17]

its application to actual components started in 1999 [15]. Automobile parts to which hydroforming is applied often have complex cross-sections and even complex shapes in the axial direction of the tube. Therefore, they often undergo pre-processing such as bending and crushing before being subjected to internal pressure. This pre-processing is referred to as pre-forming. Additionally, it is possible to create holes in the tube while it is under internal pressure. The post-processing, such as piercing holes, is referred to as post-forming. The overall process, including these pre-forming and post-forming operations, is sometimes referred to as hydroforming in a broad sense [16]. A comparison between traditional bulge forming and hydroforming is summarized in Fig. 4.1 [17]. As shown, hydroforming has a significantly broader range and higher level of complexity compared to bulge forming in the previous edition. This chapter will explain hydroforming in detail based on these advancements.

4.1.2 Theory

As mentioned in the previous section, the complex shapes of hydroformed parts applied to automotive components make it difficult to perform precise theoretical analysis. However, the fundamental deformation can be approximated as free bulge. Therefore, we will introduce the basic theory of free bulge [18]. Understanding the stress, strain, expansion limit, and burst pressure that occur in the tube during free bulge enhances our understanding of hydroforming and can be applied to the

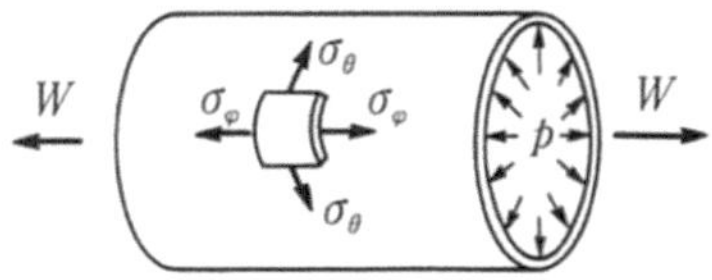

Fig. 4.2 Infinite length tube subject to internal pressure and axial force (closed ends)

development of processing methods and component design. As mentioned in the classification of processing methods discussed later, free bulge is a process in which the tube is freely expanded without using a die on the outside of the tube. Here, we consider an infinitely long cylindrical tube subjected to internal pressure p and axial load W (positive for tension and negative for compression) with both ends closed by caps (Fig. 4.2). It should be noted that in hydroforming, pushing the tube into the die from the end is referred to as axial feed. In this context, axial load refers to the load in the axial direction of the tube at the cross-section. The thickness t is assumed to be sufficiently thin compared to the radius R (= $D/2$, where D is the diameter). The deformation is assumed to occur uniformly along the axis of the tube. Based on these assumptions, it can be assumed that there is no bending moment acting on the tube. The material of the tube is assumed to be isotropic and follows the power hardening rule as follows:

$$\sigma_{eq} = C\varepsilon_{eq}^{n} \tag{4.1}$$

where σ_{eq} and $\varepsilon_{\mathrm{eq}}$ are the equivalent stress and equivalent strain, C is the plastic coefficient, and n is the strain-hardening exponent.

Let R_0 and R be the initial and final radii of the tube, and t_0 and t be the initial and final thicknesses, respectively. From the axial force equilibrium, the axial stress σ_φ can be expressed as in Eq. (4.2).

$$\sigma_\varphi = \frac{pR}{2t} + \frac{W}{2\pi Rt} \tag{4.2}$$

From the circumferential force equilibrium, the circumferential stress σ_θ can be expressed as in Eq. (4.3).

$$\sigma_\theta = \frac{pR}{t} \tag{4.3}$$

From Eqs. (4.2) and (4.3), Eq. (4.4) can be derived.

$$\frac{\sigma_\varphi}{\sigma_\theta} = \frac{1}{2} + \frac{W}{2p\pi R^2} \tag{4.4}$$

The thickness stress $\sigma_t = -p$ on the inner surface of the tube and $\sigma_t = 0$ on the outer surface of the tube. Taking the average value as the value at the center of the thickness, $\sigma_t = -p/2$. . Based on the assumption of thin walled ($t \ll R$), we have $|\sigma_t/\sigma_\theta| = t/2R \ll 1$, allowing us to neglect σ_t compared to σ_θ and σ_φ. Therefore, Eq. (4.5) can be obtained.

$$\sigma_t \fallingdotseq 0 \tag{4.5}$$

From this, it can be considered that the cylindrical tube is in a state of plane stress (biaxial stress state).

The strain is defined as follows:

$$\text{Circumferential strain } \varepsilon_\theta = ln(R/R_0) \tag{4.6}$$

$$\text{Thickness strain } \varepsilon_t = ln(t/t_0) \tag{4.7}$$

$$\text{Volume constant condition } \varepsilon_\varphi + \varepsilon_\theta + \varepsilon_t = 0 \tag{4.8}$$

$$\text{Axial strain } \varepsilon_\varphi = -\varepsilon_\theta - \varepsilon_t \tag{4.9}$$

(1) **Load with internal pressure only**

Since $W = 0$, Eqs. (4.2)–(4.4) lead to Eq. (4.10).

$$\sigma_\varphi = \frac{pR}{2t}, \sigma_\theta = \frac{pR}{t}, \frac{\sigma_\varphi}{\sigma_\theta} = \frac{1}{2} \tag{4.10}$$

The stress–strain relationship in a plane stress field is given by Eq. (4.11).

$$\varepsilon_\varphi = \left(\frac{\varepsilon_{eq}}{\sigma_{eq}}\right)\left(\sigma_\varphi - \frac{\sigma_\theta}{2}\right), \varepsilon_\theta = \left(\frac{\varepsilon_{eq}}{\sigma_{eq}}\right)\left(\sigma_\theta - \frac{\sigma_\varphi}{2}\right) \tag{4.11}$$

The equivalent stress and equivalent strain are expressed as follows when the tube material follows the yield condition of von Mises:

$$\sigma_{eq} = \sqrt{\sigma_\varphi^2 - \sigma_\varphi\sigma_\theta + \sigma_\theta^2}, \varepsilon_{eq} = \frac{2}{\sqrt{3}}\sqrt{\varepsilon_\varphi^2 + \varepsilon_\varphi\varepsilon_\theta + \varepsilon_\theta^2} \tag{4.12}$$

From Eqs. (4.10) and (4.11), Eq. (4.13) can be derived.

$$\varepsilon_\varphi = 0 \tag{4.13}$$

From Eqs. (4.5) and (4.13), it can be understood that an infinitely long cylindrical tube with both ends closed, subjected to internal pressure only, is in a state of plane stress and a plane strain (biaxial strain) state. From Eqs. (4.8) and (4.13), Eq. (4.14) can be obtained.

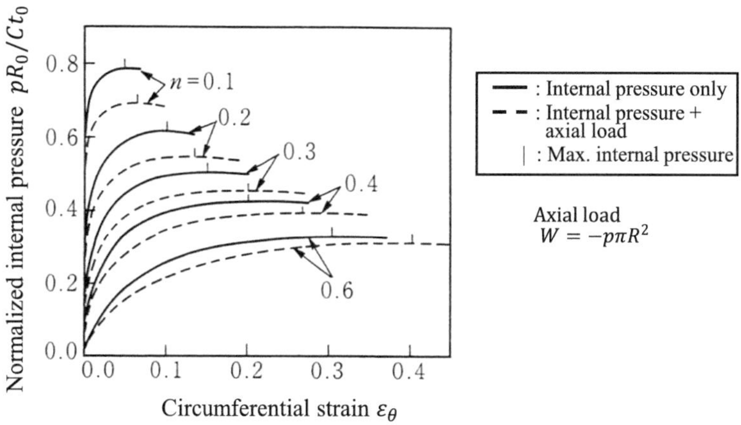

Fig. 4.3 Relation of normalized internal pressure and circumferential strain (infinite length circular tube)

$$\varepsilon_t = -\varepsilon_\theta \tag{4.14}$$

From Eq. (4.14) and Eqs. (4.6) and (4.7), it follows that $t = R_0\ t_0/R$, which means that the thickness decreases in inverse proportion to the bulge radius. The relationship between internal pressure and bulge amount (circumferential strain) is given by Eq. (4.15).

$$p = \frac{Ct_0}{R_0}\left(\frac{2}{\sqrt{3}}\right)^{n+1}\varepsilon_\theta^n \cdot \exp(-2\varepsilon_\theta) \tag{4.15}$$

Figure 4.3 shows the relationship (solid line) between the normalized internal pressure pR_0/Ct_0 and circumferential strain ε_θ. The maximum internal pressure p^* is given by Eq. (4.16).

$$p^* = \frac{Ct_0}{R_0}\frac{2}{\sqrt{3}^{n+1}}\left(\frac{n}{e}\right)^n \tag{4.16}$$

From this equation, knowing the initial dimensions of the tube R_0, t_0, and the material properties n and C allows us to determine the maximum internal pressure. Figure 4.4 shows the relationship (solid line) between the normalized maximum internal pressure $p^*R_0\ /\ C\ t_0$ and the n-value. The circumferential strain and radius at the maximum internal pressure are given by Eq. (4.17).

$$\varepsilon_\theta^* = \frac{n}{2},\ R^* = R_0 \cdot \exp\left(\frac{n}{2}\right) \tag{4.17}$$

From Eq. (4.17), it can be observed that the circumferential strain (or bulge radius) at the maximum internal pressure increases with higher n-values. However, it should

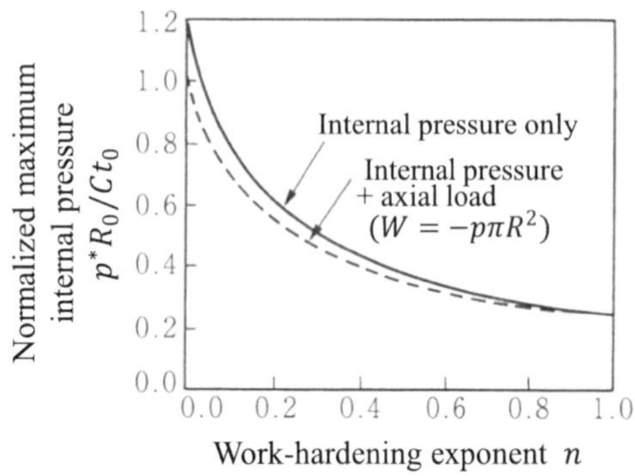

Fig. 4.4 Effect of n-value on maximum internal pressure (infinite length circular tube) [18]

be noted that this value is only half the value of the strain ($\varepsilon = n$) at the maximum load in uniaxial tensile test along the axial direction of the tube.

(2) **Simultaneous loading with internal pressure p and axial load** $W = -prR^2$

From Eq. (4.2), Eq. (4.18) can be derived.

$$\sigma_\varphi = 0 \tag{4.18}$$

Since σ_t is also 0 from Eq. (4.5), the tube is in a uniaxial stress state in the circumferential direction where only σ_θ acts. From Eqs. (4.8) and (4.11), Eq. (4.19) is obtained.

$$\varepsilon_\varphi = \varepsilon_t = -\frac{1}{2}\varepsilon_\theta \tag{4.19}$$

The relationship between internal pressure and circumferential strain is given by Eq. (4.20). Figure 4.3 shows the relationship (dashed line) between the normalized internal pressure pR_0/Ct_0 and circumferential strain ε_θ in this case. The maximum internal pressure is given by Eq. (4.21).

$$p = \frac{Ct_0}{R_0}\varepsilon_\theta^n \cdot \exp\left(-\frac{3}{2}\varepsilon_\theta\right) \tag{4.20}$$

$$p^* = \frac{Ct_0}{R_0}\left(\frac{2n}{3e}\right)^n \tag{4.21}$$

Figure 4.4 shows the relationship (dashed line) between the normalized maximum internal pressure p^*R_0/Ct_0 and the n-value in this case. The circumferential strain and radius at the maximum internal pressure are given by Eq. (4.22).

$$\varepsilon_\theta^* = \frac{2n}{3},\ R^* = R_0 \cdot \exp\left(\frac{2n}{3}\right) \tag{4.22}$$

Comparing this equation with Eq. (4.17) when only internal pressure is applied, it can be observed that the circumferential strain (bulge radius) at the maximum internal pressure is larger when the axial load $W = -prR^2$ is applied simultaneously with the internal pressure. The above describes the fundamental theory of free bulge in an infinitely long cylindrical tube. However, the theory becomes more complicated for cases where the length of the tube is finite or for cases involving die bulge using dies, so they are omitted in this book. For explanations of those theories, please refer to "Tube Hydroforming" edited by the Japan Society for Technology of Plasticity [18].

4.2 Processing Methods

Chapter 3 discussed various bending processes, including their theory and tooling. In contrast, hydroforming, in principle, involves the common process of expanding a tube by applying internal pressure. As a result, the number of hydroforming methods is not as extensive as bending processes. However, due to the variation in conditions, it is necessary to classify these methods. This section provides an explanation of the classification of hydroforming based on the following four perspectives:

1. Classification based on deformation constraint.
2. Classification based on loading conditions.
3. Classification based on processing temperature.
4. Classification based on pressure medium.

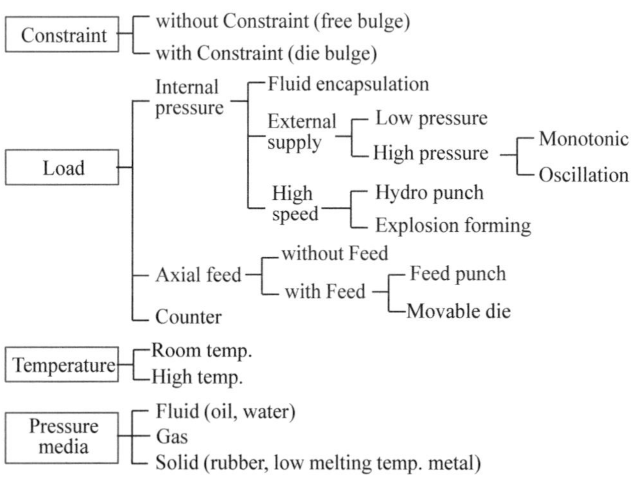

Fig. 4.5 Classification of hydroforming

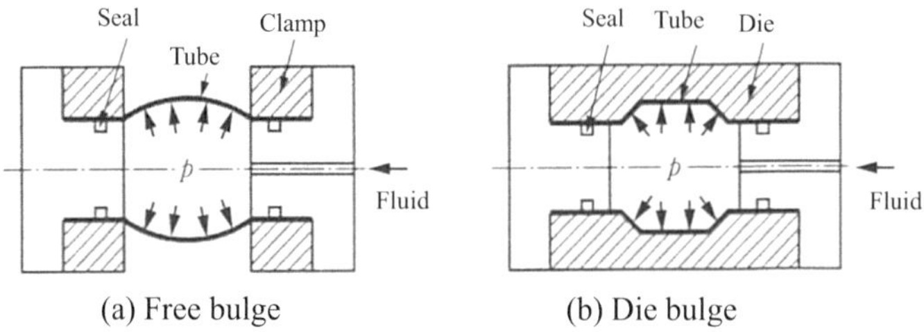

Fig. 4.6 Free bulge and die bulge

Figure 4.5 illustrates the classification of hydroforming based on these perspectives. Subsequently, the classification of the methods will be explained according to each perspective.

4.2.1 Classification Based on Deformation Constraint

(1) **Free bulge**

Free bulge is a process in which the tube is deformed freely without the use of a die on the outer side of the tube, as shown in Fig. 4.6a. This method minimizes the constraint imposed by the die and reduces the influence of friction during deformation. It is rarely used in the production of actual products but is employed to examine stress and strain states in the tube as well as the deformation limits of the material.

(2) **Die bulge**

Die bulge is a process in which the tube, placed inside a die, is expanded by applying internal pressure or axial thrust to achieve deformation according to the shape of the die, as shown in Fig. 4.6b. This method is commonly used in the production of actual products, and the subsequent explanation of various classification methods assumes the die bulge process. It should be noted that in the die bulge process, the outer surface of the tube comes into contact with the inner surface of the die, introducing friction that significantly affects formability. Further details will be discussed later.

4.2.2 Classification Based on Load Conditions

As a typical example of hydroforming, Fig. 4.7 illustrates the die structure for forming a tube into a T-shape. First, the tube is mounted on the lower die, and then the upper die descends to clamp the tube. Additionally, to prevent the die opening, a clamping force is applied, which does not significantly affect the deformation of the tube during the process if the clamping force is sufficiently large. The two main load factors acting on

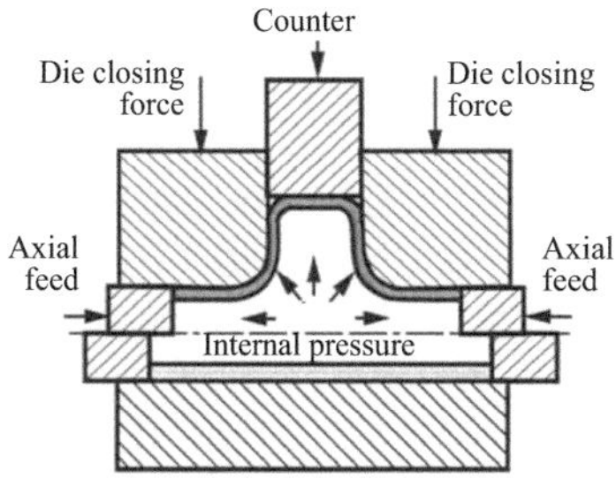

Fig. 4.7 Die structure and loads in T-forming [18]

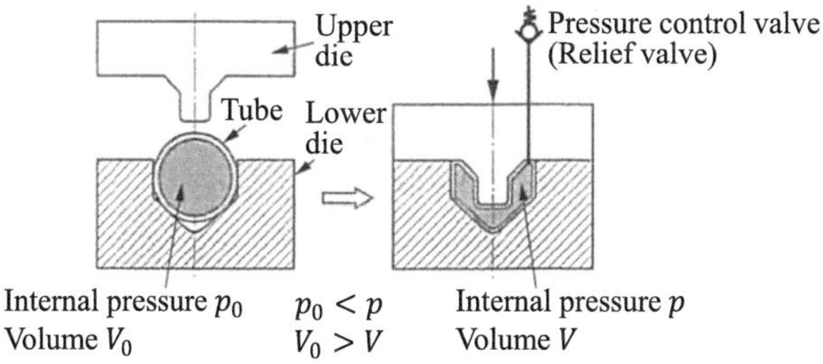

Fig. 4.8 Sealed fluid forming [19]

the tube deformation are the internal pressure applied from the inside of the tube and the axial feed exerted by the punch from the tube ends. Furthermore, in cases where a partial die is retracted during the forming process, as in the T-forming shown in Fig. 4.7, it is called "counter". Subsequently, various classifications will be explained for each load factor: internal pressure, axial feed, and counter.

(1) **Internal pressure**
(a) **Internal pressure generated inside (sealed fluid forming)**

This method involves reducing the volume of the tube by applying external force to crush the tube while it contains a sealed liquid, thereby increasing the internal pressure of the tube. It is also known as sealed fluid forming and is used as a convenient method that does not require hydraulic pumps. Although precise control of the internal pressure is not possible, by incorporating a relief valve and adjusting it to a constant pressure or below, the risk of excessive internal pressure leading to tube burst can be prevented. Figure 4.8 shows a schematic illustration [19].

(b) **Internal pressure by external supply**

This method involves increasing the internal pressure using a pump or hydraulic cylinder, allowing for precise control of the internal pressure. If high pressure is required, a pressure intensifier is used. Furthermore, the magnitude of the internal pressure by external supply distinguishes between high-pressure and low-pressure methods [15]. This classification is relative, with examples categorizing pressure above approximately 100 MPa as high-pressure and pressure below as low-pressure. The processing methods also differ slightly. Figure 4.9 shows a comparison of the forming processes, and Fig. 4.10 illustrates a comparison of cycle times. In the high-pressure method, the internal pressure is applied when die is fully closed, while in the low-pressure method, pressure is applied before the die is completely closed.

In hydroforming, the corners are formed into sharp radii by increasing internal pressure. In the low-pressure method, pressure is applied before the die fully closes, allowing for some corner shaping by the time the die is closed. Therefore, it becomes possible to achieve sharp corners even with relatively low internal pressure. Additionally, because of the low internal pressure, cycle times can be shorter, and there is also the advantage of reduced investment in machinery. However, low-pressure methods are challenging for forming complex sections or shapes with significant changes in perimeter. In such cases, high-pressure methods are more suitable. In the 1990s, when hydroforming was first applied to automotive parts, low-pressure

Fig. 4.9 High-pressure method and low-pressure method of hydroforming [15]

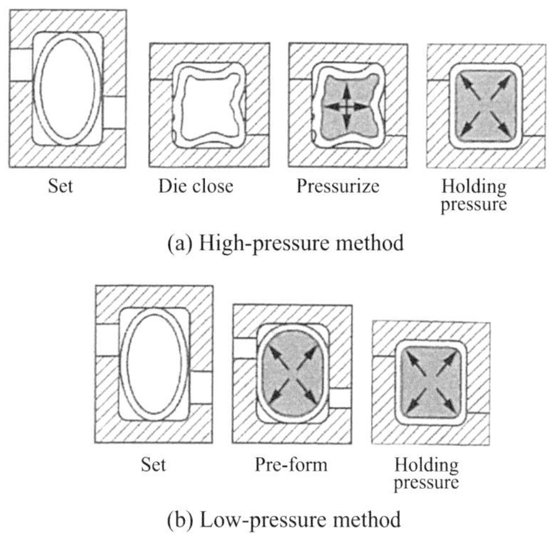

Fig. 4.10 Process diagram of high-pressure and low-pressure methods [15]

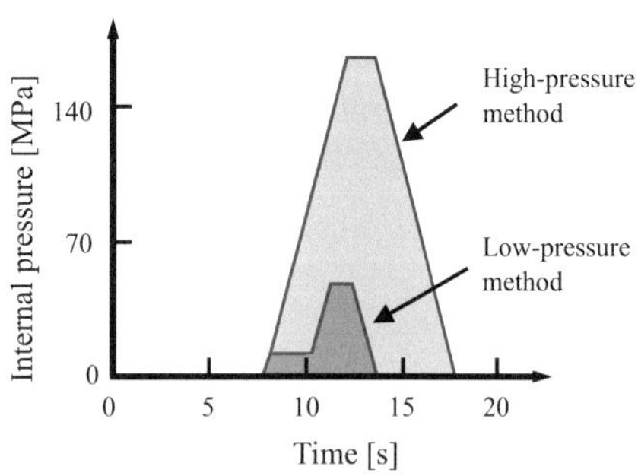

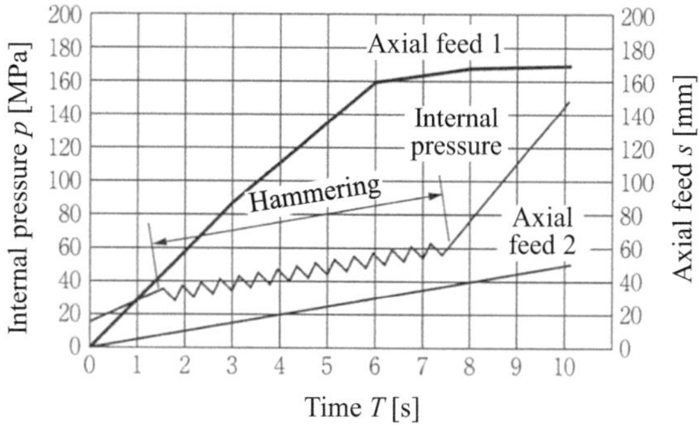

Fig. 4.11 Loading path example of hammering control method [20]

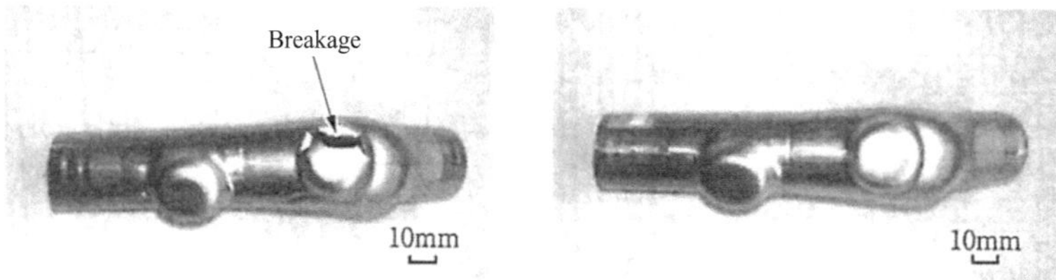

(a) without Hammering control (b) with Hammering control

Fig. 4.12 Effect of hammering control [20]

methods were also used, but recently, high-pressure methods with greater freedom in shape complexity have become more prevalent.

The internal pressure is controlled with respect to time and axial feed, and there are also developed processing methods that incorporate vibrations, known as hammering control [20]. Figure 4.11 shows an example of load paths. Oscillating frequency components in the range of 1 to several Hz are superimposed on the increase of the internal pressure. Figure 4.12 shows a comparison example of processed products with and without hammering, demonstrating the improved formability of hydroforming through hammering. This effect is attributed to the periodic reduction of the constraint of the die's inner surface on the outer surface of the tube, promoting material flow. In other words, it has a similar effect to improving lubrication conditions.

(c) **High-speed pressurization**

One method of rapidly increasing internal pressure is hydro-punching [21]. This method utilizes the pressure generated by striking the enclosed liquid inside the tube with a drop hammer. Other research examples include processing using discharge in a liquid medium [22] and explosion crimping that utilizes momentary high energy from an explosion [23].

(2) **Axial feed**

(a) **Without axial feed (expansion by internal pressure only)**

In hydroforming, the pressing of the tube from the end is generally referred to as axial feed. Axial feed promotes expansion of the tube, but the details will be explained in the next section. When only internal pressure is applied without axial feed, the tube undergoes expansion in the plane strain state, and a high expansion ratio (defined in the next section) is challenging. In addition to cases where axial feed cannot be applied due to equipment configuration, when there are shapes such as bends or collapses along the axial direction of the tube, axial feed does not transmit to those areas, creating conditions equivalent to without axial feed.

(b) **Axial feed by axial compression punch**

Typically, axial feed is achieved by using an axial compression punch, as shown in Fig. 4.7. Axial feed facilitates material flow in the axial direction of the tube and can suppress wall thickness reduction. However, as mentioned earlier, if there are shapes such as bends or collapses along the tube, axial feed is less effective. Additionally, it may be difficult to transmit axial feed when there is significant friction resistance between the die and the tube.

(c) **Axial feed by movable dies**

When material flow from the tube end alone is difficult, movable dies can be used. By allowing the surrounding dies to move in the axial direction, the friction resistance at those locations is eliminated, enabling better material flow into the internal cavity of the die. Figure 4.13 shows an example of achieving high branch tube extension using movable dies [24]. By pressing the shoulder area of the expanding section, material flow into the branch tube is promoted. The mechanism of movable dies has long been used in bellows forming [25]. As shown in Fig. 4.14, finely divided dies move accordion-like in the axial direction of the tube, allowing uniform material flow into various positions of the corrugated shape.

(3) **Counter**

In some cases, when forming out the branch tube as shown in Fig. 4.7 in a T-forming process, an additional die is provided at the top of the branch tube. This is called a counter, and the punch used for it is called a counter punch. The counter is moved in the outward direction of the die as the extension of the branch tube progresses. The role of the counter is to maintain contact between the tube and the counter at the bulge point, thereby suppressing the occurrence of bursts or wrinkles. The movement of the counter is often controlled by load rather than position control because load control ensures stable contact between the tube and the counter. For example, by applying a constant load in the inward direction of the die to the counter, the counter moves in the outward direction along with the extension deformation while maintaining contact with the top of the branch tube. There are also examples that use specialized

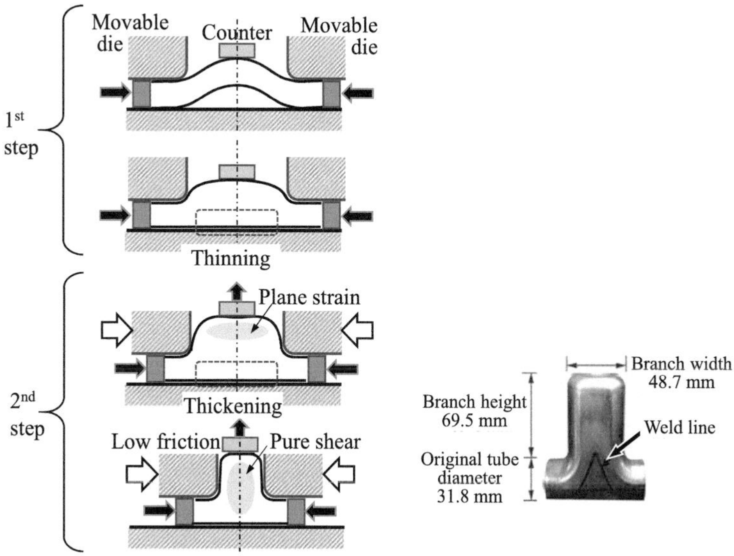

(a) Forming process (b) Formed part

Fig. 4.13 High branch forming by movable die [24]

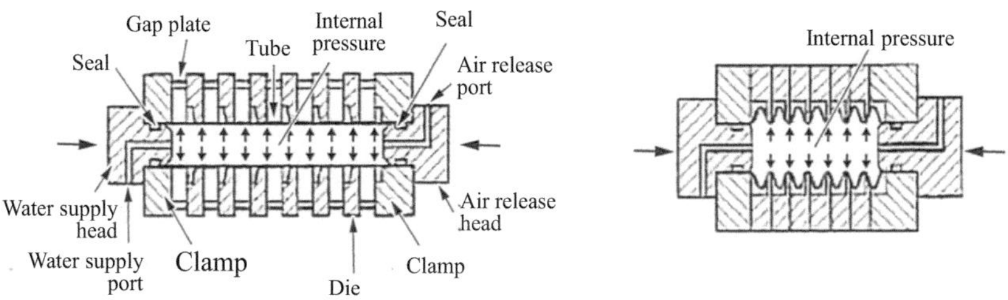

(a) Initial stage (b) Finished stage

Fig. 4.14 Forming of bellows [25]

dies that move while intersecting with movable dies and counters [26]. As shown in Fig. 4.15, the movable die has a groove, and a wide crossing counter punch moves within that groove. With this structure, the tube is constrained by the die from the beginning to the end of the process, preventing bursts and wrinkles, and enabling significant deformation. In this example, the tube is expanded to over 2.5 times its original circumference.

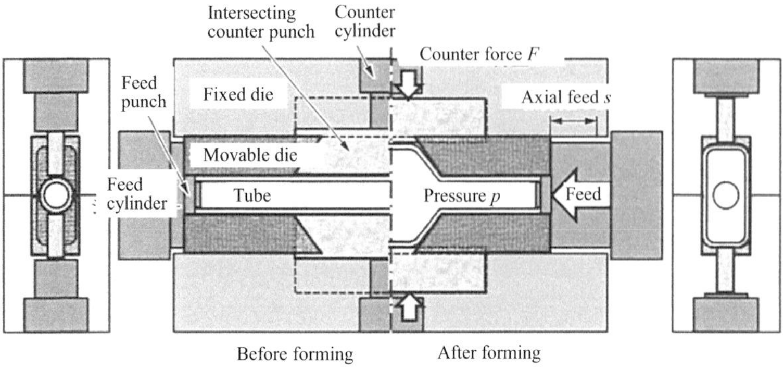

Fig. 4.15 Die structure with movable die and counter die intersection [26]

4.2.3 Classification by Processing Temperature

(1) **Room-temperature processing**

Hydroforming is generally carried out at room temperature. Although there may be a slight increase in temperature due to heat generated by plastic deformation, the internal medium cools down the tube at a temperature that can be touched immediately after processing.

(2) **High-temperature processing (hot bulge)**

High-temperature processing is employed when working with tubes that exhibit better formability at elevated temperatures compared to room temperature or when producing complex-shaped products. Gaseous media such as air or liquid media like silicone oil are used as pressure media. Solid lubricants such as molybdenum disulfide or graphite are employed as lubricants. Practical examples of aluminum alloy tube processing are shown in Fig. 4.16 [27]. This allows for the processing of products with complex cross-sectional shapes that cannot be formed at room temperature. Additionally, high-temperature processing is being investigated for magnesium tubes due to their improved formability at elevated temperatures [28]. Figure 4.17 [29] illustrates an example of high-temperature processing for the purpose of quenching. While the temperature for magnesium alloys and aluminum alloys is typically around 250–450 °C, when quenching steel, temperatures above the Ac_3 point (the temperature at which the transformation of ferrite to austenite is completed during heating) of the steel, approximately 950 °C, are required. The process involves heating the steel tube using direct electrical heating, then sealing air inside the tube, clamping a portion of the tube between upper and lower dies to form a flange, and finally bulging the tube by applying internal pressure and quenching the steel tube by the contact with the dies.

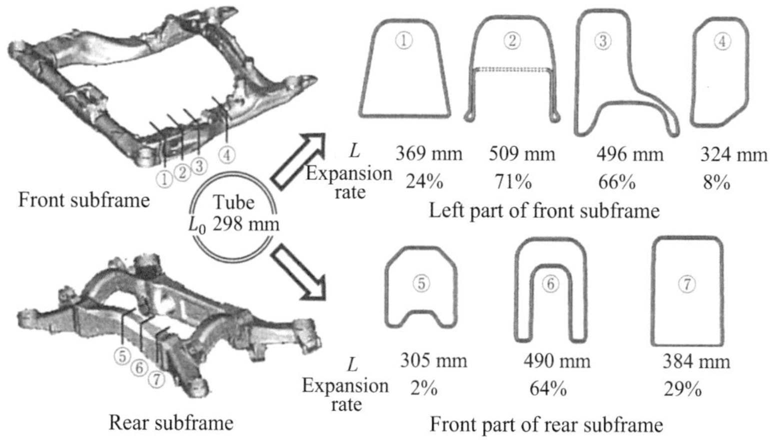

Fig. 4.16 High-temperature hydroforming of aluminum alloy tubes [27]

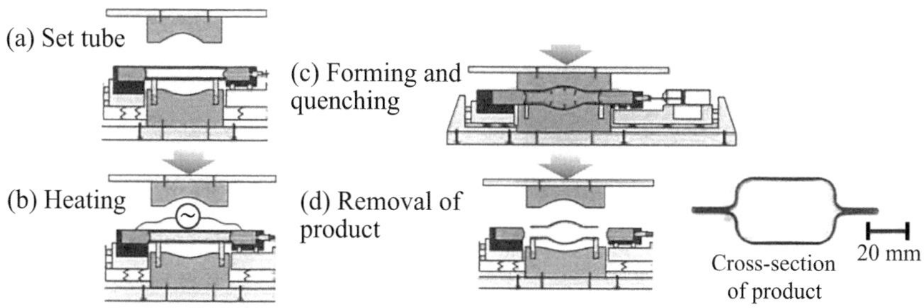

Fig. 4.17 Hot stamping of steel tube with flanges [29]

4.2.4 Classification by Pressure Medium

(1) **Liquid**

Liquid is commonly used as the pressure medium. In traditional applications such as pipe fittings, oil is typically used, but in the processing of automotive components, water is often used. In the manufacturing of automotive parts, it is common to perform degreasing and painting after processing, making water more convenient to handle as it is easily degreased.

(2) **Gas**

In the case of high-temperature hydroforming where processing is conducted above the boiling point of water, gases such as air or inert gases are used. However, in Japan, when the pressure exceeds 1 MPa for gases, it is considered as high-pressure gas production and requires that notification of the use of a high-pressure gas be

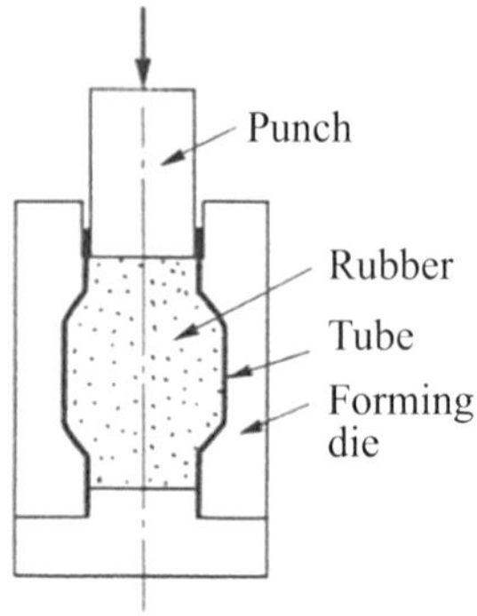

Fig. 4.18 Rubber bulge

made [30]. It should be noted that compressed air below 5 MPa is excluded from this regulation.

(3) **Solid**

While it may not strictly be classified as hydroforming, there is a processing method that uses elastic materials such as rubber as the pressure medium (Fig. 4.18), known as rubber bulge. It has been used for a long time in the processing of bicycle fittings and similar applications [31]. Compared to using liquids or gases, it offers easier sealing. However, it is important to note that the internal pressure acting on the tube may vary depending on the location. Additionally, there are experimental examples that utilize low-melting-point metals [32] or ice [33] as the pressure medium.

4.3 Processing Conditions

4.3.1 Expansion Ratio

Here, we will explain the definition of the expansion ratio in hydroforming. Generally, the expansion ratio λ is defined as the ratio of the change in the outer circumference length of the cross-section before and after hydroforming to the initial outer circumference length, usually expressed as a percentage. For example, in the case of forming a rectangular cross-section with dimensions of length a and width b, with a corner radius R, from a pipe with an outer diameter D_0, the expansion ratio λ is calculated using the following equation (usually expressed in %):

$$\lambda = \frac{2(a + b - 4R + \pi R) - \pi D_0}{\pi D_0} \tag{4.23}$$

If the cross-sectional shape varies depending on the axial position of the tube within a single product, the expansion ratio will differ at each position. It is commonly assumed that higher expansion ratios result in increased forming difficulty in hydroforming. However, it is not a definitive rule. Even with the same expansion ratio, forming difficulty increases for complex cross-sectional shapes, and positions with difficulty in material feed also contribute to increased forming difficulty.

4.3.2 *Deformation Behavior and Defects*

(1) **Deformation behavior and defects in free bulge**

First, let's explain the basic deformation behavior in free bulge. When applying only internal pressure to a tube, the central region of the tube expands. However, when axial feed is applied, the deformation behavior varies depending on the control method and load path. Figure 4.19 shows an example of a biaxial bulge test on a tube [34, 35]. As explained in Chap. 2, this test involves applying internal pressure and axial compression while controlling the circumferential and axial strains and stresses in a constant ratio. In this case, the central region of the tube expands similarly to when only internal pressure is applied. In both cases, if the internal pressure is continuously increased, the tube will eventually rupture at the central region in the axial direction.

On the other hand, in the case of die bulge, as described later, for manufacturing actual parts, axial feed is applied after loading a certain level of internal pressure. Similar deformation occurs when processing is performed using a similar load path in free bulge, as shown in Fig. 4.20 [36]. When the initial internal pressure is relatively high, the central region of the tube continues to expand along with the progression of axial feed without increasing the internal pressure, and eventually bursts at that central region. When the initial internal pressure is low, as axial feed progresses, the

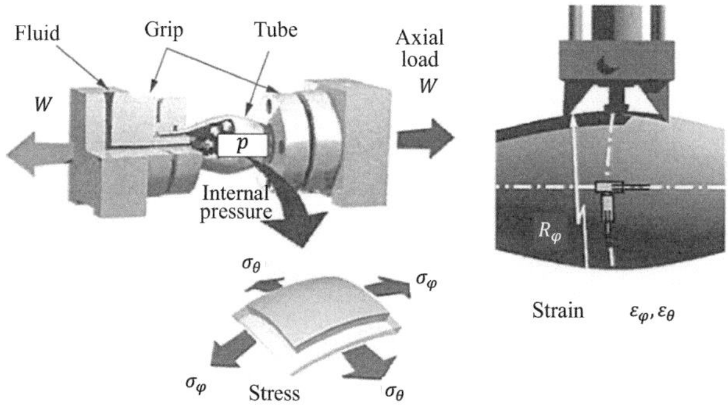

Fig. 4.19 Testing method of biaxial bulge [35]

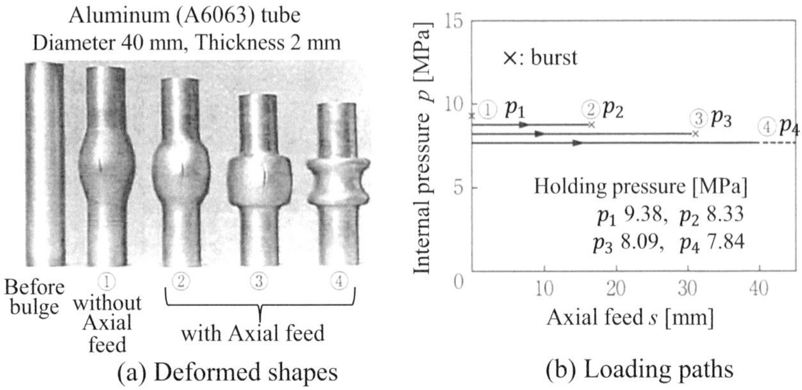

Fig. 4.20 Loading path and deformed shape in free bulge

region closer to both ends also expands, and in extreme cases, it takes on a gourd-like shape with two bulges. Ultimately, it either bursts at the most expanded position or remains intact in a biconvex shape (hyperboloid).

(2) Deformation behavior and defects in die bulge

Let's explain the typical deformation behavior of die bulge forming, specifically the example of expanding a round section straight tube into a uniform rectangular cross-section. Figure 4.21 illustrates the load paths of internal pressure and axial feed, along with the accompanying deformation behavior [37].

(Step 1) First, an initial internal pressure p_1 is applied. Generally, the initial pressure is loaded without axial feed, but in many cases, a slight amount of axial feed is applied to seal the tube ends. The magnitude of the initial pressure is estimated

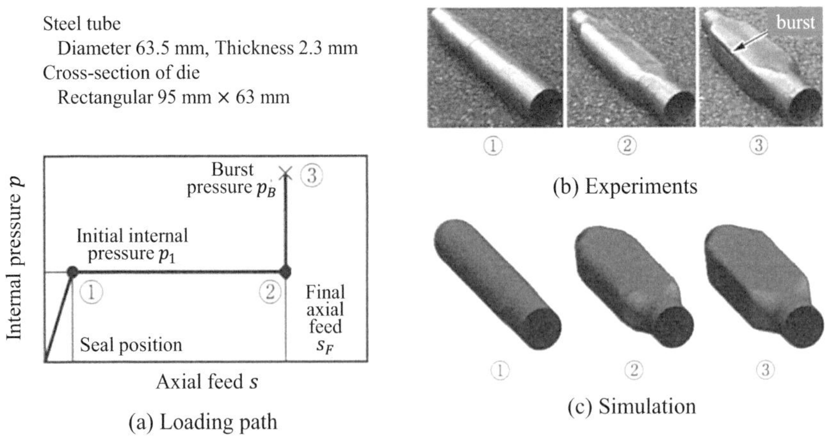

Fig. 4.21 Deformation behavior of rectangular tube expansion

based on the pressure at which the tube begins plastic deformation. At this stage, the plane strain state with minimal axial strain is considered, and p_1 can be calculated using the following equation:

$$p_1 = \frac{4}{\sqrt{3}}\sigma_y \frac{t_0}{D_0 - t_0} \tag{4.24}$$

Here, σ_y represents the yield stress of the material in uniaxial tension, and D_0 and t_0 denote the initial outer diameter and wall thickness of the tube, respectively, assuming a thin-walled circular tube. Considering the r-value of the material, Eq. (4.24) can be expressed as follows:

$$p_1 = 2\frac{1+r}{\sqrt{1+2r}}\sigma_y \frac{t_0}{D_0 - t_0} \tag{4.25}$$

It should be noted that the values obtained from the above equation serve as a rough estimate, as the appropriate values may vary depending on factors such as contact conditions with the die. As shown in Fig. 4.21, in this state, the tube is not significantly expanded.

(Step 2) Next, axial compression is applied. In some cases, internal pressure is increased while applying axial feed, but here, we will consider the simplest case where the internal pressure is maintained while increasing the axial feed to s_F. In this state, even without increasing the internal pressure, the tube undergoes significant expansion due to the axial feed. The application of axial feed induces axial compressive strain in the tube, promoting plastic deformation.

(Step 3) Finally, axial feed is stopped, and only the internal pressure is increased. This process is referred to as "calibration" and is performed to shape the corners (R) of the cross-section sharply. In the example shown in Fig. 4.21, the load path used in expressing the forming feasible range reaches the burst pressure p_B. However, in actual production, a final internal pressure p_f lower than the burst pressure p_B is set. The final internal pressure p_f is determined by corner R of the target cross-section, the wall thickness t_0, and the tensile strength σ_B of the mother tube, as expressed by the following equation:

$$p_f = \sigma_B \frac{t_0}{R - t_0} \tag{4.26}$$

However, depending on the friction conditions with the die, higher pressures may be required. This outlines the general deformation behavior of bulge forming. Next, defects in die bulge are discussed. The two main defects that can occur: burst and wrinkle.

There are two types of bursts, occurring either relatively early or late in the process. The former occurs during the aforementioned steps 1 and 2 in an area that has not yet come into contact with the die or in the central part of the contacted surfaces

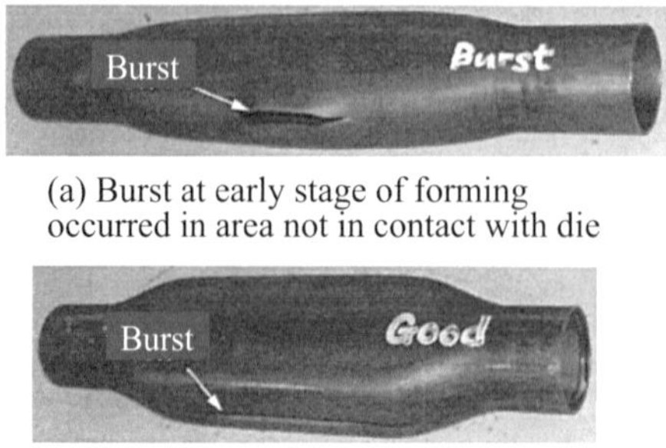

(a) Burst at early stage of forming occurred in area not in contact with die

(b) Burst at late stage of forming occurred at boundary of corner radius and flat area

Fig. 4.22 Burst in hydroforming [38]

(Fig. 4.22a) [38]. The latter occurs during step 3 at the boundary between the corner *R* and the flat section (Fig. 4.22b).

Wrinkles can be broadly classified into two types, as shown in Fig. 4.23. Figure 4.23a represents the case where wrinkling occurs when the tube is placed within the upper and lower dies, causing an inward dent due to excess material in the circumferential direction [18]. Figure 4.23b represents the case where wrinkling occurs in the axial direction of the tube during the axial feed in step 2. However, in the case of hydroforming, not all wrinkles are considered defects. If the wrinkles are eliminated during the final pressure increase in step 3, the product can be considered acceps [37]. Sometimes intentionally inducing mild wrinkles during step 2 and then eliminating them in step 3 can be advantageous for achieving a favorable wall thickness distribution in the final product. Other types of defects include the inability to shape the corner *R* to the target value, or defects where the inner side of the tube protrudes at the axial feed area of the tube end, as shown in Fig. 4.24 [37]. Additionally, springback, which can be problematic in bending processes, is less of

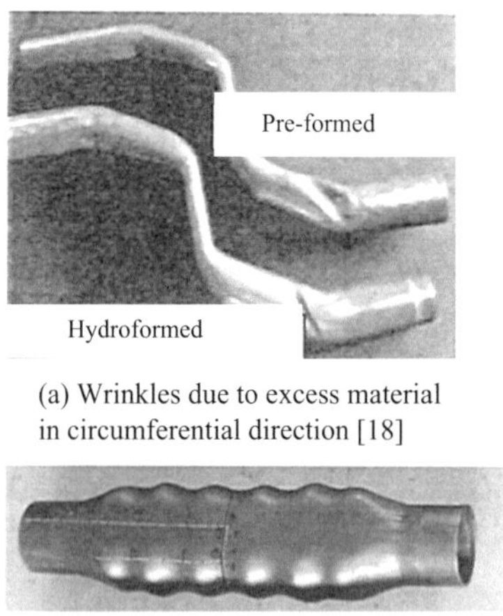

(a) Wrinkles due to excess material in circumferential direction [18]

(b) Wrinkles due to axial feed [37]

Fig. 4.23 Wrinkle in hydroforming

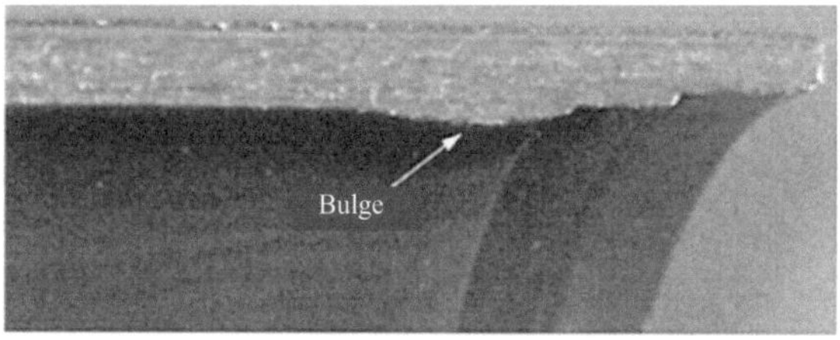

Fig. 4.24 Internal circumferential bulge due to excess axial feed [37]

an issue in hydroforming. Due to the overall circumferential plastic deformation, the amount of springback is minimal even when the load is removed. This is also why step 3 is referred to as calibration. However, when processing materials with higher strength or smaller Young's modulus, the tendency for springback increases, so it is necessary to consider springback when working with such materials

4.3.3 Influence of Load Paths

One of the factors that significantly affect the success of hydroforming is the load paths of axial feed and internal pressure. Figure 4.25 shows an example [39] where the forming conditions of hydroforming are represented by load paths. Generally, a map is drawn where bursts occur in the high internal pressure and low axial feed region, while wrinkles occur in the low internal pressure and high axial feed region.

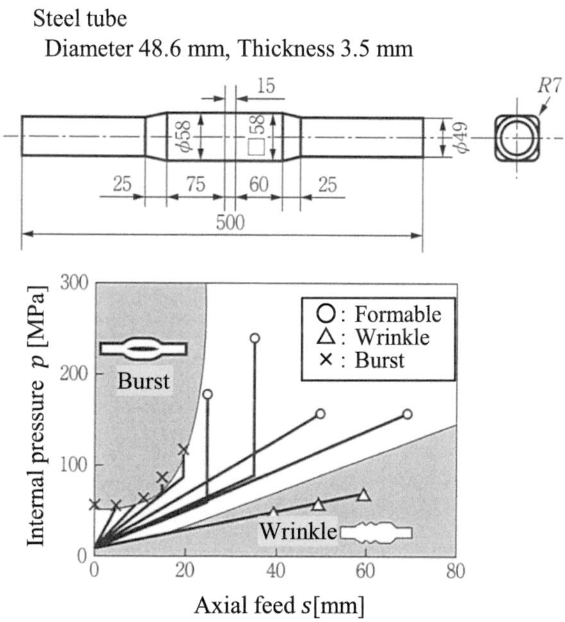

Fig. 4.25 Example of expression method for formability range in hydroforming [39]

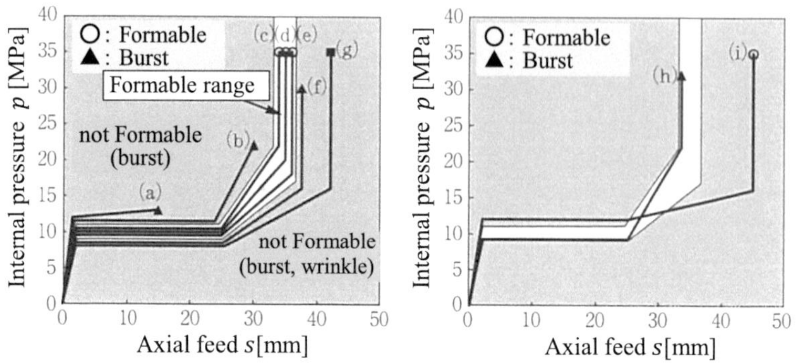

(a) Formability range determined by simply shifting loading paths

(b) Example of loading paths that cannot be explained by the (a) method

Fig. 4.26 Issues in typical expression method for formability range in hydroforming [40]

However, the representation of the forming range in figures like Fig. 4.25 is not precise. For example, there are cases (Fig. 4.26) where bursts occur even when following a path within the formable range (path h in the figure). Conversely, there

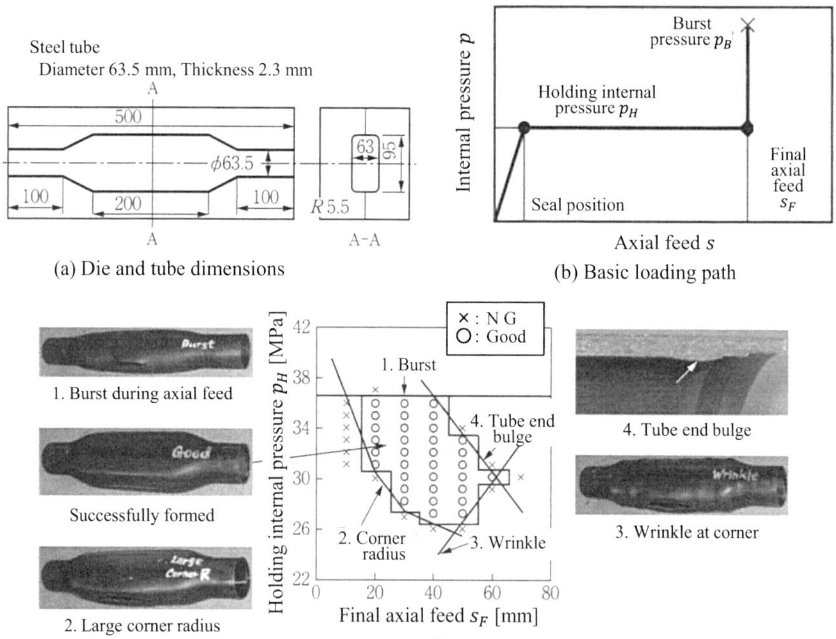

(a) Die and tube dimensions

(b) Basic loading path

(c) Formable range

Fig. 4.27 Example of expression method for formability range by final axial feed and holding pressure in hydroforming [38]

are cases where wrinkles occur initially but are eliminated during the final pressure increase (path i in the figure) [40]. Therefore, an example that represents the forming range in terms of the amount of final axial feed and holding pressure, rather than the load path itself, is shown in Fig. 4.27 [38]. In this example, for the sake of simplification, after applying initial internal pressure, internal pressure is kept constant during axial feed, and that constant value is referred to as the holding pressure. The value of the final internal pressure is not predetermined; rather, the desired shape is achieved within the range of pressures that prevent bursts. Four forming limits are considered: the corner radius limit, the wrinkle limit, the burst limit during axial feed, and the tube-end bulge limit. These four limit curves define the formable region in hydroforming. This organization not only provides guidelines for selecting appropriate internal pressure and axial feed for hydroforming a specific part but also allows for evaluating the forming margin of the tube material by comparing the area of the forming range.

The influence of the load path on the success or failure of hydroforming is highly complex, and finding the appropriate load path requires trial and error, making it a major challenge for process engineers. While simulations are valuable tools for visualizing deformation states that cannot be observed in experiments, trial and error is still necessary, similar to experimental approaches, until the appropriate load path is

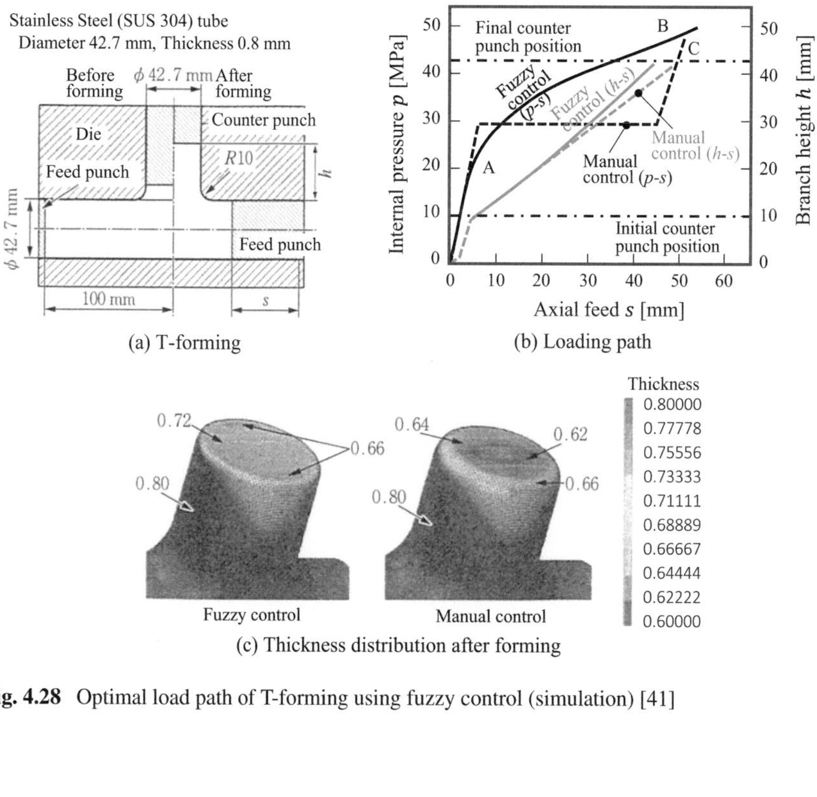

Fig. 4.28 Optimal load path of T-forming using fuzzy control (simulation) [41]

obtained. Therefore, there have been attempts to automatically determine the appropriate load path. Figure 4.28 illustrates an example of obtaining the appropriate load path for T-forming using fuzzy control [41]. Further advancements in such research are expected in order to promote the widespread use of hydroforming in the future.

4.3.4 Influence of Material Properties and Die Lubrication

Among the factors that influence the success or failure of hydroforming, material properties and die lubrication are notable. Several examples have focused on the influence of material properties, primarily the n-value and r-value, while die lubrication examines the impact through the consideration of the friction coefficient, μ.

Generally, it is believed that hydroforming formability improves with higher n-values and r-values, and lower friction coefficients, μ. Simulation results of expanding a circular tube into a square tube, as shown in Fig. 4.29, demonstrate that higher n-values, higher r-values, and lower μ lead to a more uniform distribution of wall thickness after forming [42, 43].

Figure 4.30 presents simulation results of the forming margin introduced in Sect. 4.3.2. From this figure, it can be observed that higher n-values, higher r-values, and lower μ result in an increased forming margin [44–46]. However, it is important

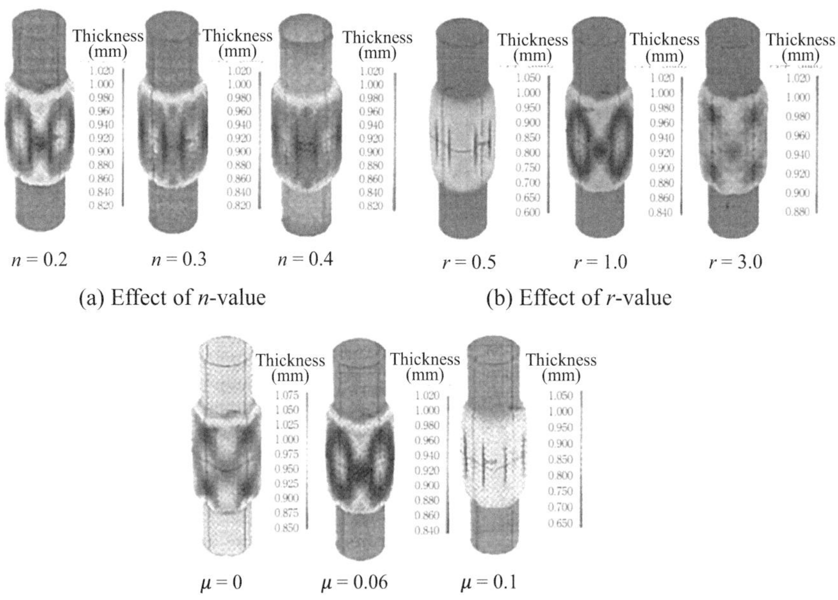

Fig. 4.29 Effect of various factors on thickness deviation in expansion into square tube by hydroforming (simulation) [42, 43]

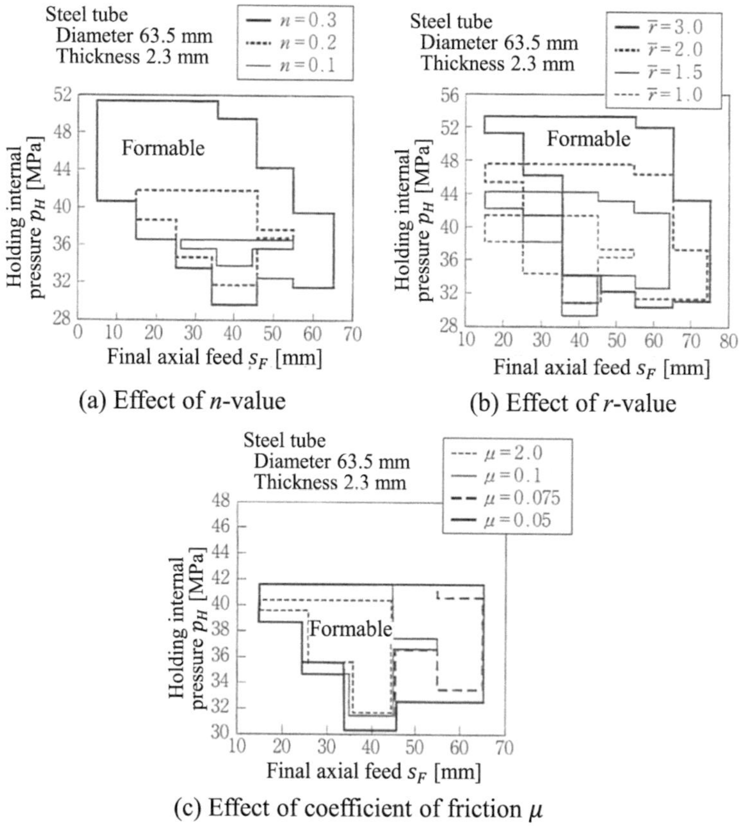

Fig. 4.30 Effect of various factors on formability range in expansion into rectangular tube by hydroforming (simulation) [44–46]

to note that while *n*-values and μ contribute to the expansion of the forming range in nearly all directions, *r*-values cause the forming range to expand while shifting towards the high-pressure side. Therefore, if a material with a higher *r*-value is used, the forming range of hydroforming indeed expands and becomes advantageous, but without adjusting the load path to increase internal pressure, there is a possibility of encountering processing defects.

Furthermore, the effects can vary depending on the shape being formed. Figure 4.31 illustrates a simulation example of forming an automotive center pillar reinforcement [47], where the effects of various factors differ between cross-section C and cross-section D. Cross-section C, being relatively close to the tube end, is more influenced by axial feed, while the effects of material properties such as *n*-value and *r*-value are minimal. On the other hand, in cross-section D, which is farther from the tube end, the effect of axial feed is less significant, and the influence of material properties becomes more prominent. Similarly, regarding the influence of the friction coefficient μ, even a slight reduction from 0.15 to 0.12 can have a significant impact

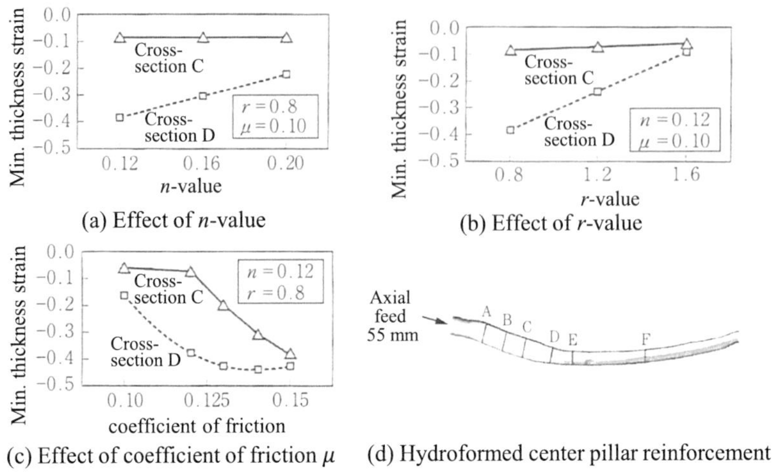

Fig. 4.31 Effect of various factors on thickness variation in hydroforming of center pillar reinforcement (simulation) [47]

in cross-section C, while a substantial decrease in μ is necessary to observe its effects in cross-section D.

Therefore, careful consideration of the shape being formed is crucial, along with the implementation of measures such as material properties and friction conditions.

4.3.5 Processing of Special Tubular Materials

In recent years, further light weighting and increased complexity of product shapes have been demanded in the automotive industry. As a result, the application of tubes with different cross-sections along the axial direction has been explored, going beyond the use of tubes with uniform cross-sections in hydroforming. In this context, we will discuss the processing of "tailored tube", which is composed of tubes with different wall thicknesses, and "tapered tube", where the diameter varies along the axial direction.

The challenging aspect of hydroforming tailored tubes lies in determining the internal pressure conditions. While it is relatively easier to form products with low expansion ratios, caution is required when forming products with higher expansion ratios. In other words, both thin-walled and thick-walled tubes are subjected to the same internal pressure, causing the thin-walled tube to bulge first, as expected. Consequently, if the initial internal pressure calculated for the thin-walled tube using Eq. (4.24) or Eq. (4.25) is set, the thick-walled tube will experience folding without bulging. Conversely, setting the internal pressure for the thick-walled tube may lead to bursts in the thin-walled tube. To address this issue, a suitable load path for tailored tubes, as shown in Fig. 4.32, has been proposed [48].

Firstly, only the thin-walled tube side is subjected to axial feed, while the initial internal pressure is set for the-thin walled tube. This allows the thin-walled tube to

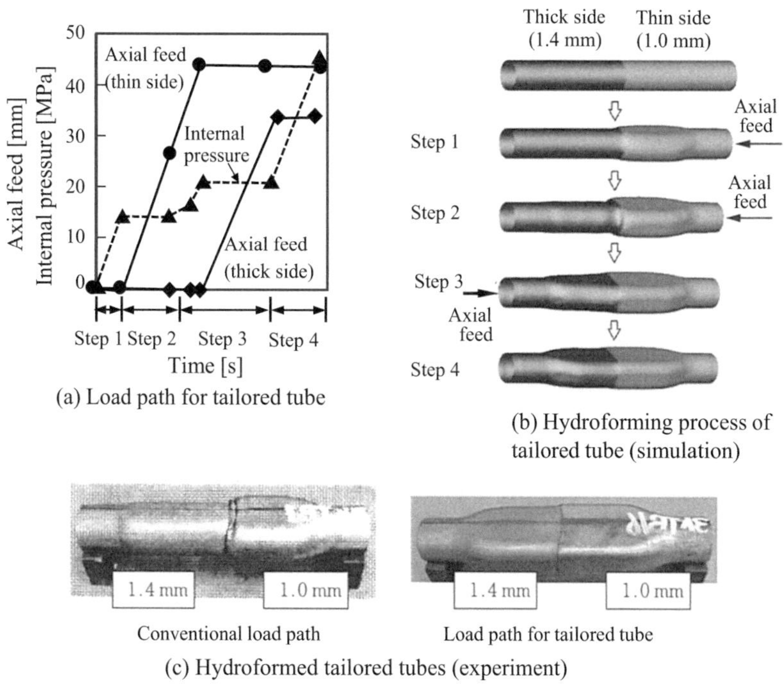

Fig. 4.32 Loading path suitable for tailored tube hydroforming

expand first. Although there is some effect of axial feed on the thick-walled tube, it is not significant due to the frictional resistance between the thin-walled tube and the die. Next, the internal pressure is gradually increased to the appropriate level for the thick-walled tube. By this point, the thin-walled tube has already undergone significant expansion and is supported by the surrounding die, preventing it from bursting. Finally, axial feed is applied to the end of the thick-walled tube, allowing the thick walled tube to expand as well. Through this load path, hydroforming of tailored tubes becomes achievable

On the other hand, one challenge in hydroforming tapered tubes is the axial feed method.

In conventional axial feed, a tube with a uniform cross-section slides through the die cavity with a uniform cross-section. However, in the case of tapered tubes, the cross-section varies along the axial direction, making it impossible to perform axial feed using the same structure. Therefore, a method using a movable die, as shown in Fig. 4.33, has been proposed [49, 50]. The larger diameter end of the tapered tube is restrained by the movable die, and the material is pushed inwards together with this movable die. By adopting this axial feed structure, hydroforming of tapered tubes becomes feasible.

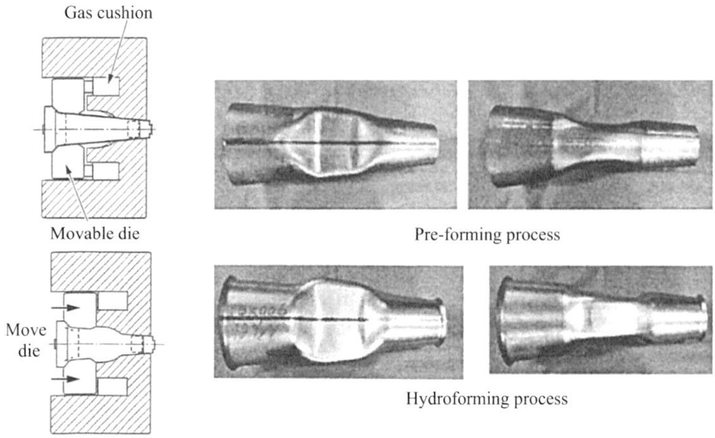

(a) Experimental setup (b) Experimental result

Fig. 4.33 Example of hydroforming of tapered tube

4.4 Pre-forming and Post-forming

4.4.1 Pre-forming

As mentioned in Sect. 4.1.1, pre-forming is often conducted as a preliminary step before hydroforming complex-shaped products such as automotive components. The quality of pre-forming significantly influences the feasibility of subsequent hydroforming processes. The main pre-forming techniques include expanding and reducing the tube diameter, bending, and crushing. The primary objective of pre-forming is to fit the tube into the hydroforming die; thus, it is crucial to prevent pinching when closing the upper and lower dies. Pinching refers to the phenomenon of the tube being caught between the dies. However, simply fitting the tube into the hydroforming die is not enough, as the feasibility of hydroforming is affected by the position and shape of the tube within the die.

For instance, let's consider an example [51] where the bending shape affects the deformation behavior in hydroforming. While the details of bending methods and deformation behavior were discussed in Chap. 3 (omitted here), it is generally observed that the tube undergoes thinning on the outer side of the bend and thickening on the inner side of the bend. Consequently, if a bent tube is placed in the hydroforming die without axial feed and solely expanded using internal pressure, there is a high risk of burst in the outer side of the bend. However, in the case of hydroforming with axial feed, as shown in Fig. 4.34, the outer side of the bend makes contact with the die first, increasing the gap between the inner side of the bend and the die. Subsequently, the inner side of the bend expands, resulting in a risk of burst in the inner side of the bend. In such cases, it is preferable to adjust the fit between the inner side of the bend and the die when placing the tube in the hydroforming die.

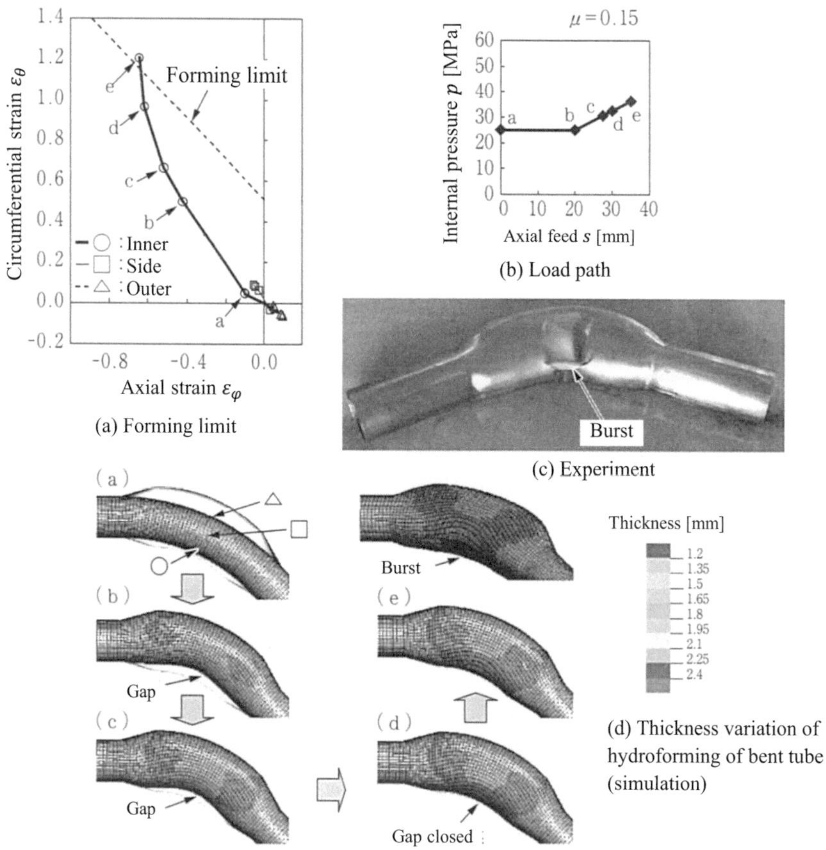

Fig. 4.34 Burst at inner side of bend in bent tube hydroforming

In crushing operations as well, the shape within the hydroforming die is crucial. Although certain levels of denting and wrinkling can be alleviated through the application of internal pressure during the hydroforming, significant excess material and severe wrinkles cannot be alleviated even during final pressurization (refer to Fig. 4.23a). Therefore, measures are taken to minimize denting and wrinkling during the crushing process using movable dies, as depicted in Fig. 4.35 [18]. Additionally, sealed fluid forming, as discussed in Fig. 4.8, is sometimes employed as a pre-forming method. By closing the die with internal pressure, it is possible to prevent denting and wrinkling compared to cases without internal pressure.

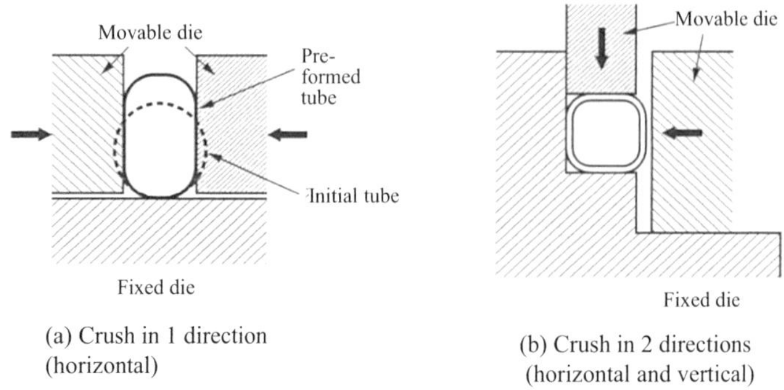

Fig. 4.35 Improvement of die structure for pre-forming

4.4.2 Post-forming

Post-forming refers to the process other than deformation using the internal pressure of hydroforming. One commonly used post-forming technique is hydro-piercing, which involves creating holes [11]. Hydro-piercing can be performed continuously after the hydroforming process within the die, resulting in reduced manufacturing steps. This process enables high accuracy by piercing holes before the tube is removed from the die, and is commonly applied for positioning holes.

Hydro-piercing can be classified into two types, as shown in Fig. 4.36: inward piercing, where holes are created towards the inside of the tube, and outward piercing,

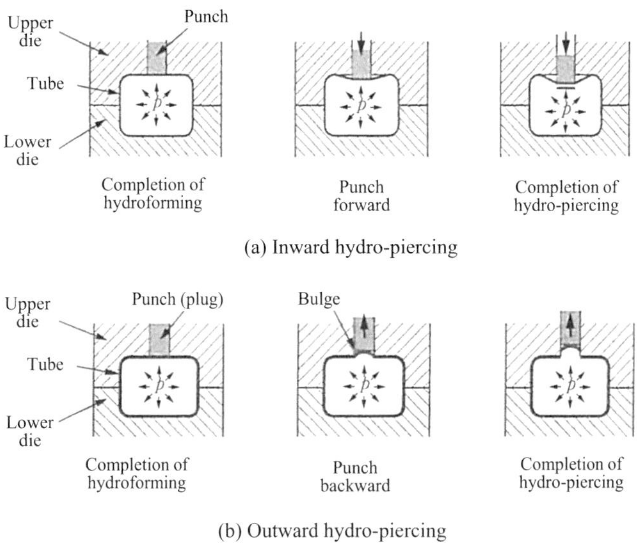

Fig. 4.36 Hydro-piercing [18]

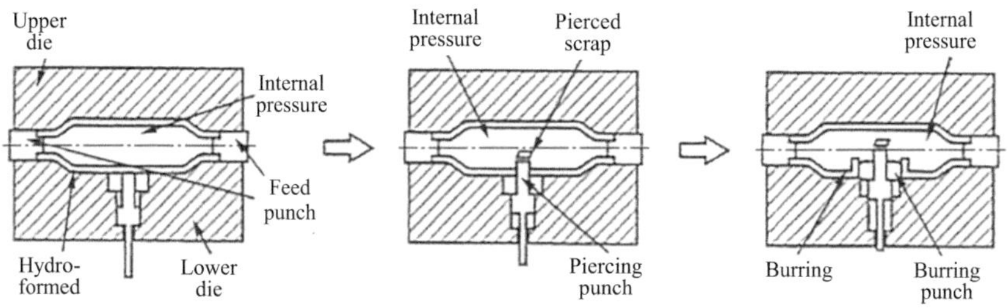

Fig. 4.37 Hydro-burring [53]

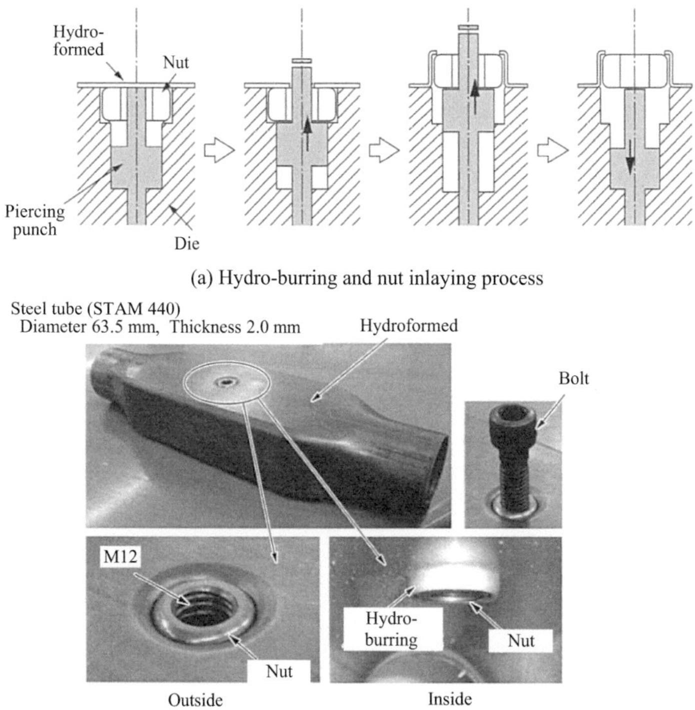

Fig. 4.38 Nut inlaying in hydroforming process [54]

where holes are created towards the outside of the tube [18]. In general, inward piercing is more commonly used. The approximate punch load, F_{pi}, required for hole creation can be estimated using the following equation.

$$F_{pi} = \frac{1}{\sqrt{3}}\sigma_B t_0 L_{pi} + p_{pi} A_{pi} \tag{4.27}$$

Here, σ_B represents the tensile strength of the tube, t_0 is the thickness of the tube, L_{pi} is the punch circumference, p_{pi} is the internal pressure during piercing (typically using the final internal pressure, p_f), and A_{pi} denotes the cross-sectional area of the punch end. The first term on the right side represents the punching load, and the second term represents the resistance to internal pressure.

It is worth noting that at the moment of hole creation and afterward, sealing continues between the punch and the pierced surface, as well as between the die and the product surface around the hole, resulting in almost no water leakage. Therefore, it is possible to create multiple holes simultaneously [52]. In cases where handling the punched-out tabs is troublesome, some part of the punch tip is rounded to suppress fracture and allow bending tabs inward, similar to the pull tabs of beverage cans [11].

Other post-forming techniques include hydro-burring, as shown in Fig. 4.37, where the hole created by hydro-piercing is further expanded towards the inside of the tube [53]. Additionally, developments such as nut embedding, as depicted in Fig. 4.38, utilizing hydro-burring, have been made [54]. Furthermore, techniques such as hydro-flanging [55], where flanges are formed by partially clamping the material, and hydro trimming [56], used for cutting the tube, as shown in Fig. 4.39, are also available.

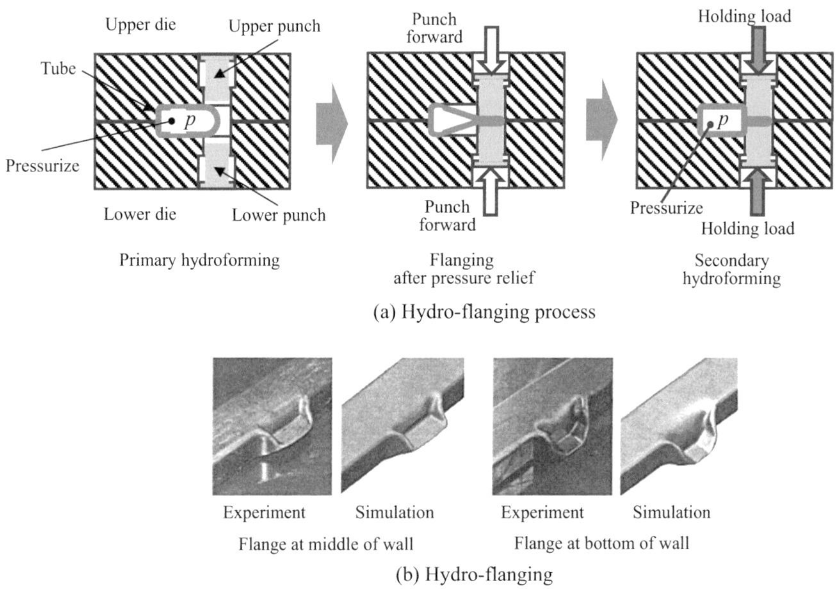

Fig. 4.39 Hydro-flanging [55]

4.5 Die Design

4.5.1 Upper and Lower Dies

The dies used in hydroforming differ from those used in sheet stamping, as there is only a female die (die) and no male die (punch). However, due to the necessity of putting it in and taking it out of the tube and the formed product, the die is divided into two parts: the upper and lower dies. The division plane between the upper and lower dies, called the parting plane, is set at the outermost position in the top view to allow the removal of the formed product after hydroforming (refer to Fig. 4.40). However, if the position of the parting plane varies between the left and right sections, thrust forces act on the die. The thrust force is not only determined by the offset within the cross-section but also influenced by the axial offset and the position of the internal parting line within the die (as shown in Fig. 4.41) [57].

Fig. 4.40 Die parting surface and resulting thrust force in hydroforming

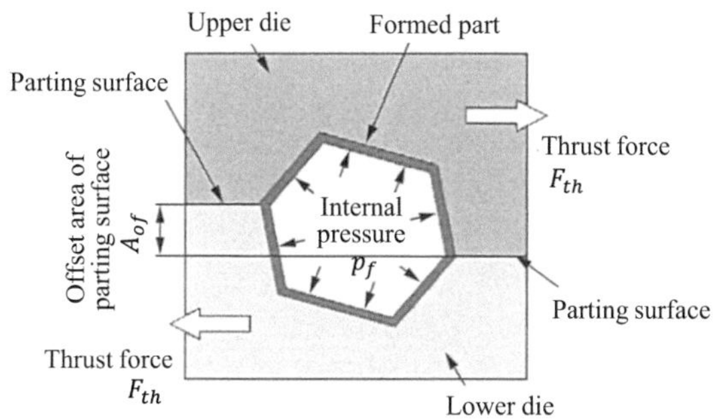

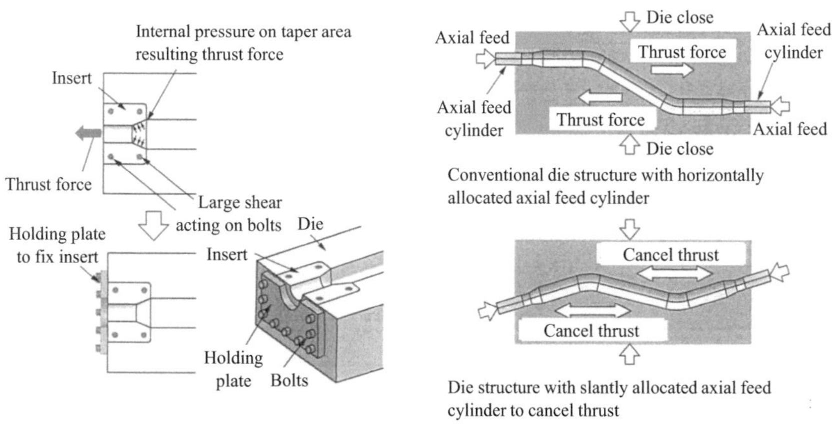

Fig. 4.41 Examples of measures against thrust force in hydroforming [57]

The thrust force F_{th} is calculated using the offset area A_{of} and the final internal pressure p_f obtained from Eq. (4.26) as follows:

$$F_{th} = p_f A_{0f} \tag{4.28}$$

Depending on the offset amount and internal pressure, the thrust force can become quite significant. Therefore, it is necessary to consider the use of thrust blocks in the die or arrange the die cavity to cancel out the thrust force. In the example shown in Fig. 4.41b, the thrust force is canceled by tilting the S-shaped formed product. The clamping force F_{cl} required to maintain the upper and lower dies in the closed position is calculated based on the projected area A_{pr} in the top view of the formed product and the final internal pressure p_f obtained from Eq. (4.26) as follows:

$$F_{cl} = p_f A_{pr} \tag{4.29}$$

Additionally, when the corner radius R of the product is small, not only does the final internal pressure p_f increase according to Eq. (4.26), but also the corner radius R in the die cavity becomes smaller, leading to increased stress concentration by internal pressure during processing. In case the caused stress is high, consideration of die thickness and die material is necessary to avoid die fracture during processing. The stress caused by thrust depends on the cross-sectional and overall shape of the product and is desirable to be estimated by finite element analysis and other analysis.

4.5.2 Axial Feed Punch

The force applied to the axial feed punch, F_x, can be decomposed into three forces as shown in Fig. 4.42:

$$F_x = F_p + F_\mu + F_f \tag{4.30}$$

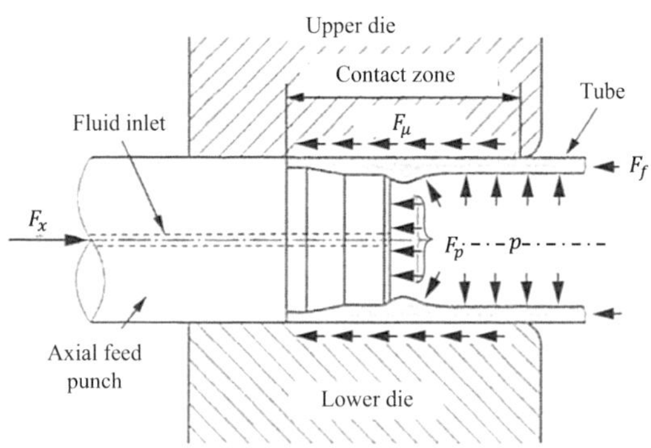

Fig. 4.42 Forces acting on axial feed punch [18]

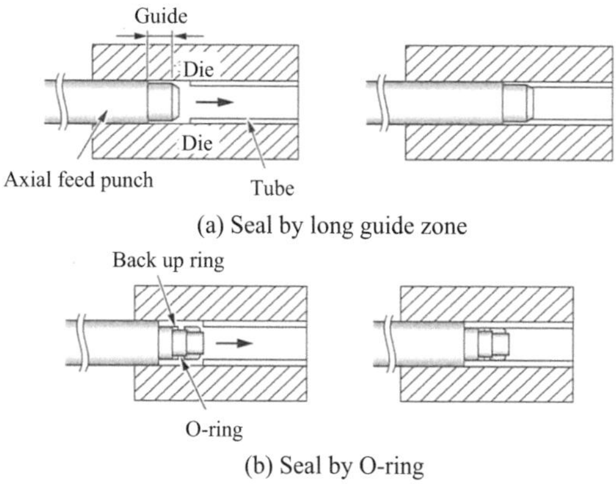

Fig. 4.43 Examples of sealing for axial feed punch

Here, F_p represents the reaction force against the internal pressure, generally calculated by multiplying the cross-sectional area of the axial feed punch end by the final internal pressure p_f. F_μ represents the frictional force between the tube and the die, and F_f represents the axial force related to the plastic deformation of the tube. In the load path explained in Fig. 4.21 for hydroforming, the values of F_μ and F_f are significant in step 2, while the value of F_p is significant in step 3. Depending on the shape being formed, and the strength and dimensions of the tube, generally F_p in step 3 tends to be dominant and easily calculated. Therefore, it is recommended to use this value for designing the specifications of the axial feed cylinder and punch.

Another important aspect in the design of the axial feed punch is its sealing capability. It is required to seal the pressure inside the tube and allow easy separation from the tube end after processing. An example of the sealing part of the axial feed punch is shown in Fig. 4.43 [17]. O-rings are sometimes used to facilitate initial water filling, but they alone cannot withstand high pressures. Therefore, metal seals, such as those achieved by plastic deformation of the tube end, are generally used. For example, the punch is recessed into the tube end by creating steps [57]. It is also common to provide a guide at the tip to prevent wrinkles on the tube end and ensure a reliable seal with the metal seal section.

4.6 Hydroforming Machine

4.6.1 Hydroforming System

Figure 4.44 shows a typical hydroforming system [58]. The machines used and the production speed varies for each process. In particular, in the process of narrowly defined hydroforming, which involves applying internal pressure, a significant time is required to fill the pressure medium internally, resulting in longer processing time (cycle time). Depending on the size of the product and the method of water

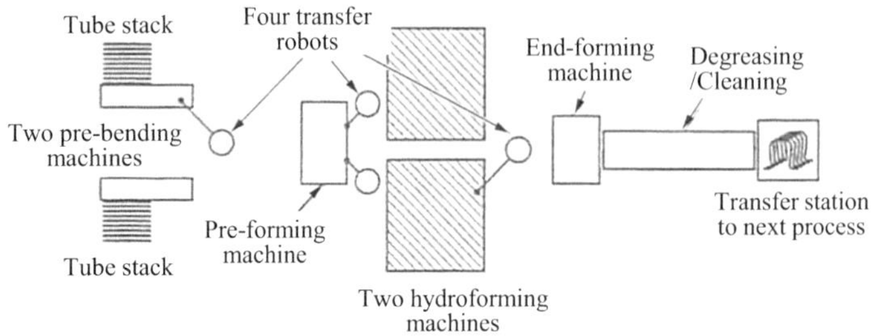

Fig. 4.44 Layout example of full hydroforming process [58]

filling, the cycle time typically ranges from 20 to 60 s. Therefore, in the example shown in the figure, two bending machines and two hydroforming machines are installed to balance the production speed. Additionally, to avoid excessive investment in equipment, it is common to process multiple products simultaneously within a single die.

4.6.2 Equipment Configuration

Figure 4.45 shows a typical equipment configuration for hydroforming [59]. The piping system consists of an oil circulation system and a water circulation system. Oil is used for operating the clamping device, high-pressure generation device, and axial feed device. Water is filled into the tube for hydroforming processing, then recovered, passed through a filter, and reused. The clamping device is required to perform opening and closing for inserting and removing the raw tube and finished product, as well as to prevent opening during processing under internal pressure. There are three types of clamping mechanisms shown in Fig. 4.46 that satisfy these two functions:

1. **Top Drive Method**: Similar to a conventional hydraulic press, it performs die opening, die closing, and clamping using a slide.
2. **Under Drive Method**: It performs die opening with a slide, then inserts clamping blocks between the crown and slide to bear the load, and clamps it from below using cylinders.
3. **Mechanical Lock Method**: It performs die opening with a slide and clamps it using a taper ring or taper clamps.

Additionally, there are specialized processing machines, such as the one shown in Fig. 4.47, which separates die opening/closing and mechanical locking to achieve compactness [60]. It involves sliding the closed upper and lower dies into the rear C-frame and locking them using a mechanical lock mechanism.

The high-pressure generation system, as shown in Fig. 4.48, is composed of a primary-side hydraulic chamber and a secondary-side water chamber. The pressure

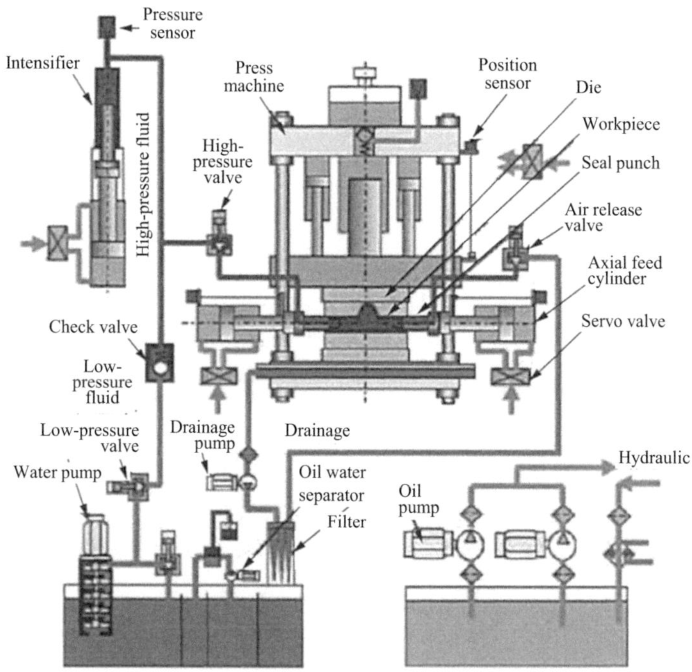

Fig. 4.45 Hydroforming machine configuration [59]

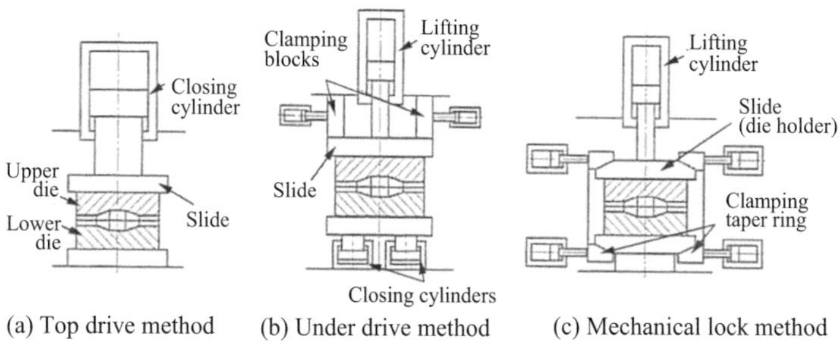

Fig. 4.46 Die closing methods for hydroforming [18]

generated is governed by the ratio of their respective cross-sectional areas [18]. While the single-action type is common as shown in Fig. 4.48a, there are also double-action types like in Fig. 4.48b. Furthermore, since it is necessary to achieve a high-precision load path between the internal pressure and the axial feed, servo valves are used for control. Additionally, since the processing water is reused, equipment for recovery and circulation is required. To prevent corrosion and bacterial growth in the machine, an emulsion solution containing antiseptic and corrosion inhibitors is mixed with the processing water.

Fig. 4.47 Compact hydroforming machine [60]

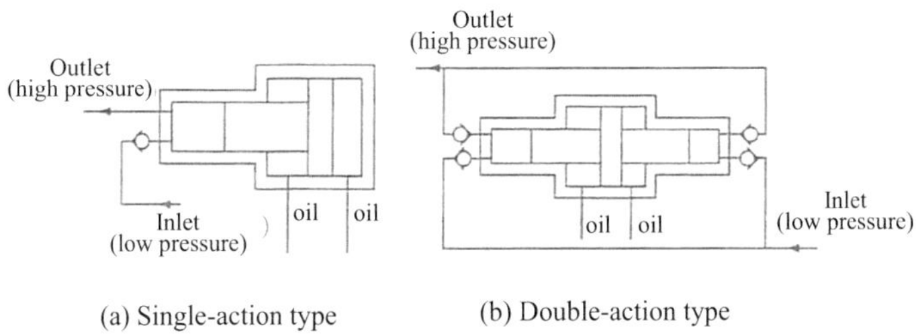

Fig. 4.48 Intensifier for hydroforming [18]

Axial feed devices also require high-precision control, similar to internal pressure, and are controlled using servo valves. Axial feed involves both load control and position control, but in actual production, position control, which provides dimensional accuracy of the product, is often preferred.

4.7 Examples of Hydroformed Products

4.7.1 Automotive Components

As mentioned in Sect. 4.1.1, hydroforming has experienced significant growth in recent years due to its expanded application in automotive components. Chassis parts application is most common and dominant with subframes being particularly prevalent, as shown in Fig. 4.49. Examples include engine cradles, hydroformed from a single tube in a U-shape [61], and a rectangular frame composed of hydroformed tubular beams [62].

Additionally, there are numerous examples of applying hydroforming to create torsion beams by collapsing the central portion of a tube into a U-shape or V-shape,

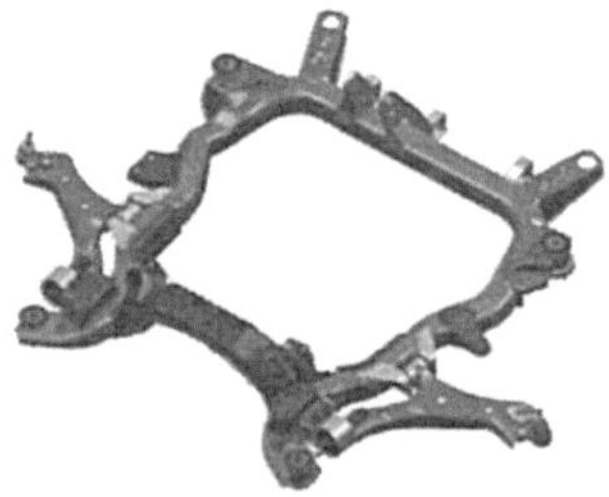

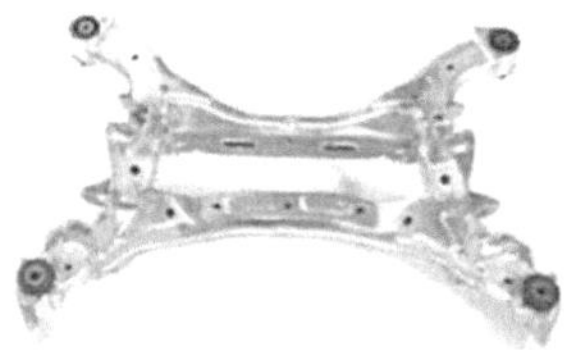

(a) Steel engine cradle (b) Aluminum alloy rear subframe

Fig. 4.49 Application examples of hydroforming to subframes

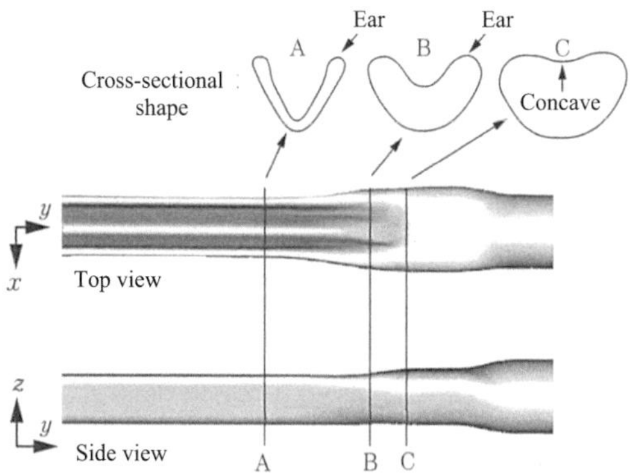

Fig. 4.50 Application example of hydroforming to torsion beam [63]

as depicted in Fig. 4.50 [63]. Furthermore, there is a case of developing axle housing that expand the original tube to three times its original circumference (expansion ratio of 200%) using only cold forming techniques, as shown in Fig. 4.51.

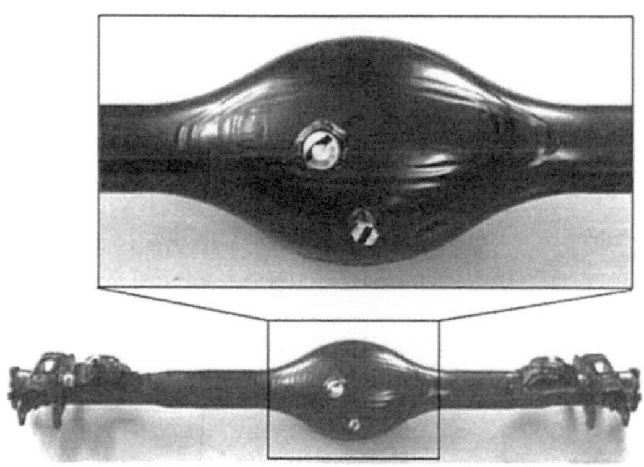

Fig. 4.51 Hydroformed axle housing expanded to three times original tube circumference

Hydroforming has also found multiple applications in the field of exhaust systems. It is applied to various branch pipes, such as the exhaust manifolds depicted in Fig. 4.52 [64]. The application of hydroforming for the body parts is also progressing, with examples including front pillars [65] as shown in Fig. 4.53 and center pillars [66] as depicted in Fig. 4.54. In other areas, hydroforming has been employed in steering columns for the steering system [67], as illustrated in Fig. 4.55.

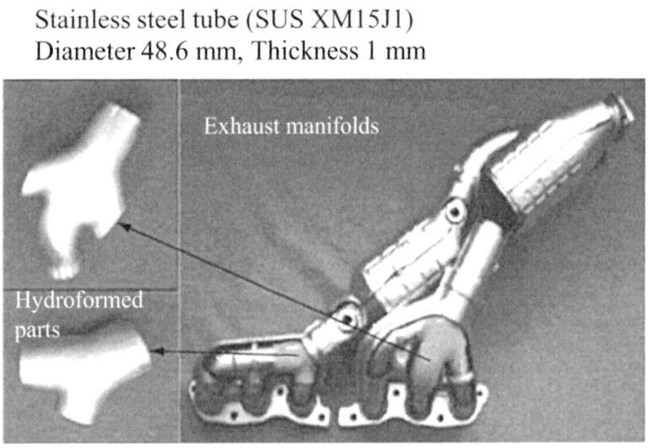

Fig. 4.52 Application example of hydroforming to exhaust manifolds [64]

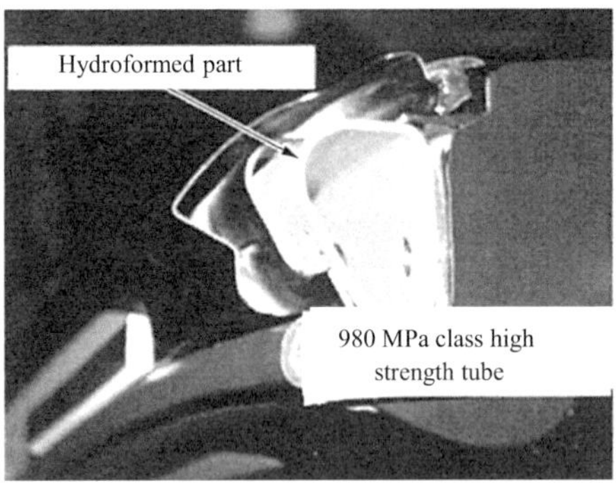

Fig. 4.53 Application example of hydroforming to front pillar reinforcement [65]

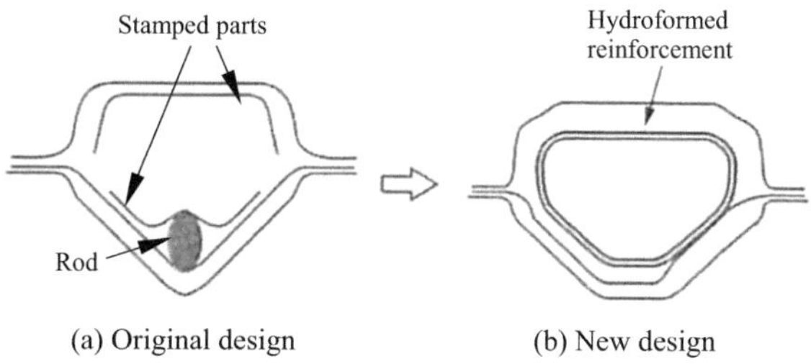

Fig. 4.54 Application example of hydroforming to center pillar reinforcement [66]

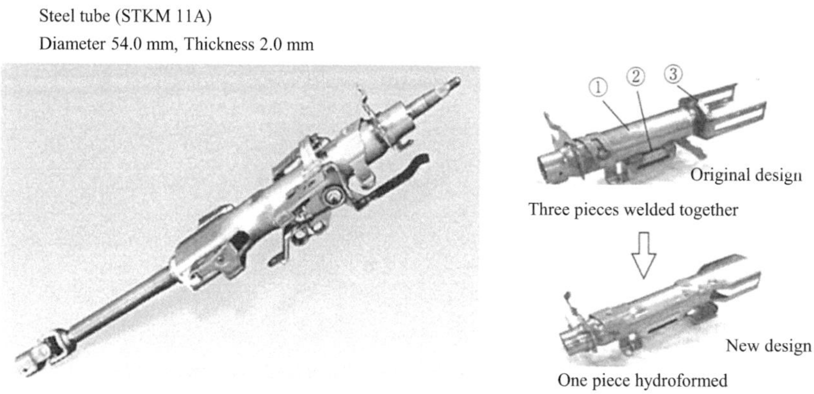

Fig. 4.55 Application example of hydroforming to steering column [67]

4.7.2 *Other Components*

One commonly adopted application since the era known as bulge forming is in pipe fittings [8], bicycle joints [9], and musical instruments. Examples include pipe fittings for piping systems as shown in Fig. 4.56 [68], bicycle joints with multiple-branch pipes as shown in Fig. 4.57, and horns as shown in Fig. 4.58 [69]. Furthermore, we will introduce examples of recent expanded applications. It has been applied not only to four-wheeled automobiles but also to motorcycles, as seen in the application to swing arms shown in Fig. 4.59 [70]. Figure 4.60 shows an example of application to a gas turbine combustor using super heat-resistant alloy pipes [71]. Additionally, the expansion of applications to smaller sizes is being considered. For the purpose of applying to components such as eyeglasses, there are studies on branch pipe forming using pipes with a diameter of less than 10 mm, as shown in Fig. 4.61 [72]. Application examples with diameters of less than 1 mm are also being considered for the medical field and the communication-electronics field. For more details on such microfabrication, please refer to Chap. 9.

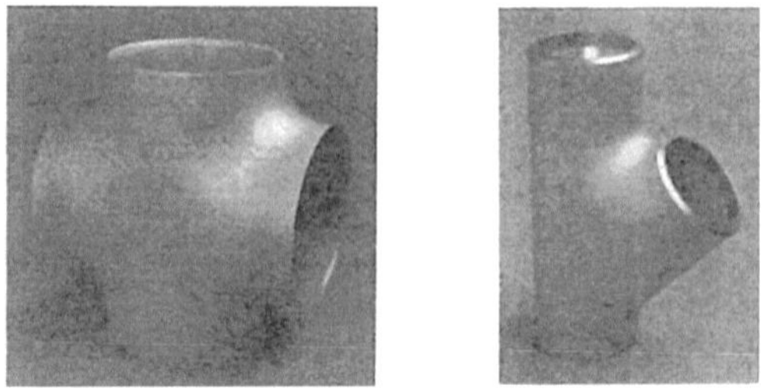

(a) Cross-fitting (b) Y-fitting

Fig. 4.56 Application examples of hydroforming to pipe fittings [68]

(a) Hunger lug (b) Head lug

Fig. 4.57 Application examples of hydroforming to bicycle joints

Fig. 4.58 Application examples of hydroforming to brass instrument (horn) [69]

Fig. 4.59 Application examples of hydroforming to motorcycle [70]

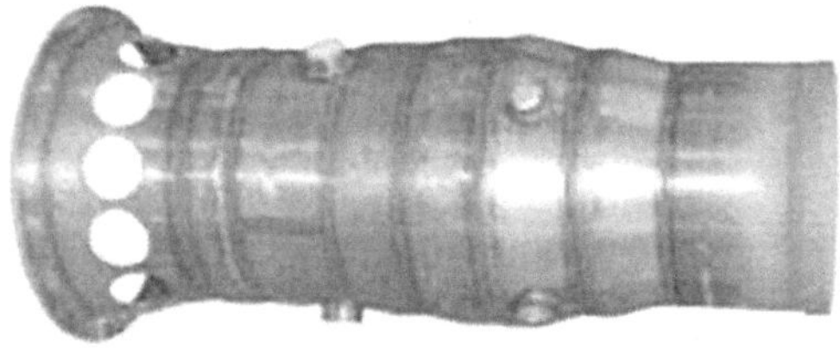

Fig. 4.60 Application examples of hydroforming to gas turbine combustor [71]

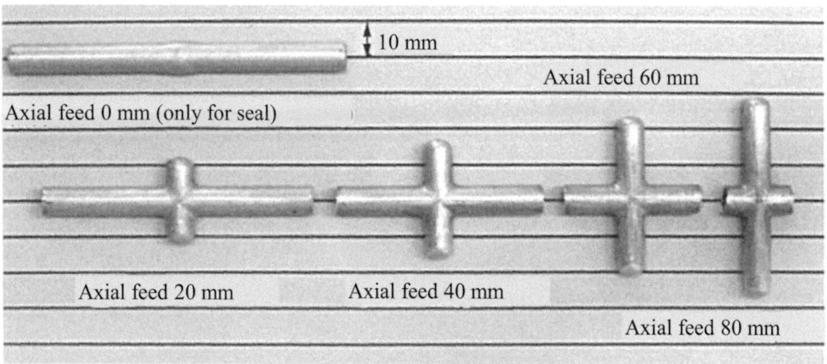

Initial diameter 8.0 mm, thickness 0.5 mm

Fig. 4.61 Cross forming of small aluminum alloy tube by hydroforming [72]

References

1. US Patent : 2203868
2. Takagi, R.: Report on Research and Guidance in Bicycle Production Technology. Japan Bicycle Industry Association (1957)
3. Ogura, T., Ueda, T.: Reports of Government Industrial Research Institute. Nagoya **7-2**, 89–94 (1958)
4. Ogura, T. Ueda, T.: Rep. Gov. Ind. Res. Inst., Nagoya **7-10**, 719–723 (1958)
5. Ogura, T. Ueda, T., Ishikawa, M.: Rep. Gov. Ind. Res. Inst., Nagoya **8-9**, 603–608 (1959)
6. Ogura, T. Ueda, T., Ishikawa, M.: Rep. Gov. Ind. Res. Inst., Nagoya **11-13**, 131–136 (1962)
7. Ogura, T. Ueda, T., Ishikawa, M.: Rep. Gov. Ind. Res. Inst., Nagoya **11-9**, 524–528 (1962)
8. Kimura, J.: Proceedings of the 75th Symposium on the Technology of Plasticity, JSTP, pp. 52–59 (1981)
9. Takagi, R.: J. JSTP **12-120**, 59–66 (1971)
10. Manabe, K.: J. JSTP **39-453**, 99–1004 (1998)
11. Leitlof, F.U., Geisweid, S.: J. JSTP **39-453**, 1045–1049 (1998)
12. Shibata, M.: Proceedings of the 180th Symposium on the Technology of Plasticity, JSTP, pp. 9–16 (1998)
13. Manabe, K.: Proc. ICEM-98 119–125 (1998)
14. Kuriyama. Y.: J. JSTP **39-453**,1009–1013 (1998)
15. Hamada, M.: Nikkei Mech. **539**, 32–46. Nikkei Business Publications (1999)
16. Mizumura, M., Yoshida, T.: Mach. Tool **46-10**, 46–51. Japan Industrial Publishing (2002)
17. Abe, H., Sonobe, O.: Press Work. **39-7**, 24–27. Nikkan Kogyo Shimbun (2001)
18. The Japan Society for Technology of Plasticity (Ed.): Tube Hydroforming. Morikita Publishing (2015)

19. Manabe, K., Fuchizawa, S.: J. JSTP **52-600**, 36–41 (2011)
20. Yogo, T., Ito, M., Mizuno, T.: J. JSTP **45-527**, 1026–1030 (2004)
21. Tominaga, H., Takamatsu, M.: J. JSTP **7-69**, 548–555 (1966)
22. Nakazawa, K., Yaji, S., Kaga, H.: J. JSTP **16-171**, 306–312 (1975)
23. Hokamoto, K., Kakimoto, E.: J. JSTP **44-512**, 906–910 (2003)
24. Sato, K., Hiramatsu, K., Hiroshige, I., Mano, K.: Proceedings of the 2010 Spring Meeting, JSTP, pp. 1–2 (2006)
25. Ukita, H.: J. JSTP **32-36**, 818–824 (1991)
26. Wada, M., Kaneda, H., Miaumura, M., Iguchi, K.: J. JSTP **55-647**, 1102–1106 (2014)
27. Kiyama., K., Kitano, Y., Nakao, K.: J. JILM **56-1**, 63–67 (2006)
28. Jinishi, T., Naganuma, T., et al.: Proceedings of the 2010 Spring Meeting, JSTP, pp. 189–190 (2010)
29. Nogiwa, K., Ishizuka, M., et al.: Proceedings of the 50th Joint Conference on the Technology of Plasticity, JSTP, pp. 41–412 (2017)
30. The High Pressure Gas Safety Institute of Japan (Ed.): High Pressure Gas Safety Regulations, 16th edn. The High Pressure Gas Safety Institute of Japan (2016)
31. Kanbara, H., Yoshitomi, Y., Shimaguchi, T., Nomura, H.: J. JSTP **23-255**, 315–320 (1982)
32. Ohashi, T., Hayashi, K., Saotome, Y., Amada, S.: Proceedings of the 2000 Spring Meeting, JSTP, pp. 261–262 (2000)
33. Ohashi, T., Matsui, K., Saotome, Y.: Proceedings of the 52nd Joint Conference on the Technology of Plasticity, JSTP, pp. 27–28 (2001)
34. Kuwabara, T.: J. JSTP **54-624**, 18–24 (2013)
35. Mihara. Y.: J. JSTP **53-614**, 183–187 (2012)
36. Fuchizawa, S.: J. JSTP **41-478**, 1075–1081 (2000)
37. Mizumura, M., Kuriyama, Y.: J. JSTP **44-508**, 524–529 (2003)
38. Mizumura, M., Honda, O., Yoshida, T., Iguchi, K., Kuriyama, Y.: Nippon Steel Technical Report, vol. 380, pp. 101–105 (2004)
39. Kimura, T., Hidaka, T., Yasutomo, T., et al.: Proceedings of the 51st Joint Conference on the Technology of Plasticity, JSTP, pp. 347–348 (2000)
40. Mizumura, M.: Doctoral thesis, The University of Tokyo (2006)
41. Manabe, K., Miyamoto, S., Koyama, H.: Proceedings of the 2002 Spring Meeting, JSTP, 259–260 (2002)
42. Amino, M., Manabe, K.: Proceedings of the 48th Joint Conference on the Technology of Plasticity, JSTP, pp. 373–374 (1997)
43. Manabe, K., Amino, M.: Proceedings of the 49th Joint Conference on the Technology of Plasticity, JSTP, pp. 303–304 (1998)
44. Mizumura, M., Kuriyama, Y.: J. JSTP **45-516**, 60–64 (2004)
45. Mizumura, M., Kuriyama, Y.: J. JSTP **45-517**, 103–107 (2004)
46. Mizumura, M., Kuriyama, Y.: J. JSTP **45-525**, 817–821 (2004)
47. Yoshida, T., Kuriyama, Y., Sumimoto, D., et al.: Proceedings of the 2000 Spring Meeting, JSTP, pp. 425–426 (2000)
48. Iguchi, K., Mizumura, M., Niwa, T., Kuriyama, Y., Sawada, S.: CAMP-ISIJ 20–977 (2007)
49. Tomizawa, A., Yasuyama, M., Kameoka, N.: Proceedings of the 56th Joint Conference on the Technology of Plasticity, JSTP, pp. 197–198 (2005)
50. Tomizawa, A., Kameoka, N.: Proceedings of the 56th Joint Conference on the Technology of Plasticity, JSTP, pp. 199–200 (2005)
51. Mizumura, M., Kuriyama, Y.: J. JSTP **53-614**, 225–230 (2012)
52. Watanabe, Y., Abe, S., Tomizawa, A., et al.: CAMP-ISIJ **20**, 966–969 (2007)
53. Mizumura, M., Kuriyama, Y., Iguchi, K.: Proceedings of the 56th Joint Conference on the Technology of Plasticity, JSTP, pp. 189–190 (2005)
54. Mizumura, M., Sato, K., Kuriyama. Y.: Proceedings of the 2008 Spring Meeting, JSTP, pp. 231–232 (2008)
55. Uchida, T., Kojima, M., Tomizawa, A., et al.: Proceedings of the 2010 Spring Meeting, JSTP, pp. 181–182 (2010)

56. Liewald, M., Wagner, S.: Proc. TUBEHYDRO **2007**, 19–26 (2007)
57. Arita, H.: J. JSTP **57-671**, 1142–1143 (2016)
58. Abe, H.: J. JSTP **45-525**, 809–813 (2004)
59. Fukumura, T.: J. JSTP **53-614**, 199–203 (2012)
60. Hiramatsu, K., Ishihara, S., Kadoma, Y., Haeno, T., Honda, S., Sato, K.: J. JSTP **46-539**, 1147–1150 (2005)
61. Kikuchi, F., Oyama, K., Tashiro, K.: J. JSTP **47-550**, 1064–1068 (2006)
62. Nasu, Y.: Press Work. **41-43**, 38–41. Nikkan Kogyo Shimbun (2003)
63. Iguchi, K., Wada, M., Mizumura, M., Suehiro, M.: Proceedings of the 2010 Spring Meeting, JSTP, pp. 301–302 (2014)
64. Hamanishi, Y., Kato, K.: Proceedings of the 322nd Symposium on the Technology of Plasticity, JSTP, pp. 37–43 (2001)
65. Nikkei Automotive Technology (Ed.): Nikkei Automotive Technology, vol. 1, p. 31. Nikkei Business Publications (2009)
66. Majima, S., Sumimoto, D., Sugiyama, T., Yoshida., T., Terada, K., Kuriyama. Y., Takahashi, S.: Proceedings of the 2000 Spring Meeting, JSTP, pp. 427–428 (2000)
67. Watanabe, Y., Abe, S., Shimoi, K., Tomizawa, A., Uchida, T., Kojima, Y., Kikuchi, F.: Proceedings of the 2007 Spring Meeting, JSTP, pp 255–256 (2007)
68. Kimura, J.: Press Work. 27–11, 81–86. Nikkan Kogyo Shimbun (1989)
69. Ouchi, T.: J. JSTP **19-210**, 573–578 (1978)
70. Kimura, T., Yasutomo, T., Kawasaki, T., Kanamaru, A., Araki, T.: KHI Tech. Rev. **151**, 60–62. Kawasaki Heavy Industries (2002)
71. Ikeuchi, Y., Kitano, H., Kimura, T., Fujiwara, S., et al.: KHI Tech. Rev. **161**, 49. Kawasaki Heavy Industries (2006)
72. Shirayori, A.: J. Mater. Process Technol. Center **5-11**, 22–27 (2014)

Chapter 5
Tube End-Forming

5.1 Fundamentals

5.1.1 Classification and Overview [1, 2]

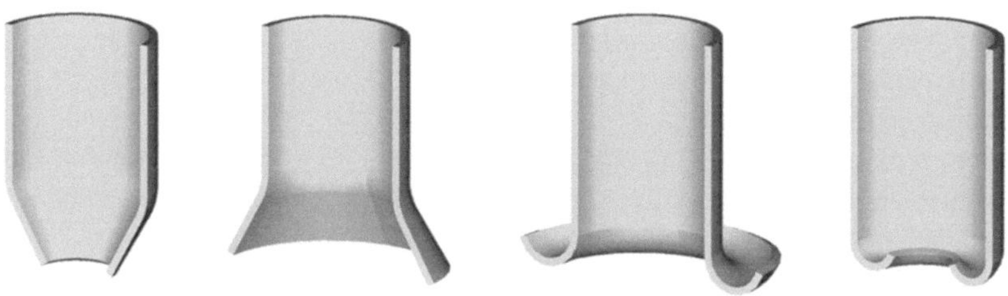

Fig. 5.1 Schematics of tube end-formed product

Tube end-forming includes processes such as necking, flaring, curling, and reverse forming. Figure 5.1 illustrates schematics of formed products for each type of process. Necking is a process that compresses the end of a tube in the circumferential direction using a rigid tool from the outside of the tube. In necking, compressive stress is also induced in the axial direction of the tube due to axial loading and frictional forces. Flaring is a process that expands the tube end significantly from the inside to the outside using tools or other means. Similar to other processes of tube end-forming, this process can be categorized into methods such as press forming, rotary forming, bulging using internal pressure, and high-energy rate forming techniques like electromagnetic forming. Curling involves bending the edge portion of the tube

Y. Kuriyama et al., *Tube Forming*, https://doi.org/10.1007/978-981-95-3421-0_5

end or deep-drawn container outwards or inwards to enhance product handling safety and reinforce the end. Reverse forming is an advanced technique that takes curling further by completely flipping the inner or outer surface of the tube, creating a new cylindrical tube that aligns with the axial center of the original tube.

To understand the material deformation and processing defects in tube end-forming, the stress states for each part of necking and flaring, including the changes during the process, are examined as shown in Fig. 5.2. In both processes, a plane stress field is assumed based on the assumption zero stress in the thickness direction for thin-walled cylindrical tubes. The numerals 1–3 represent different parts, with the tube end as 1, the boundary between the taper part and parallel part as 3, and the intermediate part between 1 and 3 as 2. Capital letters as subscripts indicate the processing stages, where S represents the initial stage, M represents the intermediate stage, and E represents the final stage.

The stress state in necking, assuming a plane stress field (for thin-walled cylindrical tubes) with zero stress in the thickness direction, results in a compressive-compressive stress field in the third quadrant of Fig. 5.2. The stress history varies depending on the part of the tube. Here, we assume that circumferential wrinkles do not occur and only axisymmetric buckling takes place for this process. Part 1, where there is no material in front, is always in the state of 1_{SME}. In part 3, where buckling occurs, it transitions to the state of 3_E near the bending part, and deformation progresses in the direction of the normal vector n_3, strain increment direction is normal to yield locus, increasing the circumference and causing buckling. When the intermediate part 2 moves along the die slope, the stress state of 2 moves on the yield surface from 2_M to 2_E.

From a different perspective, the stress state changes as necking progresses. In the mid-stage (M) of necking, at the parallel part just before the bending part which is boundary of parallel part and taper part, the state is elastic, denoted as 2_{Ma}, while at the bending part, it is denoted as 2_{Mb}, and at the tube end, it is denoted as 1_M [3]. In the later stage of necking, during the occurrence of buckling (E), at the parallel part just before the bending part (denoted as 3_E) the state is plastic, and it transitions to 2_E at the bending part. In the third quadrant that represents necking process, there are no processing limits due to fracture or rupture. However, the processing limits are determined by various types of defects, including buckling (refer to Table 5.1).

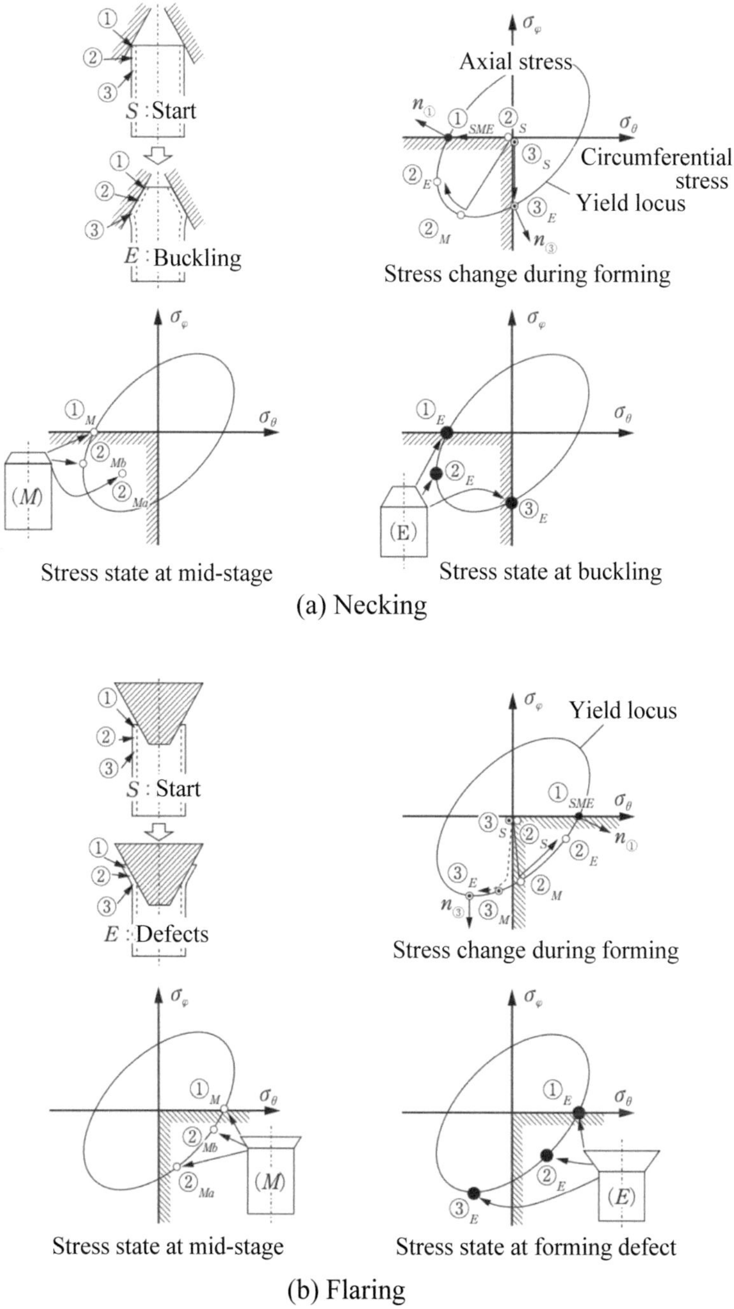

Fig. 5.2 State of stress during end-forming

Table 5.1 Buckling and forming defects in conical necking [2]

Zone	Forming Part			
Mode	1	2	3	4
Appearance	Half wave length Buckling 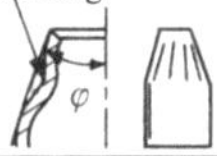	Quarter wave length Buckling 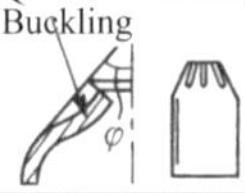	Wavy deformation on die; Round tube 	Curling deformation 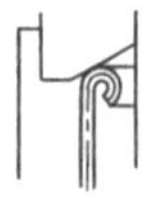
Condition	$\varphi < \sim 20°$	$\sim 30° < \varphi$	$\sim 30° < \varphi$	$\sim 60° < \varphi$ outer constraint, well lubrication
Material	Hard (small n-value) small t_0/d_0 *	Hard (small n-value) small t_0/d_0	–	–

* In the case of high deformation, buckling occurs even with large t_0/d_0

Zone	Cylindrical Part	
Mode	5	6
Appearance	Axisymmetric Buckling 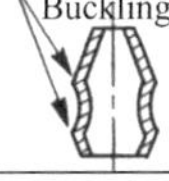	Asymmetric Buckling 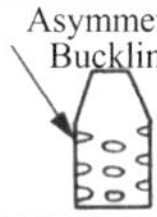
Condition	Free Necking (without constraint tool)	Outside constraint necking
Material	Soft (large n-value)	Soft (large n-value)

The stress state in flaring is primarily in a tension–compression stress field in the fourth quadrant of Fig. 5.2. The stress history varies depending on the part of the tube. In part 1, where there is no material in front, it is always in the state of 1_{SME}. At the bending part 3, it is in the state of 3_E. When the intermediate part 2 moves along the die slope, the stress state of 2 moves on the yield surface from 2_M to 2_E. From a different perspective, the stress state changes as flaring progresses. In the mid-stage (M) of flaring, the entry of bending part, which is the boundary between the parallel part and the taper part, is closer to a plastic state 2_{Ma}, while at the bending part, it is denoted as 2_{Mb}, and at the tube end, it is denoted as 1_M. In the later stage (E) with the occurrence of defect formation, the entry side of the bending part becomes 3. In the parallel part, buckling occurs, and at the tube end where σ_θ becomes maximum, the stress state is 1_E, which may lead to crack formation.

There are two types of curling: outward curling and inward curling. In the case of outward curling, the deformation experienced by the tube mainly consists of circumferential elongation, radial bending, and thickness direction shear, which are mixed together. In the reverse process after the curling tip is oriented in the opposite direction to the axial direction of the original tube, additional bending-back occurs, and even after curling is completed, buckling may occur without reverse deformation.

Inward curling and inward reverse are similar to outward curling and outward reverse, but the formable range of tube dimensions and the suitable conditions on the tool are narrower compared to outward process.

Among tube end-forming, necking is the most widely used. Necking can be achieved through (1) press forming, (2) rotary forming, and (3) ironing (refer to Sect. 6.3). Press forming (1) allows for high-efficiency production through continuous processing. However, when the forming force becomes significant, the process needs to be divided and the degree of deformation controlled to prevent tube buckling. Rotary forming techniques (2) include spinning (refer to Sect. 6.2.2), forming with roller tools [4], forming with planetary ball dies [5], and orbital rotary flaring (refer to Sect. 6.4.4). Additionally, rotary nosing using an asymmetric die has been proposed (refer to Sect. 6.4.2) [6]. The forming surface of the asymmetric die consists of a portion of the conical surface in contact with the material and a non-contacting relief area. In terms of contact area, rotary nosing with asymmetric die falls between press forming and spinning. Although its productivity is lower compared to press forming, it offers excellent flexibility and can achieve higher forming limits if suitable processing conditions are selected. Ironing machines (3) are commonly used for reducing the diameter of pipes with small taper angles. In necking processes, there are cases where the open end is completely closed, similar to the closure of high-pressure cylinders or metal bats (refer to Sect. 5.2), to create a final product. However, in general, necking is frequently performed as a pre-processing step for joining tubes and components in applications such as piping components and structural members.

5.1.2 Theory of Necking Process [2, 7]

Necking processes encompass various techniques, but they share a common characteristic of material forming occurring in a compression-compression stress field, as mentioned earlier. Here, we present the theory for tapering thin-walled circular tubes using a press, which is the most fundamental necking process (Fig. 5.3). The target thin-walled circular tube is assumed to be an *n*-power hardening material with planar anisotropy. We apply membrane theory and total strain theory, ignoring stress in the thickness direction, bending moments, and shear deformation.

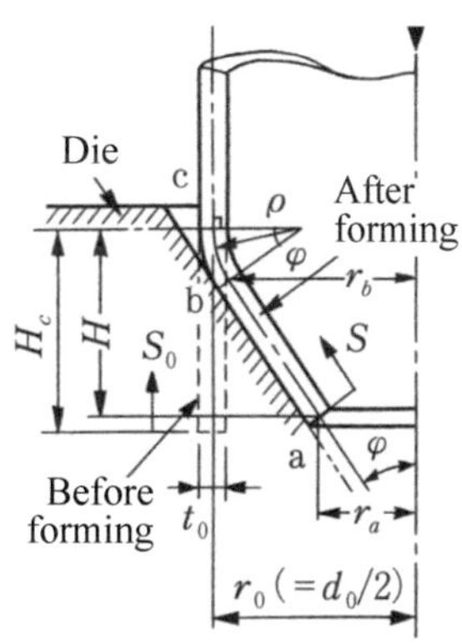

Fig. 5.3 Deformation in necking process

(1) **Bending radius ρ**

The radius ρ of the bending part formed at the die entrance (between points b and c in Fig. 5.3) is given by the following equation. When the tube end starts to conform to the die or when bending alone occurs, such as in curling deformation, the equation is expressed as Eq. (5.1). In a typical necking process where a cone shape is formed and bending and unbending are involved, Eq. (5.2) is applicable [7–9].

$$\rho = \sqrt{\frac{t_0 r_0}{4(1 - \cos\varphi)}} \tag{5.1}$$

$$\rho = \sqrt{\frac{t_0 r_0}{2(1 - \cos\varphi)}} \tag{5.2}$$

(2) **Fundamental equations**

Below, we present the fundamental equations.

(a) **Strain and *r*-value**

The strains in the circumferential direction θ, thickness direction t, and meridional direction φ are expressed as Eq. (5.3). The plastic anisotropy coefficient is given by Eq. (5.4), where F, G, and H are Hill's anisotropy parameters.

$$\varepsilon_\theta = \ln(r/r_0),\ \varepsilon_t = \ln(t/t_0),\ \varepsilon_\varphi = -(\varepsilon_\theta + \varepsilon_t) \tag{5.3}$$

$$r_\varphi = (\varepsilon_\theta/\varepsilon_t)_{\sigma_\theta=0} = H/G,\quad r_\theta = \left(\varepsilon_\varphi/\varepsilon_t\right)_{\sigma_\varphi=0} = H/F \tag{5.4}$$

(b) **Stress–Strain relationship**

Applying Hill's anisotropy theory, the stress–strain relationship in the total strain theory is expressed as Eq. (5.5), with Eqs. (5.6) and (5.7).

$$\sigma_\varphi = -\frac{1}{\overline{\lambda}}\left\{\frac{r_\varphi}{r_\theta}\varepsilon_\theta + \left(r_\varphi + \frac{r_\varphi}{r_\theta}\right)\varepsilon_t\right\},\ \sigma_\theta = \frac{1}{\overline{\lambda}}\left(\varepsilon_\theta - r_\varphi\varepsilon_t\right) \tag{5.5}$$

where $\overline{\lambda} = \dfrac{3r_\varphi\{1+(1/r_\theta)+(r_\varphi/r_\theta)\}}{2\{1+(r_\varphi/r_\theta)+r_\varphi\}} \cdot \dfrac{\varepsilon_{eq}}{\sigma_{eq}}$

$$\left.\begin{aligned} \sigma_{eq} &= \sqrt{\frac{3}{2}\cdot\frac{1}{r_\varphi + r_\theta + r_\theta r_\varphi}\left\{r_\theta\left(1 + r_\varphi\right)\sigma_\varphi^2 + r_\varphi(1 + r_\theta)\sigma_\theta^2 - 2r_\varphi r_\theta\sigma_\varphi\sigma_\theta\right\}} \\ \varepsilon_{eq} &= \sqrt{\frac{2}{3}\cdot\frac{r_\varphi + r_\theta + r_\theta r_\varphi}{r_\varphi\left(1 + r_\theta + r_\varphi\right)}\left\{\frac{r_\varphi}{r_\theta}(\varepsilon_\theta + \varepsilon_t)^2 + \varepsilon_\theta^2 + r_\varphi\varepsilon_t^2\right\}} \end{aligned}\right\} \tag{5.6}$$

$$\sigma_{eq} = C\varepsilon_{eq}^n \tag{5.7}$$

(c) **Equilibrium equations**

In the cone forming part:

$$\frac{d(r\sigma_\varphi t)}{dr} - \sigma_\varphi t(1 + \mu\cos\varphi) = 0 \tag{5.8}$$

In the bending deformation part:

$$\frac{d(r\sigma_\varphi t)}{d\varphi} + \rho\sigma_\theta t\sin\varphi = 0 \tag{5.9}$$

(d) **Forming load**

Ignoring bending stress:

$$W = 2\pi r_0 t_c \sigma_{\varphi c} \tag{5.10}$$

Considering bending stress:

$$W = 2\pi r_0 t_c \left\{ \sigma_{\varphi c} + C\left(\frac{t_b + t_c}{4\rho}\right)^{1+n} \right\} \tag{5.11}$$

Note that the above equations take into account the unbending at point b in Fig. 5.3.

(e) **Dimensionless quantities**

By dimensionless stress, forming load, and bending radius using the following equation, it becomes possible to conduct general discussions:

$$\left.\begin{array}{l} \overline{\sigma}_{eq} = \dfrac{\sigma_{eq}}{C}, \overline{\sigma}_\varphi = \dfrac{\sigma_\varphi}{C}, \overline{\sigma}_\theta = \dfrac{\sigma_\theta}{C} \\ \overline{W} = \dfrac{W}{C r_0 t_0}, \overline{\rho} = \dfrac{\rho}{\sqrt{d_0 t_0}} \end{array}\right\} \tag{5.12}$$

5.1.3 *Theory of Flaring Process*

Here, we present the theory and results of the flaring process, which involves pressing a tapered punch into the workpiece to achieve flaring (Fig. 5.4). Similar to necking, we assume an n-power hardening material and apply membrane theory and total strain theory.

In the flaring process, the fundamental equations such as bending radius, equilibrium equation, and forming load, as well as the concept of dimensionless quantities,

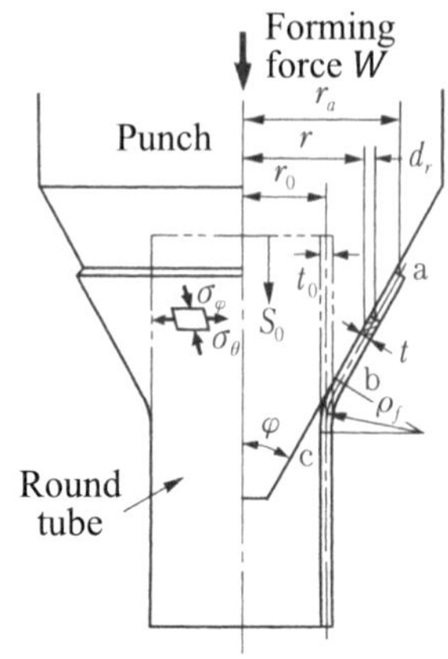

Fig. 5.4 Deformation in flaring process

are similar to those in the necking process. The degree of deformation is measured by the flaring ratio = $(r_a/r_0) - 1$.

Representative examples of theoretical analysis results are shown below [10, 11]. Figure 5.5 illustrates an example of stress and strain distribution obtained through theoretical analysis considering the bending deformation at the punch entrance. The influence of the friction coefficient μ between the punch and the tube increases the meridional stress σ_φ at the punch contact boundary side, but on the other hand decreases the circumferential stress σ_θ.

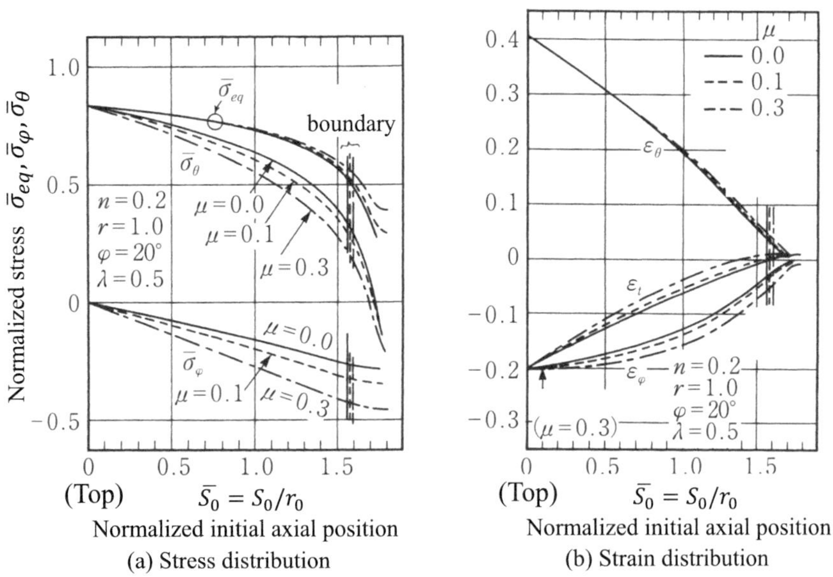

Fig. 5.5 Stress and strain distribution in flaring process

Furthermore, the punch half-angle φ and μ as processing conditions show qualitatively similar effects on frictional forces. As φ decreases, the contact length of the punch increases, resulting in higher frictional forces. When μ is large, similar results are observed.

Consequently, the strain distribution corresponds to this, with higher μ and smaller φ, meridional strain ε_φ (compression) and thickness strain ε_t at the punch contact boundary side increases.

5.1.4 *Theory of Curling and Reverse Forming*

The deformation mechanism of curling and the forming force in reverse forming have been studied using the energy method, assuming that the thickness of the circular tube, t_0, is sufficiently small compared to the mean radius, d_0, and there is no friction between the material and the tool surface. The results are presented below.

(1) **Deformation mechanism of curling** [12–14]

Whether the tube end unbends to the tool surface and forms flaring or unbending does not occur and curling continues depends on t_0, d_0, and the tool conditions (punch half-angle φ and the radius of curvature at the entrance). Consider the state where the tube is curled along the entrance with a radius ρ, as shown in Fig. 5.6a. Point P represents the intersection between the arc section and the conical surface. Analytical studies have shown that curling persists when the energy increment of the curling mode is smaller than the energy increment of the unbending mode for the material in front of the point P to the tube end [12]. According to this analysis, the following condition must be satisfied for curling: the curvature radius ρ of the tube formed in the arc section must be smaller than a value ρ_m determined by the following Eq. (5.13).

$$\rho_m = \sqrt{\frac{t_0 d_0}{8\{2\cos\varphi - (\pi - 2\varphi)\}\sin\varphi}} \tag{5.13}$$

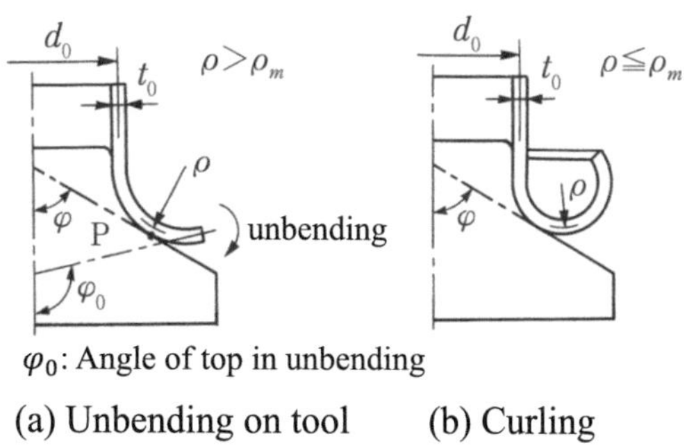

Fig. 5.6 Unbending and curling in flaring process

When normalized by $\sqrt{t_0 d_0}$, it becomes Eq. (5.14).

$$\overline{\rho}_m = \frac{1}{\sqrt{8\{2\cos\varphi - (\pi - 2\varphi)\}\sin\varphi}} \tag{5.14}$$

By setting $\varphi = 60°$ in the equation, $\overline{\rho}_m$=1.15 is obtained, and at $\varphi_0 = 120°$, unbending occurs and transitions to flaring. Moreover, by using $\varphi_0 = 60°$ and $\overline{\rho}_m$=1 for the entrance section radius, curling can persist without unbending. As φ increases, the unbending towards the tool surface is delayed, resulting in the formation of a curled section as shown in Fig. 5.6b. Of course, at $\varphi = 90°$, $\overline{\rho}_m$ becomes infinite, and curling always occurs unless the tube buckles or the edge breaks.

On the other hand, for the curling mode by a conical tool without an entrance section, the dimensionless representation of the tube's bending radius ρ_0 obtained by analysis is given by the following Eq. (5.15) [8, 11].

$$\overline{\rho}_0 = \frac{1}{\sqrt{8(1 - \cos\varphi)}} \tag{5.15}$$

Figure 5.7 [12] shows the curling region and the flaring region based on Eqs. (5.14) and (5.15). The intersection of $\overline{\rho}_m$ and $\overline{\rho}_0$ curves represents $\varphi = \varphi_S$ (approximately 45°), which is the starting angle of curling with a conical tool.

When using a tool with $\varphi = 60°$, for example, even if an entrance section with a smaller radius is provided, it deforms according to $\overline{\rho}_0$ (refer to Eq. (5.15)), and a gap is formed between the entrance section and the material.

(2) **Processing force in reverse forming with arc tool** [15]

In Fig. 5.8, set $\varphi = 90°$ and the radius of the arc section as$r_d = \rho + t_0/2$. The material's flow stress is expressed as, where σ_{eq}, ε_{eq}, n, C, and ε_0 represent the equivalent stress, equivalent strain, work-hardening exponent, strength coefficient, and initial strain, respectively. The value of $\overline{\rho}_0$ that minimizes the sum of the expansion deformation energy from diameter d_0 to d_1 of a circular tube with a constant initial thickness t_0 and the bending and unbending is given by Eq. (5.16), and the corresponding minimum processing force W_{min} is given by equation (Al-Hassani-Johnson equation) (5.17).

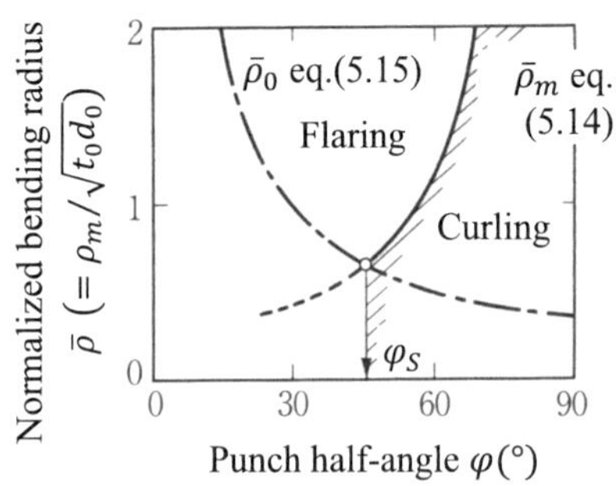

Fig. 5.7 Starting angle of curling with conical tool

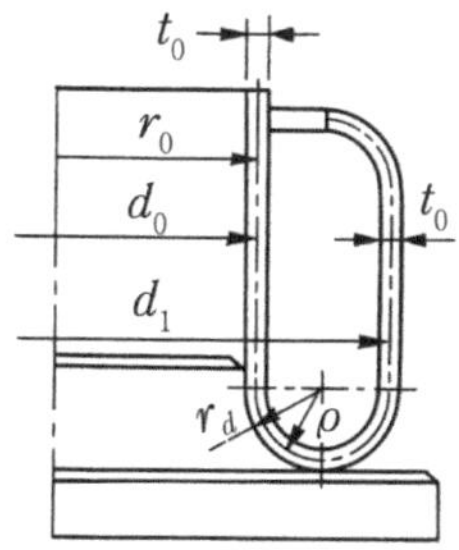

Fig. 5.8 Reverse forming with arc tool

$$\overline{\rho}_0 = \frac{1}{2\sqrt{2}} \tag{5.16}$$

$$W_{\min} = \frac{2\pi d_0 t_0 C}{n+1}\left\{\varepsilon_0 + \frac{2}{\sqrt{3}}\ln\left(1+\sqrt{\frac{2t_0}{d_0}}\right)\right\}^{n+1} \tag{5.17}$$

The analysis of reverse forming [16–19] is based on the same principles as the previous analysis, using the total strain theory and the energy method, and there are no significant differences in $\overline{\rho}_0$ and $W_{\min}$ obtained from these analyses.

5.2 Processing Methods [20]

Figure 5.9 shows the types of end-forming methods for tubes. These include flaring, nosing (necking), curling, and reverse forming, as well as processes such as upset forming, which changes the thickness of the tube through axial, radial, and lateral compression deformation, and crushing of pipe sections.

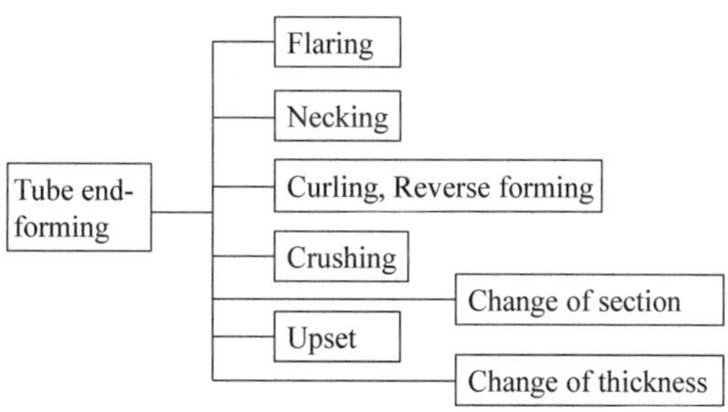

Fig. 5.9 End forming methods for tubes

Figure 5.10 illustrates the classification of end-forming based on the shape [20]. For nosing (necking), there are tapering, flanging, and necking processes, with a unique process called closing where the opening end is fully closed. For flaring, there are flaring, flanging, step forming, and other processes. For both nosing (necking) and flaring, there are eccentric forming and curling processes.

The representative methods include press forming, but there are also rotary forming (spinning, full-contact rotary forming, and orbital rotary forming), as well as ironing by rotary ironing. Figure 5.11 shows the representative processing methods [20]. For information on rotary forming and rotary ironing, please refer to Sect. 6.3. Here, we will discuss the processing methods, machinery, and examples of processed products for tube end-forming using presses.

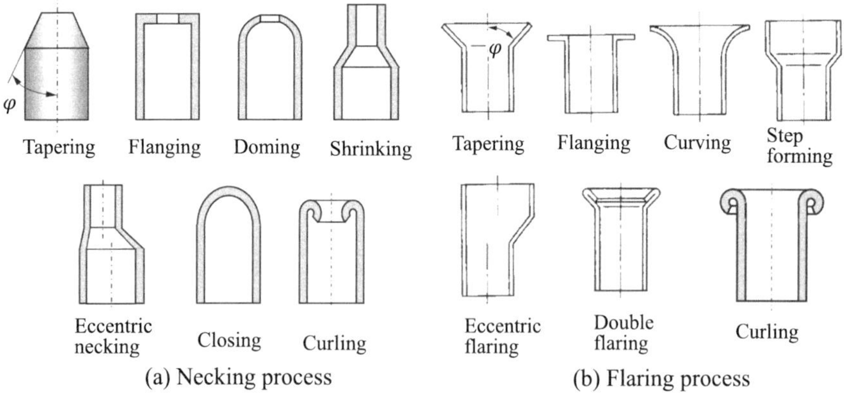

Fig. 5.10 Classification of end-forming based on the shape [20]

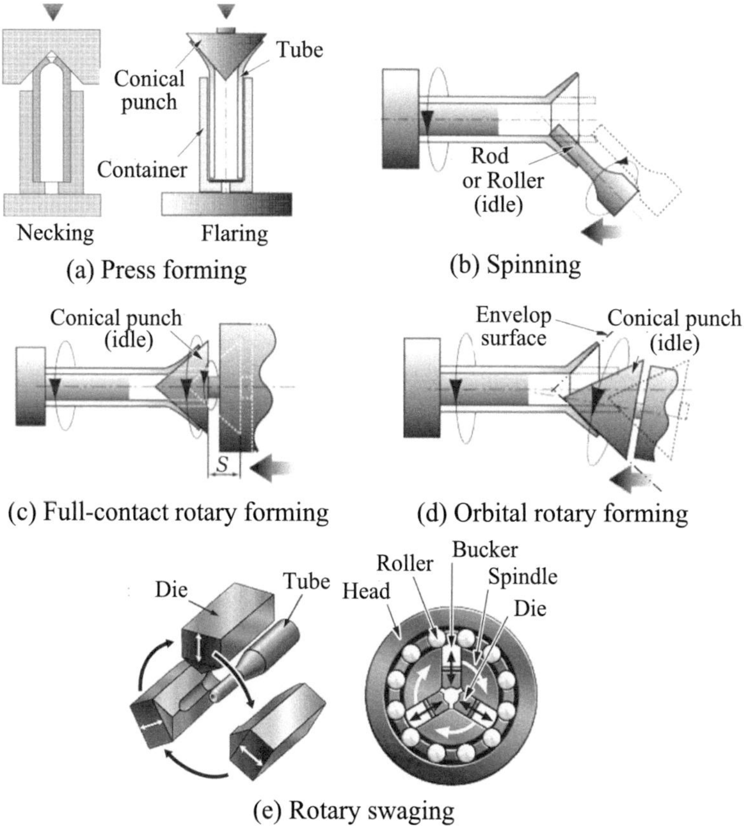

Fig. 5.11 Representative tube end-forming methods [20]

5.2.1 *Nosing Process by Press*

(1) **Overview**

As shown in Fig. 5.12, this is a processing method that uses a press to push the pipe material into a die and shape it into the die's form. The equipment is cost-effective and highly productive compared to other methods. The processing shape can be tapered, narrowed, eccentrically necked, or curled by changing the die, making it versatile.

Flanging and curved surface processing are also possible, but there are challenges in surface accuracy.

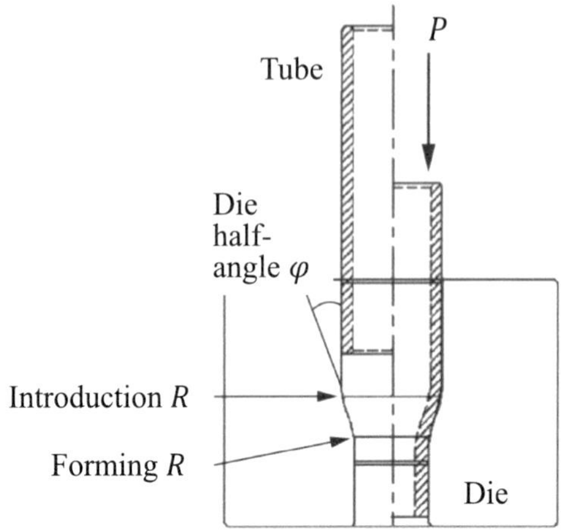

Fig. 5.12 Nosing process by press

(2) **Application guidelines**

The taper angle (half-angle φ) for tapering and narrowing processes should be 5° or more, with 22–30° being suitable. The achievable reduction ratio ($(d_0 - d_1)/d_0$, where d_0 is the initial outer diameter and d_1 is the final outer diameter) in a single process is typically around 20–28%, depending on the processing speed. If a higher reduction ratio is required, the process may be divided into multiple steps. In such cases, as the processing force increases due to work-hardening, the reduction ratio should be gradually reduced.

Flanging, as shown in Fig. 5.13, involves tapering first (Fig. 5.13b) and then flatting (Fig. 5.13c, c′) to create the flange. The use of a mandrel in Fig. 5.13c prevents the flange from curling inwards [21].

When forming a curved surface using a press, as shown in Fig. 5.10, a significant reduction ratio can be achieved by restraining the entire tube to prevent buckling during forming. Eccentric necking, as shown in Fig. 5.10, involves necking the tube with an eccentric die (a die with the same shape on the entrance and exit sides but with the center offset).

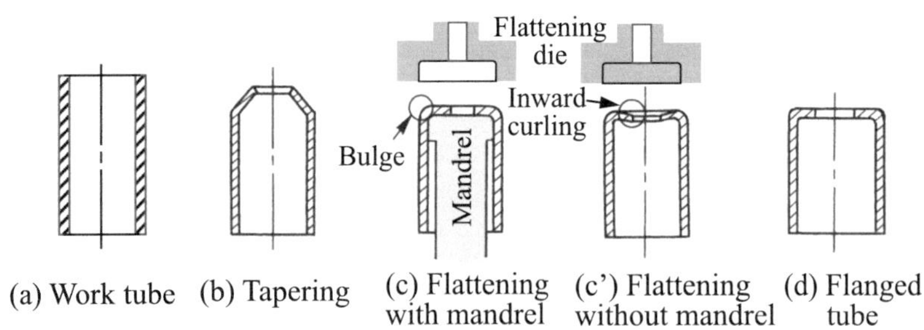

Fig. 5.13 Flanging process [21]

(a) General vertical press (b) Horizontal press with clamping

Fig. 5.14 Examples of press machines [22]

(3) **Processing machinery**

Figure 5.14 shows examples of press machines. Figure 5.14a is a photo of a general vertical press. If the tube is long, it may be challenging to process it using a commercially available general-purpose press, so custom-made machines are often used to match the product's processing specifications. The machine shown in Fig. 5.14b is a horizontal forming press with functions for clamping the tube and forming (pressing) the tube end [22]. In the necking process, hydraulic presses may be used instead of mechanical presses to alleviate material buckling, depending on practical considerations.

5.2.2 *Flaring Process by Press*

(1) **Overview**

The flaring process by press involves using a punch or similar tool instead of a die like used in necking to expand the end of a pipe from the inside to the outside, as shown in Fig. 5.15. Figure 5.15a illustrates flaring (tapering) [21], and Fig. 5.15b demonstrates stepped flaring [21, 23]. Other processing shapes include flanging, curved surface processing, eccentric flaring, and double flaring, as shown in Fig. 5.10.

(2) **Application guidelines**

For flaring processes shown in Fig. 5.10 (flaring, flanging, etc.), punches that match the desired forming shape are used for the process. The punch half-angle for flaring and stepped flaring is optimal around 20–25°, and the expansion ratio per process ($(d_0 - d_1)/d_0$, where d_0 is the initial outer diameter and d_1 is the final outer diameter)

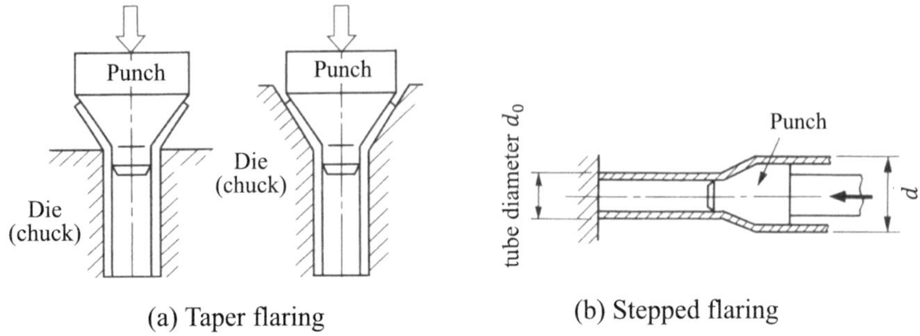

Fig. 5.15 Flaring process by press [21, 23]

is around 20–25%, depending on the material. In multi-step processes, gradually increasing the punch half-angle and reducing the expansion ratio per process is preferable. If cylindrical buckling occurs in the tube, restraining tools should be used.

Eccentric flaring shown in Fig. 5.10 involves expanding the end of the tube by eccentrically displacing the axis of the tube and the expanded section. An eccentric punch shown in Fig. 5.16 is used for this process. Due to the eccentric load applied to the tools, including the punch and the tube, measures should be taken to prevent their escape or rotation using guides or other means.

Figure 5.17 shows a processing method called double flaring, where the flared section of the tube is folded back to create a double layer. It is used in air and hydraulic piping joints without seals [21].

(3) **Processing machinery**

The processing machine can be either vertical or horizontal, similar to necking processes. Flaring with punches causes wavy deformation in the forming section, as shown in Fig. 5.18 [21].

If precision is required, the method (machine) shown in Fig. 5.19 is adopted. The punch is inserted with the die in place (Fig. 5.19a), the die is retracted while keeping the punch fixed at the leading end (Fig. 5.19b). When the die has fully retracted, the punch is then retracted (Fig. 5.19c). This series of movements ensures that the inner

Fig. 5.16 Eccentric flaring by eccentric punch

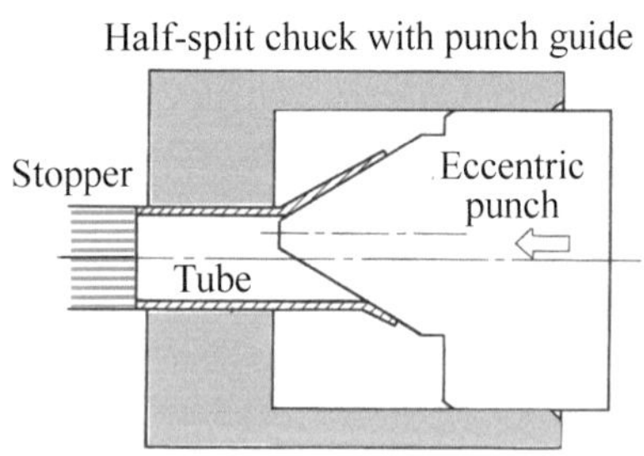

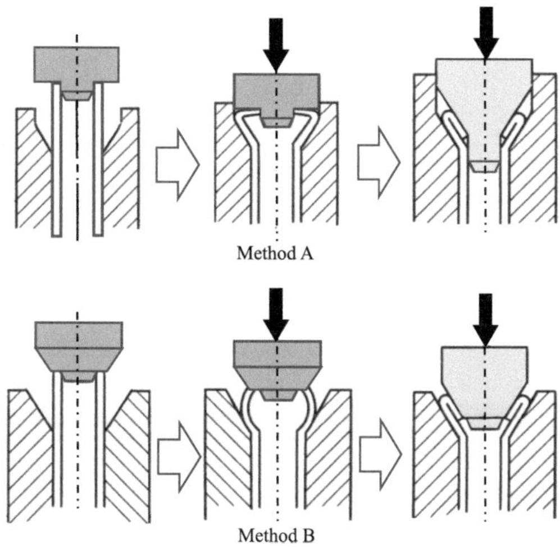

Fig. 5.17 Double flaring process

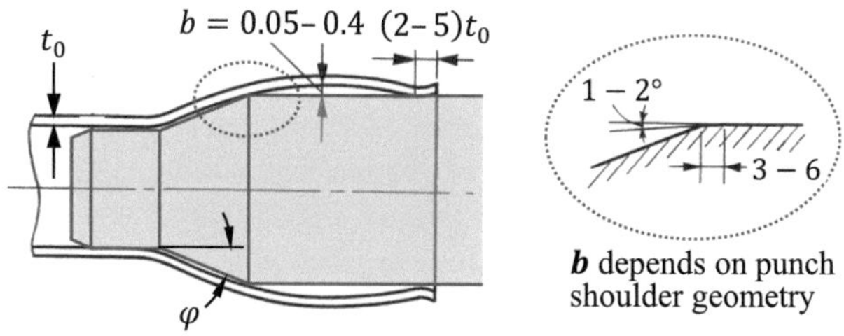

Fig. 5.18 Wavy deformation in flaring by punch [21]

and outer diameters of the expanded section are constrained and ironed with high precision. However, this process requires dedicated machines capable of performing complex operations [21] (detailed in Sect. 5.5.2).

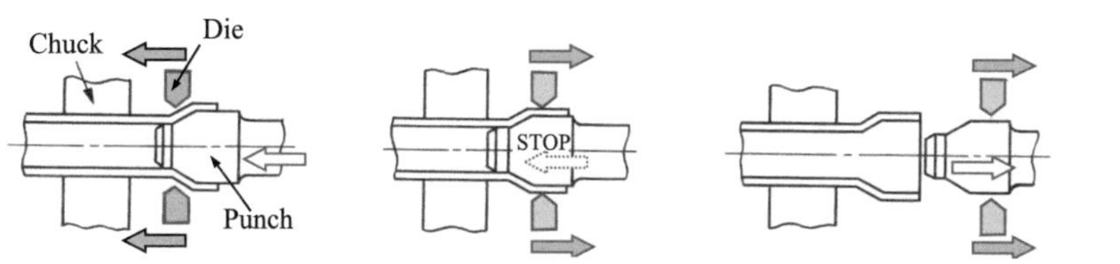

Fig. 5.19 Step forming process with ironing [21]

5.2.3 *Examples of Necking and Flaring by Press*

(1) **Summary**

There are few examples of using a single method or process in forming products through this process. Instead, they are often manufactured through a combination of multiple methods (necking and flaring) and multiple process steps.

(2) **Examples of formed products**

Figure 5.20 shows some characteristic examples of end-formed products in automotive parts. Figure 5.20a depicts an air suspension tube, and Fig. 5.20b represents an exhaust pipe made of ferritic stainless steel. These products are capable of forming complex shapes with a bulged central section by combining necking and flaring in multiple steps. In the case of Fig. 5.20b, this allows a transition from conventional flange fastening to band fastening, reducing the number of components and achieving lightweight design. Figure 5.20c illustrates a tube axle housing, and Fig. 5.20d shows a steering column tube. In these cases, the central section of the product remains as a straight tube, with the left end being necked and the right end being flared. The central section is unformed and has the same diameter of 57 mm as the original tube. The left section has a necking diameter of 38.1 mm (necking ratio of 33%), while the right section has a flaring diameter of 75 mm (flaring ratio of 32%), resulting in a double difference in diameter between the two ends. Figure 5.20e represents a tube propeller shaft, which is a necking-formed product with a die half-angle of 50° that allows deformation in the axial direction, contributing to crash safety.

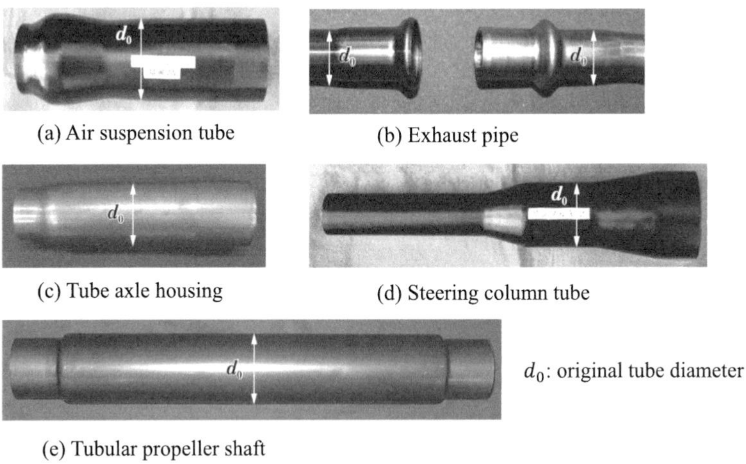

(a) Air suspension tube

(b) Exhaust pipe

(c) Tube axle housing

(d) Steering column tube

(e) Tubular propeller shaft

Fig. 5.20 Characteristic examples of end-formed products by press

5.2.4 *Curling and Reverse Forming by Press*

(1) **Overview**

Curling by axial compression utilizes arc-shaped (semicircle arc [24], quadrant arc [15–19]) or conical tools, as shown in Fig. 5.21. When conical tools [24–27] are used, external curling deformation is achieved with convex tools, while internal curling deformation is achieved with concave tools. Reverse forming is also carried out by continuing the curling process.

(2) **Application guidelines**

In curling or reverse forming, determining the radius of the tool arc is crucial. If it is too small, the cylindrical section may buckle, and if it is too large, the tip may fracture. When using conical tools, a convex tool half-angle φ (Fig. 5.21c) of approximately 60° or more enables external curling deformation, and stable outward reverse forming can be achieved within the range of 60–70°. For concave tools, a half-angle φ (Fig. 5.21f) of approximately 60° or more enables internal curling deformation, and beyond 80°, it becomes prone to buckling, so a φ within this range is recommended. Continuation of the curling process enables forming of double tubes as well. As shown in Fig. 5.22a, after flanging, the flange section is fixed, and the cylindrical tube can be reversed in the axial direction [28]. This is similar to re-draw in sheet metal press forming; however, in this case, it has long been known that reverse forming can be achieved without using a circular punch, as shown in Fig. 5.22b [29]. However, when not using a circular punch for reverse forming,

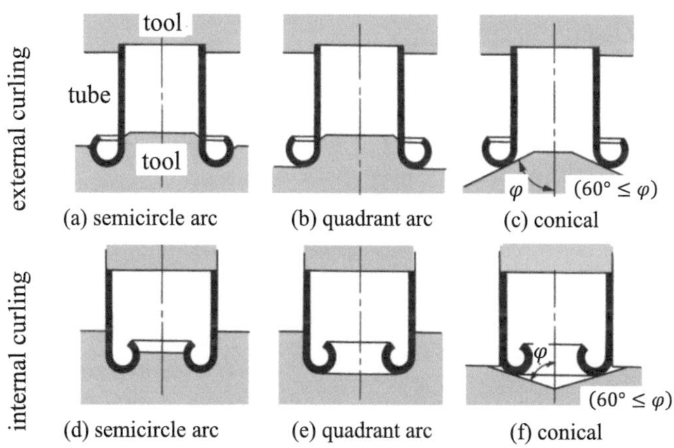

Fig. 5.21 External and internal curling by punch

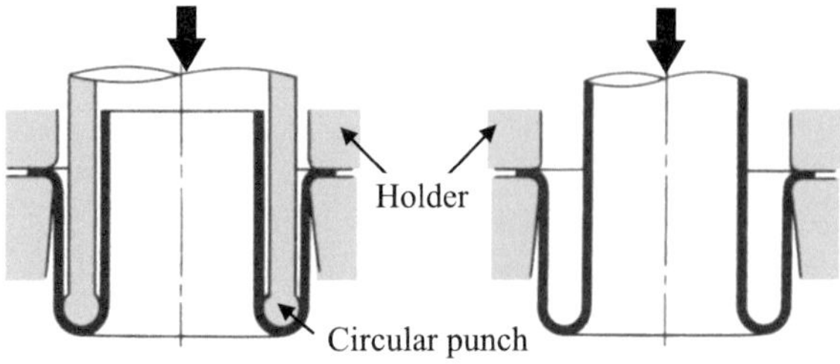

(a) Using circular punch [29] (b) Without circular punch [30]

Fig. 5.22 Examples of reverse forming by curling

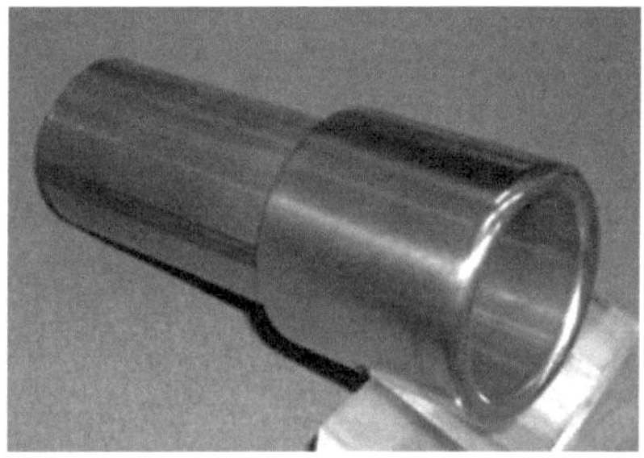

Fig. 5.23 Example of reverse forming product (integrated muffler cutter) [30]

the shape of the reversed form is determined by the diameter and thickness of the cylindrical tube, and only specific shapes can be reversed.

(3) **Processing machinery and product examples**

Vertical-type universal hydraulic presses are commonly used for this forming. Examples of processed products include the integrated muffler cutter for automotive exhaust tailpipes shown in Fig. 5.23 [30].

5.2.5 Can End Processing of Aluminum Beverage Bottle Cans

(1) **Overview**

Figure 5.24 shows aluminum beverage bottle cans[1]. Aluminum bottle cans were introduced by Japanese can manufacturers as a practical alternative to PET bottles in the year 2000. In terms of Japan's aluminum can demand in 2015, bottle cans accounted for 2.56 billion out of a total of 21.41 billion cans. Beverage cans made of aluminum account for slightly over 30% [31] of the Japanese aluminum sheet

[1] Japan Seikan Association: http://www.seikan-kyoukai.jp/process/index.html (as of February 2019).

Fig. 5.24 Aluminum beverage bottle cans

shipments, making it an extremely important field. Bottle cans are available in two types: two-piece type[2] and three-piece type[3]. The two-piece type undergoes necking down from an original diameter of 66–38 mm at the opening. The three-piece type can be necked down to 28 mm in addition to 38 mm. Complex forming processes, such as the necking-down of intricate can end sections shown in Fig. 5.25 [32], are employed. The can end (mouth) processing of bottle cans can be considered a high-volume production technique involving multiple processes.

(2) Examples of mouth shape

Figure 5.25 illustrates a side view of a three-piece type bottle can and an enlarged cross-section of its mouth. The lower opening of the main body is sealed by crimping the bottom lid. The mouth is necked down to a diameter that allows easy drinking, and the opening end features a curled shape for safety considerations. Below that, a screw thread for cap fastening and a bead connecting to the shoulder section are formed [33].

Fig. 5.25 Side view and mouth of bottle can [33]

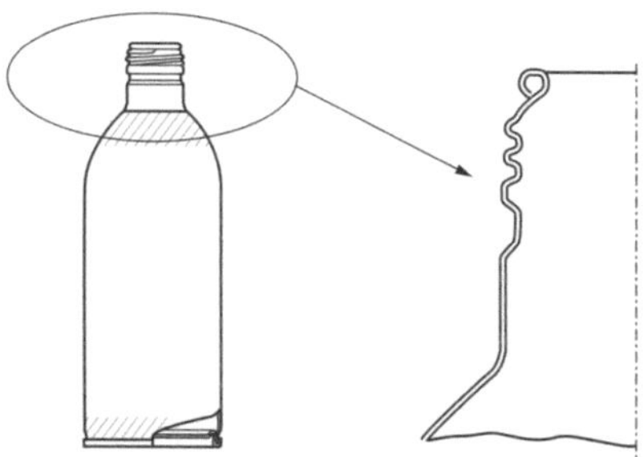

[2] Universal Can Corporation: http://www.unican.co.jp/product02.html (as of February 2019).

[3] Daiwa Can Co., Ltd.: http://www.daiwa-can.co.jp/product/drink_01.html (as of February 2019).

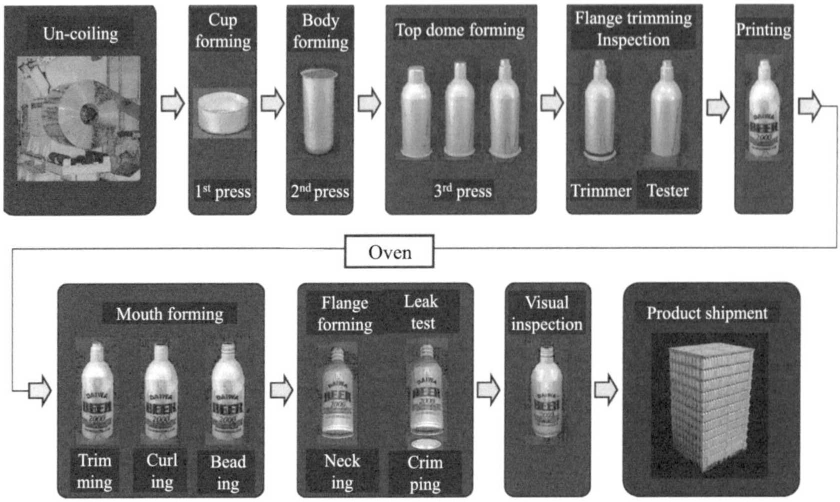

Fig. 5.26 Manufacturing process of 3-piece type bottle can [34]

(3) Processing methods and processes

Figure 5.26 provides an overview of the processing steps for the three-piece can manufacturing process. It involves a series of automatic processes, starting from cup forming and body (drawing and ironing, DI) forming, all the way to inspection. As for the can end processing, there are top dome forming and mouth forming. In the case of the three-piece type, the remaining thickness of the bottom section (of the original sheet thickness) during body forming is utilized to ensure screw formability, rigidity of the screw section, and necking down up to 28 mm.

Figure 5.27 illustrates an example of the top dome forming process. The top dome forming consists of five steps, and in the third step, efforts are made to smooth the boundary between the small-diameter cylindrical section formed by the necking process in the previous step and the shoulder section, creating a seamless dome shape. The capping (crimping) process after filling the contents also involves can end sealing. Aluminum bottle cans employ a pilfer-proof design (when the cap is twisted to open, the lower part of the cap with perforations is detached, indicating that the can has been opened), and axial load resistance and screw section rigidity are lower compared to glass bottles. Therefore, special screw shapes and strict control of crimping conditions are implemented different from glass bottles [32].

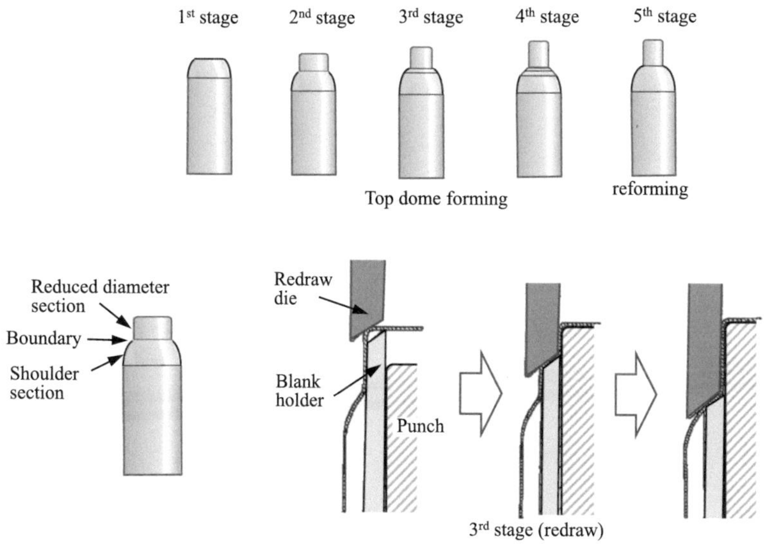

Fig. 5.27 Top dome forming process of bottle can [34]

5.3 Processing Force

5.3.1 Processing Force in Necking [7, 35]

The processing force required for necking operation using a press can be considered as the resultant of three forces: the force required for diameter reduction, the force to overcome friction between the die and the material, and the force required for bending deformation at the die entrance. These relationships are illustrated in Fig. 5.28. The force for diameter reduction (Force 1) remains constant regardless of the die half-angle φ as long as the degree of diameter reduction is the same. The force to overcome friction (Force 2) increases as φ decreases, resulting in greater contact length between the material and the die. The force for bending deformation (Force 3) increases as φ increases, indicating tighter bending deformation at the die entrance. Consequently, the overall processing force exhibits a minimum value at a certain φ. It is important to understand these relationships when considering the influence of various factors on the processing force.

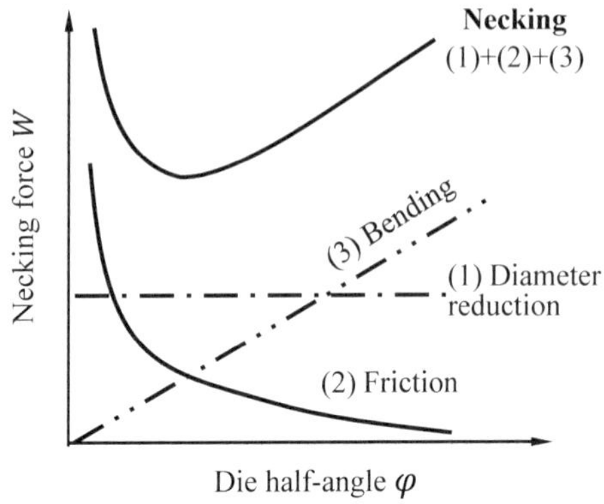

Fig. 5.28 Relationship between die half-angle and necking force

5.3.2 *Factors Influencing the Processing Force in Necking [7, 35–37]*

The processing force in necking operation is influenced by factors such as die half-angle, lubrication, tube dimensions, material properties, and tool shape. The following describes the impact of each of these factors.

(1) **Die half-angle**

Figure 5.29 is an example of a graph showing the relationship between the processing force (W) and punch stroke (L) with the die half-angle (φ) as a parameter in tapering using a necking die. When comparing at the same punch stroke, the processing force increases as the die half-angle φ becomes larger. The Δd in the figure represents the reduction in inner diameter at the top of formed section. When considering the same amount of forming (Δd = constant), there exists an optimal die half-angle, around 20–30°, where the processing force is minimal, except in the initial stage of forming. By using this optimal die half-angle for the operation, the required minimum

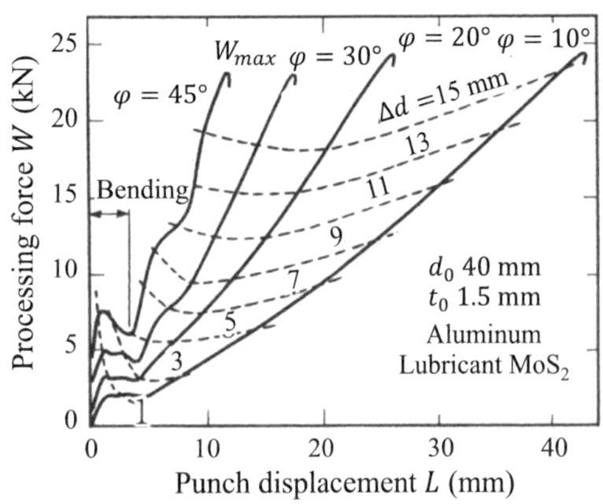

Fig. 5.29 Relationship between processing force and punch displacement with die half-angle [35]

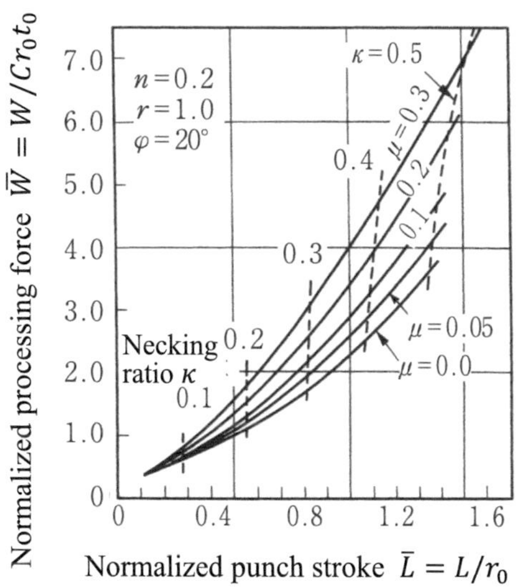

Fig. 5.30 Effect of friction coefficient on processing force [36]

processing force is achieved, and the compressive load on the cylindrical section is correspondingly reduced, resulting in less susceptibility to buckling of the cylindrical section.

(2) **Lubrication**

As shown in Fig. 5.30, a higher friction coefficient μ leads to higher processing forces. It is important to maintain good lubrication between the die and the tube to reduce the processing force. Additionally, a higher friction coefficient μ causes the optimal die half-angle φ_m to shift towards a larger value as the forming progresses (Fig. 5.31). Poor lubrication results in significant changes to the optimal die half-angle during forming, leading to unstable processing.

(3) **Tube dimensions**

As factors affecting the processing force, there are the tube diameter d_0 and the wall thickness t_0. The processing force W increases or decreases proportionally to d_0 and t_0, as shown in Eq. (5.10). In the case of thin-walled tubes ($0.02 \leq t_0/d_0 \leq 0.04$), the influence of t_0/d_0 is small, except for cases with a die half-angle $\varphi > 30°$ where the bending at the die entrance is relatively large.

(4) **Material properties**

The processing force varies proportionally with the C-value. As mentioned above, the processing force is also proportional to d_0 and t_0. Therefore, the processing force W can be expressed in dimensionless form as $\overline{W}(= W/Cr_0t_0)$. The influence of the n-value on the processing force is shown in Fig. 5.32. For the same C-value, a smaller n-value results in a larger processing force. The r-value has little influence on the processing force.

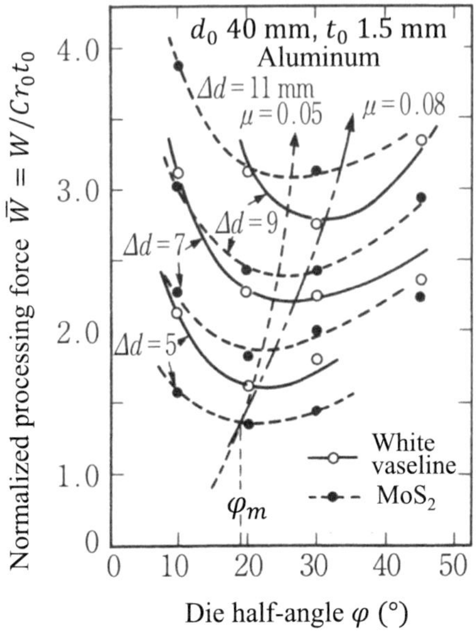

Fig. 5.31 Effect of lubricant on optimal die half-angle [35]

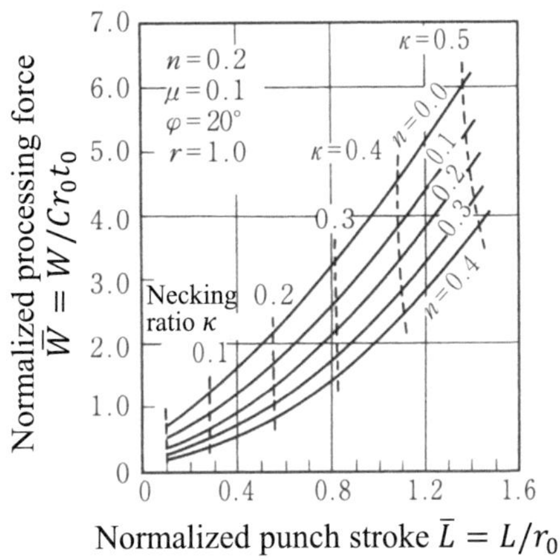

Fig. 5.32 Effect of n-value on processing force [36]

(5) Tool shape

In tapering processes, when the die half-angle φ exceeds 30°, the bending deformation at the die entrance becomes significant. Therefore, by using an arc-shaped die for curved surface forming instead of conventional tapering, a smoother processing force W-displacement L curve can be achieved, as shown in Fig. 5.33.

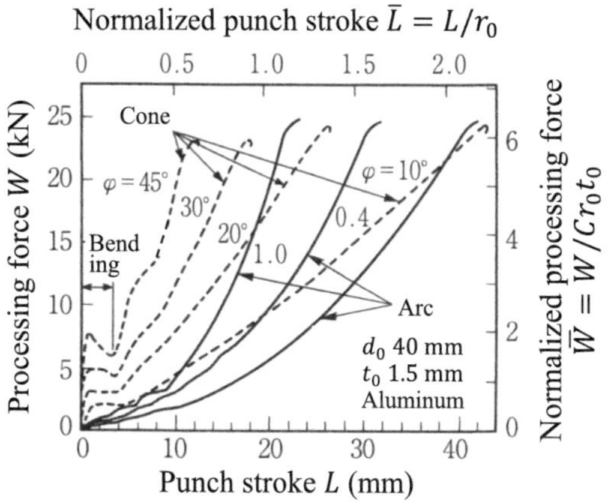

Fig. 5.33 Effect of tool shape on processing force [35]

5.3.3 Estimation of Processing Force in Necking [38]

Figure 5.34 shows a monograph for the processing force W (punch force) in taper forming using a press die for mild steel. The processing force W can be determined by following the sequence indicated by the dashed lines in the figure, based on the desired necking ratio l_1 (=r_a / r_0). In practical operations, the average value of the friction coefficient μ often ranges from 0.1 to 0.3. Assuming $\mu = 0.3$ for estimation purposes results in a slightly larger estimation of the processing force, providing a safety margin. Similar charts are also provided for aluminum and steel pipes. Additionally, the processing force in eccentric necking is calculated using the upper bound element technique (UBET) [39].

In recent years, the formed shape and processing force can be easily obtained using the FEM with general-purpose codes. Particularly for axisymmetric forming, a two-dimensional axisymmetric model can be applied, which is efficient with short computational time [40]. On the other hand, at production sites, it would be advantageous to have a method for easily determining the processing force. Equation (5.10) [35] appears relatively straightforward, but it does not consider stroke. In these equations, the stress state is assumed to be 3_E as shown in Fig. 5.2, but finite element analysis reveals it to be 2_{Mb} or 2_E [3]. Therefore, it is not easy to determine the processing force using these equations. However, considering that the circumferential stress σ_θ changes slightly and is approximately equal to the yield stress $-\sigma_y$, the processing force can be estimated by focusing on the circumferential stress σ_θ.

Using the equation for thin-walled cylindrical tubes, we have Eq. (5.18):

$$p_r = \frac{t_r}{R}\sigma_\theta \tag{5.18}$$

Here, p_r represents the die face pressure in the axial direction per unit axial length and per unit circumferential length, R is the radius of the necking section, and t_r is the radial thickness, i.e., $t/\sin\varphi$. By neglecting thickness variation and work-hardening

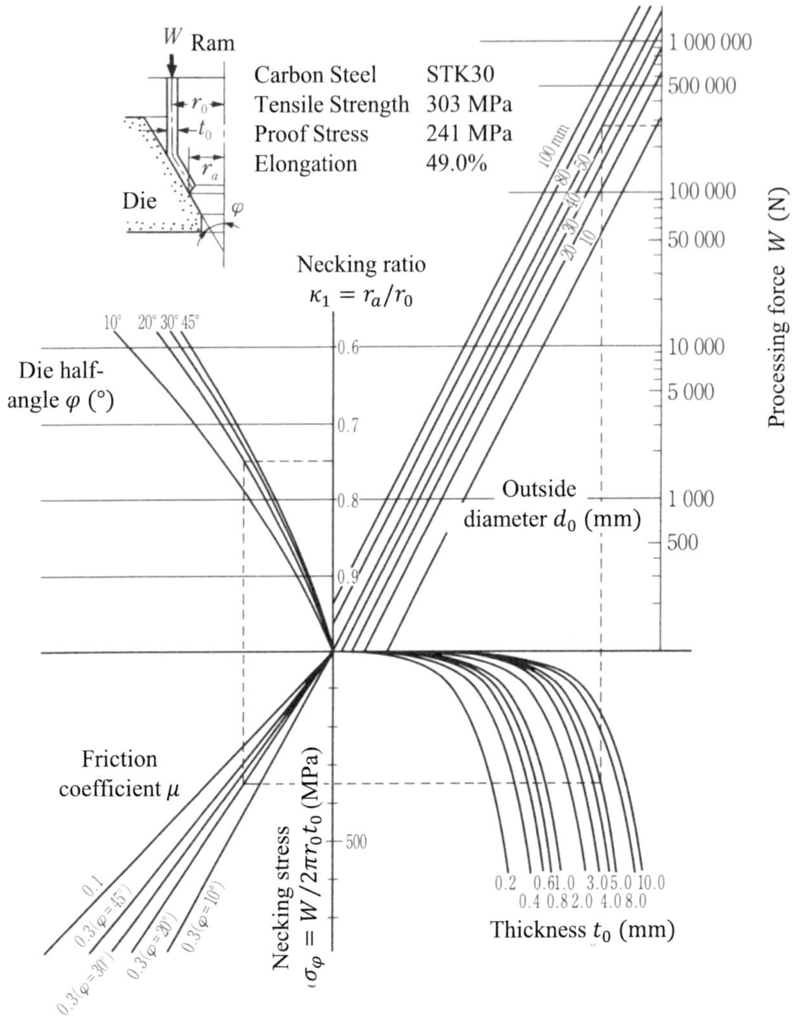

Fig. 5.34 Processing force of tapering by press [38]

and considering $\sigma_\theta = -\sigma_y$, the axial forming load W can be obtained using the following equation, Eq. (5.19):

$$W = 2\beta\pi t_0 S(\tan\varphi + \mu)\sigma_y \tag{5.19}$$

Here, μ is the friction coefficient and β is a coefficient used to compensate for neglected thickness variation and work-hardening. Figure 5.35 shows a comparison between the static implicit finite element method and the simplified Eq. (5.19). The mother tube is made of S45C material, with outer diameter 30 mm and wall thickness

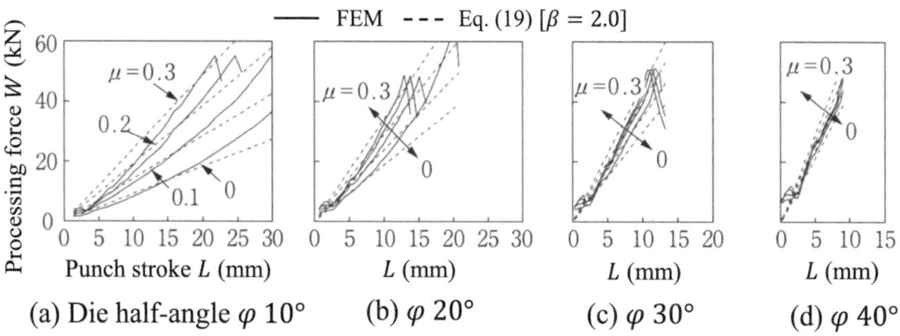

Fig. 5.35 Comparison of necking load-punch stroke diagrams using simplified equation and FEM

1.0 mm. The simple Eq. (5.19) can consider the effect of stroke and provide a rough understanding of the influence of various forming conditions on the forming load.

5.3.4 Processing Force in Flaring

In flaring processes using a press die, for the case of flare forming, the processing force is minimal at optimal punch half-angle φ_m due to the frictional force between the punch and tube and the bending deformation of the tube at the punch entrance, as well as processing force in necking as shown in Fig. 5.28 [11]. When the friction coefficient μ is large, the processing force increases in the range of smaller φ values, causing φ_m to shift to the right and become larger in the figure. In experiments with pure aluminum tubes, φ_m is approximately $15° \leq \varphi_m \leq 45°$ [10]. It should be noted that φ_m does not vary with the thickness-to-diameter ratio (t_0/d_0) if the tube is thin-walled. Furthermore, the relationship between the processing force W and μ is approximately linearly increasing as μ increases, as shown in Fig. 5.36 [10, 41]. In contrast, in curved surface forming, as shown in Fig. 5.37, the effect of bending is improved, and the processing force becomes almost constant, unaffected by the curvature of the arc (r_w/t_w) [10].

Fig. 5.36 Effect of friction on flaring force by press [10, 41]

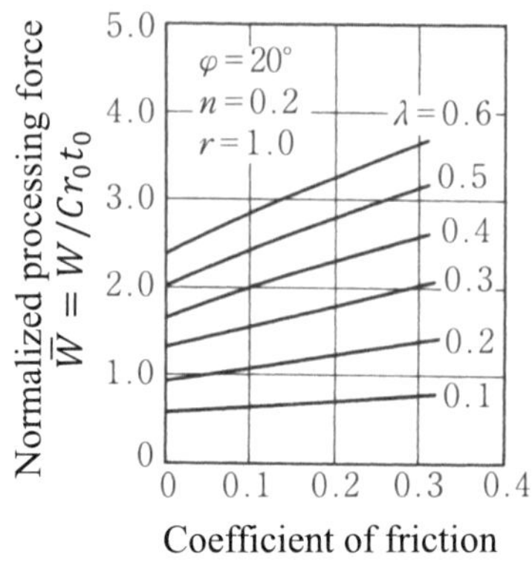

Fig. 5.37 Effect of punch curvature on curved surface forming force by press [10]

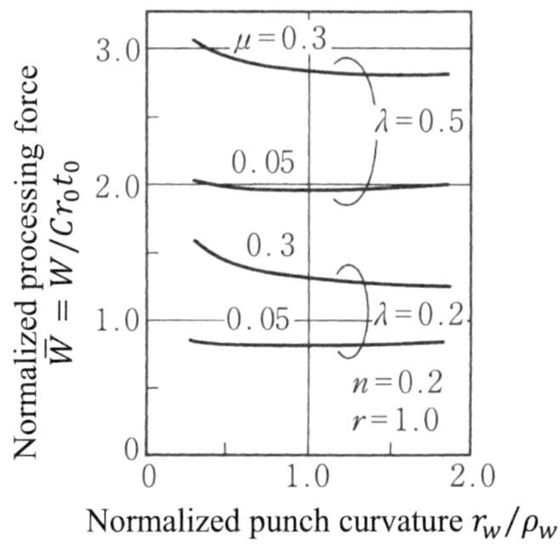

5.3.5 *Processing Force in Curling and Inverse Forming*

In reverse forming using a press die, after curling, the forming load increases and then reaches an approximately constant load value, leading to the inverse state. Therefore, the inverse load W_{inv} is larger than the curling load W_{cur}, and estimation of W_{inv} is important. When W_{inv} is smaller than the buckling load W_B of the cylindrical tube, it is possible to achieve inverse forming, while when W_{cur} is smaller than W_B, curling forming is possible. W_{cur} and W_{inv} vary depending on the material and dimensions of the cylindrical tube and tool shape. An example of W_{cur} and W_{inv} using a quadrant arc tool is shown in Fig. 5.38 [42]. Both W_{cur} and W_{inv} are minimal when the radius of the circular arc section r_d has a specific value. If the forming is done using a tool with r_d larger or smaller than this value, the forming load increases sharply. This minimum load is generally somewhat larger than the value calculated by the Al-Hassani-Johnson Eq. (5.17). The r_d at this point is also larger than the value derived from the same equation. Additionally, W_{cur} is higher for inner curling than for outer curling [17, 18], and the same for W_{inv}.

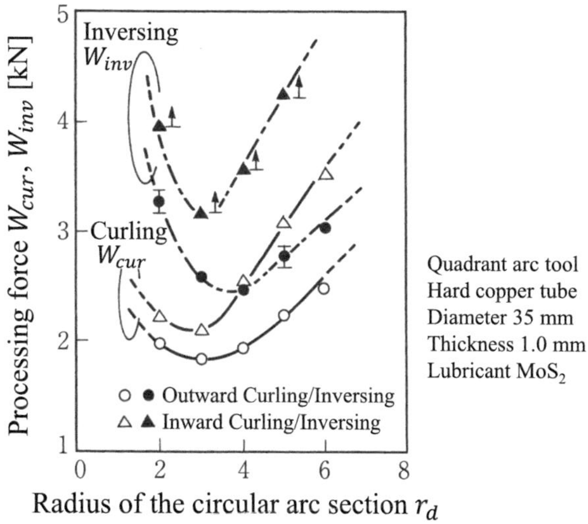

Fig. 5.38 Effect of radius of circular arc section r_d on processing force in curling and reverse forming [42]

5.4 Forming Limits

5.4.1 Forming Limits and Defective Deformations in Necking [2, 35]

The processing limits in necking are determined by buckling, wrinkling, waviness deformation, and curling deformation (Table 5.1) [35]. Each defective deformation has conditions under which it is more likely to occur. For example, two types of wrinkles (Modes 1 and 2) are more likely to occur in thin-walled tubes and in materials with smaller n-values. However, even in thick-walled tubes, if buckling and wrinkling are suppressed in some way and highly deformed, defective deformation due to the formation of shear bands can occur. It has also been pointed out that Mode 1 wrinkles in thin-walled tubes are shear bands in the thickness direction [43, 44]. Although it is difficult to determine the specific occurrence points of these defective deformations during processing, the shape changes in the force–displacement (W-L) diagram shown in Sect. 5.3.2 can provide a rough estimation of their occurrence points [35]. Mode 1 and 2 wrinkles occur before the maximum force point and occur earlier when the die half-angle is larger and t_0/d_0 is smaller. On the other hand, Mode 5, which is the axisymmetric buckling of the cylindrical section, defective deformation occurs at the same time as the maximum force.

5.4.2 *Factors Affecting the Forming Limits in Necking [35, 45, 46]*

Factors that generally affect the forming limits include the thickness-to-diameter ratio of the tube (t_0/d_0), tool shape (straight, curved, presence of guide), lubrication conditions, material properties, and the presence of restraining tools.

(1) **Thickness-to-diameter ratio**

The influence of t_0/d_0 on the relationship between buckling (wrinkling) in the forming section and the die half-angle φ is shown in Fig. 5.39. Mode 1 wrinkles indicated in Table 5.1 are more likely to occur as φ decreases, while Mode 2 wrinkles are more likely to occur as φ increases. The processing limits improve as t_0/d_0 increases. Furthermore, as t_0/d_0 increases, the range of die half-angles in which Mode 1 wrinkles occur expands.

(2) **Tool shape**

In tapering processes, the influence of bending deformation at the die entrance becomes significant when the die half-angle φ exceeds 30°. As a result, defective deformations such as waviness deformation (Mode 3), curling deformation (Mode 4), and edge buckling (Mode 2) indicated in Table 5.1 are more likely to occur. Therefore, considering the effect of arc-shaped forming shown in Sect. 5.3.2 [5], incorporating an arc-shaped guide at the die entrance can improve the forming limits. The possibility of predicting the forming limits using finite element analysis and forming limits

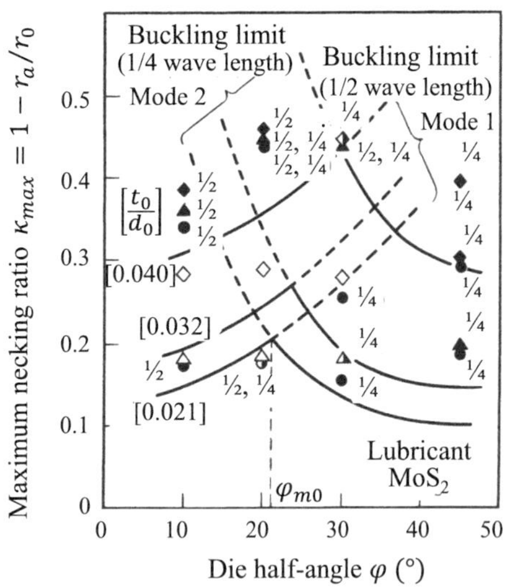

Fig. 5.39 Effect of thickness-to-diameter ratio on forming limit [35]

has been demonstrated [40]. Axisymmetric buckling (Mode 5) can be reproduced even in two-dimensional axisymmetric analysis. On the other hand, edge buckling with longitudinal wrinkles (Mode 2) involves three-dimensional deformation, and predicting it accurately with two-dimensional axisymmetric analysis is challenging. However, with a change in approach, it can be predicted to some extent.

Figure 5.40a illustrates the results of necking of an aluminum circular tube as an example. The mother tube is made of A6063 material, with outer diameter 40 mm and wall thickness 1.0 mm. When the die half-angle is small, the contact area increases, resulting in an increase in axial load and a higher likelihood of axisymmetric buckling (Mode 5) at the boundary between the taper section and the parallel section. This phenomenon can be quantitatively evaluated by the increase in the diameter of the buckling section, δ_D. On the other hand, under conditions where edge buckling with longitudinal wrinkles (Mode 2) is more likely to occur with a larger die half-angle, the analysis reveals that the tube lifts off from the inner surface of the die, as shown in Fig. 5.40b. The instability is caused by lifting off from the inner surface under the presence of circumferential compressive stress, leading to edge buckling. Therefore, the occurrence of edge buckling can be predicted using the amount of lift-off δ_W from the die surface. Figure 5.40c presents a comparison between predictions and experimental results. Even though it is a simple indicator, it enables the prediction of defective deformations assuming a friction coefficient μ of 0.25.

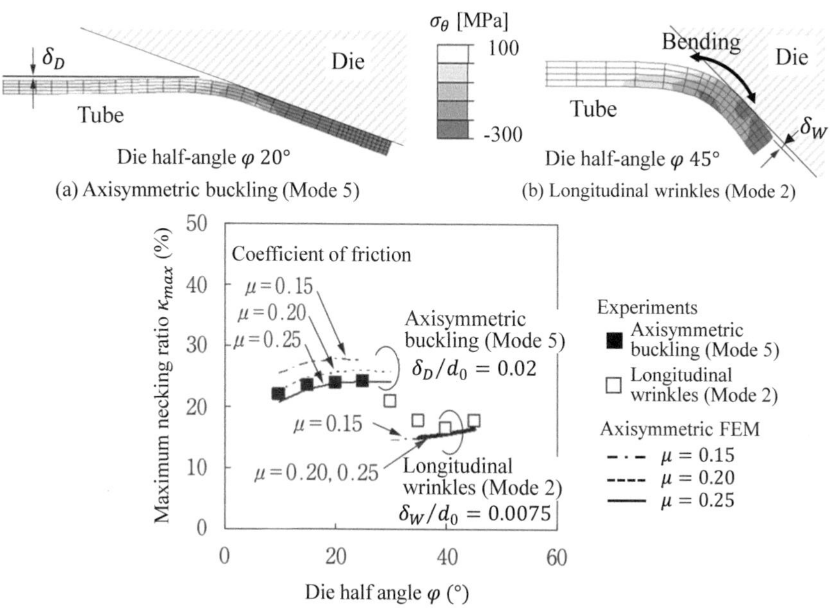

Fig. 5.40 Prediction of forming limit by axisymmetric FEM

(3) **Lubrication** [47, 48]

The effect of lubrication on forming limits has opposite effects depending on whether the forming limit is determined by cylindrical buckling or by forming section buckling. In forming conditions where cylindrical buckling (Mode 5) determines the forming limit, improving the surface finish of the die and reducing frictional resistance through forced lubrication, for example, result in a decrease in processing force (refer to Sect. 5.3.2 [2]). As a result, the load burden on the cylindrical section is reduced, leading to an improvement in the forming limit. Conversely, in conditions where forming section buckling (Modes 1 and 2) determines the forming limit, lubrication in the forming section is rather undesirable as it lowers the frictional resistance that restrains the occurrence of buckling.

(4) **Material properties**

Figure 5.41 compares the forming limit of a hard and a soft pure aluminum tube, both having a thickness of 1.5 mm. Except for the case of a $\varphi = 45°$ for the hard material, the forming limit is determined by cylindrical buckling (Mode 5) of the cylindrical section, indicating that the hard material with a higher cylindrical buckling load has a higher forming limit. However, when the thickness changes, the mode of defective deformation also changes, and there are cases where the forming limit of the soft material becomes higher than that of the hard material.

A higher n-value of the material has the effect of suppressing forming section buckling (Modes 1 and 2). Therefore, the forming limit of the forming section is higher for the soft material with a higher n-value. By locally annealing only the forming section, it is possible to improve the forming limit. An overview of the defective deformation varying with t_0/d_0 is illustrated in Fig. 5.42.

(5) **Restraint tool**

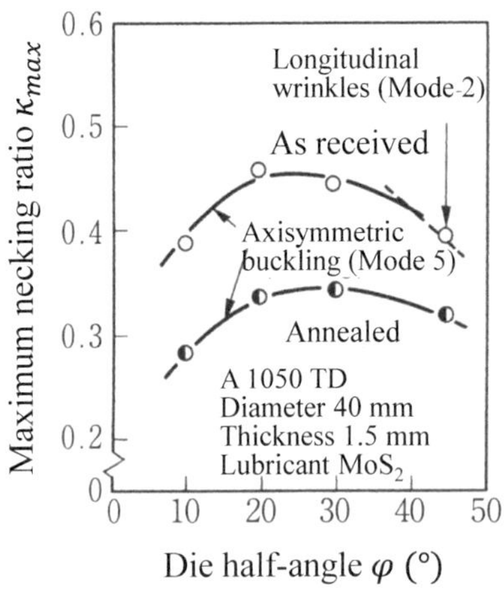

Fig. 5.41 Effect of material property on forming limit of conical necking by press [46]

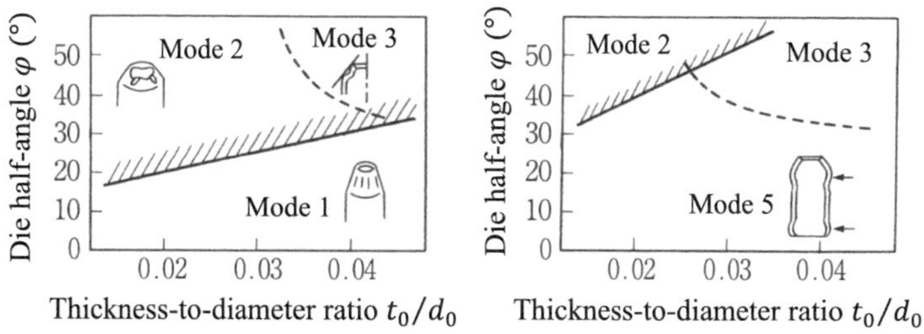

Fig. 5.42 Overview of defective deformation in necking by press varying with t_0/d_0 and die half-angle φ [37]

Figure 5.43 illustrates the effect of restraint from outside of the cylinder section on the forming limit. When a soft material undergoes the free necking process, buckling occurs at the cylindrical section (Mode 5) and at the forming section (Mode 2). However, when the soft material is necked with an external restraint, buckling Mode 5 is suppressed, and buckling Mode 6 is induced. Consequently, the forming limit is improved. External restraint is effective against cylindrical section buckling, but it has no effect on forming section buckling in hard materials or under processing conditions with a die half-angle $\varphi \geq 45°$. In such cases, it is beneficial to combine external restraint for the cylindrical section and internal restraint using a flexible tool (rubber) for the forming section [49].

Rotary necking with an asymmetric die is proposed to improve the forming limit by reducing circumferential restraint of a tubular workpiece, as shown in Fig. 5.44 [6, 50]. The asymmetric die consists of a part of a symmetric conical surface for forming and relief surface smoothly connected to the conical surface. The forming is achieved by rotating the die relative to the workpiece, and by pushing the die against

Fig. 5.43 Effect of restraint from outside of cylinder section on forming limit (pure aluminum O material) [45]

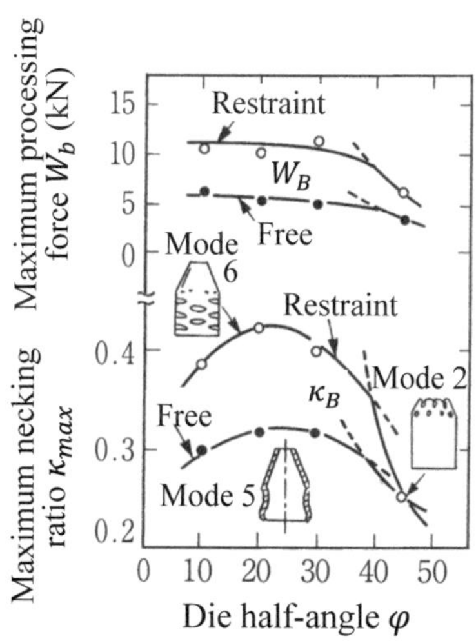

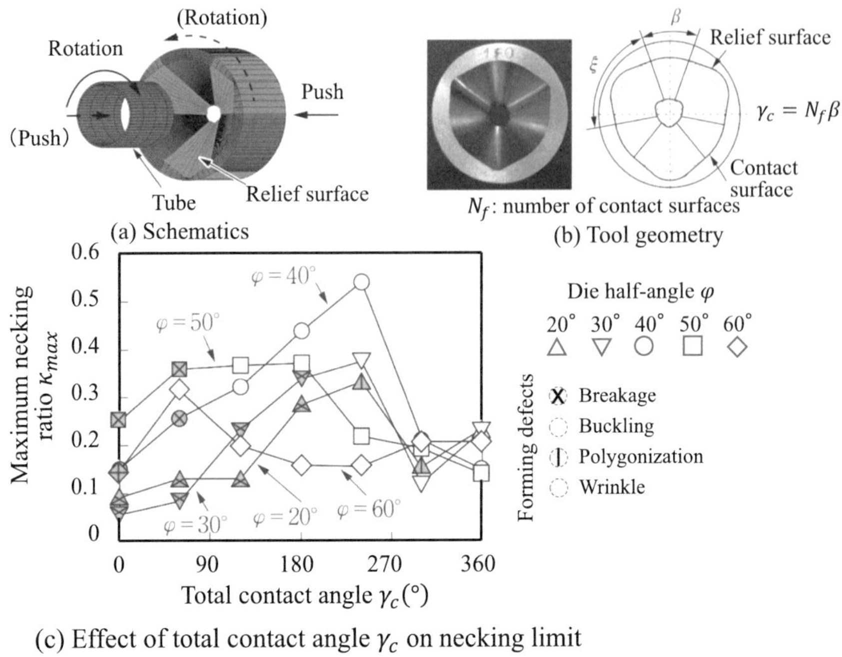

Fig. 5.44 Rotary necking with asymmetric die

the workpiece. The presence of the relief suppresses axial loads causing buckling and circumferential stress causing wrinkling, enabling high necking ratios to be achieved.

The total contact angle γ_C (Fig. 5.44b) is one of the factors that govern the type of defective deformation mode and the forming limit. The forming phenomenon is similar to press forming with a conical die when the total contact angle $\gamma_C = 360°$ and closer to spinning forming when $\gamma_C = 0°$. Figure 5.44c shows an example of necking an A6063 tube with outside diameter 30 mm and thickness 0.5 mm using an asymmetric die of $N_f = 3$ with a feed of 0.1 mm/rev. In the case of die half-angle $\varphi = 40°$, breakage occurs with smaller γ_C, while wrinkling occurs with larger γ_C. The limit necking ratio becomes maximum at $\gamma_C = 240°$. Thus, optimizing circumferential tool restraint leads to the maximum forming limit in a single feed compared to conventional press forming or spinning. However, in spinning, even if the forming amount per feed is small, a very high necking ratio can be achieved through repetitive processing.

5.4.3 *Forming Limit and Defective Deformation in Flaring Processes*

The forming limit of flaring using a press is determined by fracture at edge, unconformity to die like curling, and buckling at a cylindrical section, as shown in Table 5.2. In welded pipes, breakage may occur near the heat-affected zone or when the welded area undergoes significant hardening, resulting in a convex shape (Mode 5). The forming limit determined by fracture shows a good correlation with the strain gradient at the tube ends and is independent of the tool shape, such as cones or arcs [51]. In other words, a higher fracture limit is achieved by a conical punch with larger punch half-angle φ or an arc punch with smaller curvature radius ρ. If the buckling limit is reached in the cylindrical section before the fracture limit, the forming limit is determined by buckling Modes 6 and 7, similar to necking, as indicated in Table 5.2. In the case of conical forming, defective deformation includes waviness and curling, similar to necking. Curling deformation occurs when the punch half-angle φ exceeds a certain angle [26, 27]. Specific defects in eccentric flaring processes include breakage and extreme thinning at the large expansion region (Fig. 5.45). Additionally, the tube end may have a significant inclination, leading to larger trim length.

Table 5.2 Breakage and forming defects in flaring

Zone	Forming Part				
Mode	1	2	3	4	5
Appearance	Breakage 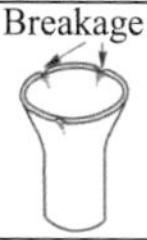	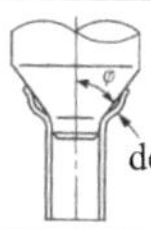 Wavy deformation	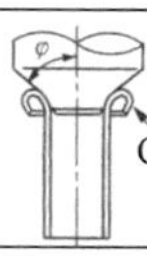 Curling	Weld line 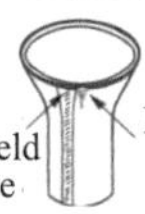 Breakage	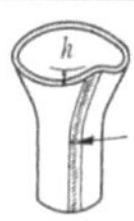 Weld line
Condition	small φ small gradient of ε_θ	~30° < φ	~60° < φ	–	Large φ
Material	Hard (small n-value) small t_0/d_0	–	–	Welded tube	Welded tube (hard weld line)

Zone	Cylindrical Part	
Mode	6	7
Appearance	Buckling 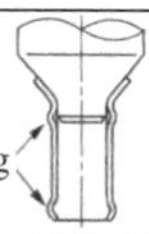	Buckling 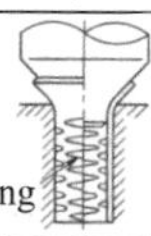
Condition	Free flaring (without constraint tool) Largeφ	Outside constraint flaring
Material	Soft (large n-value) small t_0/d_0	Soft (large n-value) small t_0/d_0

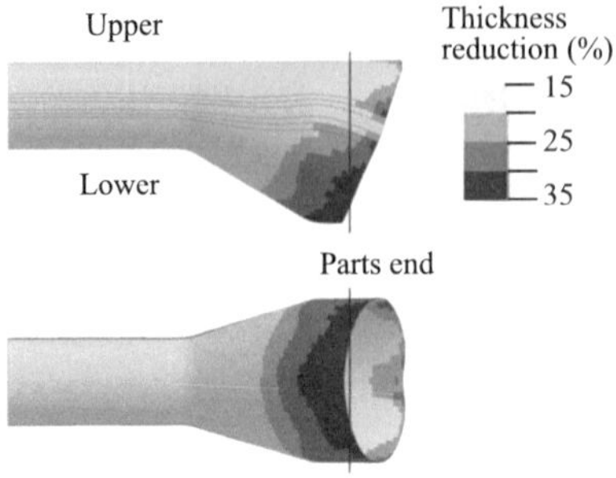

Fig. 5.45 Issues in eccentric flaring process: large tube end slant and significant wall thinning [52]

5.4.4 Forming Limit and Defective Deformation in Curling and Reversing Process

In the case of curling and reversing processes using a press, the forming limit is determined by buckling at the tool entry section or on the sidewall of the cylindrical section (Fig. 5.46a–c), as well as fracture at the tube end (Fig. 5.46d). Among these, buckling is determined by the relationship between W_{B} and W_{cur} or W_{inv} as mentioned before, and it is more likely to occur when the bending radius r_d is small. In the case of inner curling or inner reversing, W_{cur} and W_{inv} are larger than the values for outer curling and outer reversing, so special attention is required for buckling. On the other hand, when r_d is large, in outer curling tube end fracture is more likely to occur [15–18], and in inner curling, mouth edge wrinkles may occur (Fig. 5.46f). Additionally, with a large r_d, in outer curling, there are instances where mouth edge wrinkles occur at the curl edge (Fig. 5.46g) during the tube end being formed into the curl.

When curling tubes with high work-hardening, such as stainless steel pipes, buckling is likely to occur near the tool entry section of the cylindrical pipe, making curling forming difficult. In such cases, it is effective to heat the tool with high frequency [24] and locally anneal the forming section, reducing the forming loads W_{cur} and W_{inv} to become lower than the buckling load W_B. Furthermore, pre-processing the tube end into a tapered shape, as shown in Fig. 5.47, is effective to prevent buckling and to reduce the forming load [21].

In full-contact rotary forming, the local annealing effect can be expected due to frictional heat. This effect leads to a prevention of buckling on the sidewall of the cylindrical section, and there are instances when external restraint jigs to prevent buckling become unnecessary for inner curling. Generally, in rotary forming, the axial component of the forming force is smaller compared to press forming. Especially in forming processes such as spinning, eccentric rotary forming, and orbital rotary forming, the forming force can be very small with localized contact of the tool on

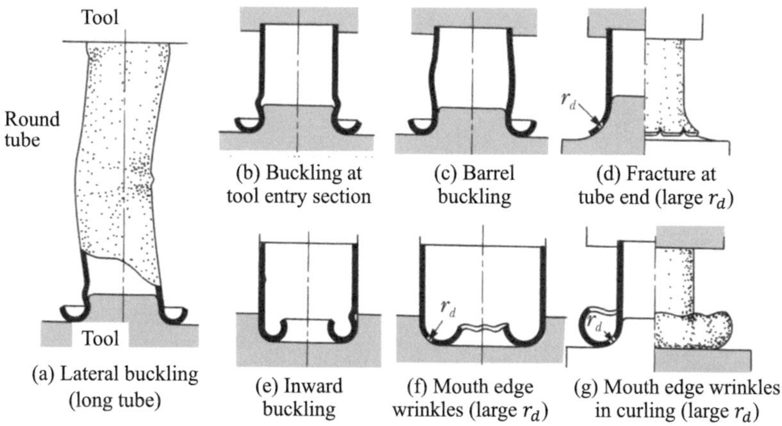

Fig. 5.46 Defects determining forming limit in curling process

Fig. 5.47 Pre-processing of tube end to prevent buckling in curling [21]

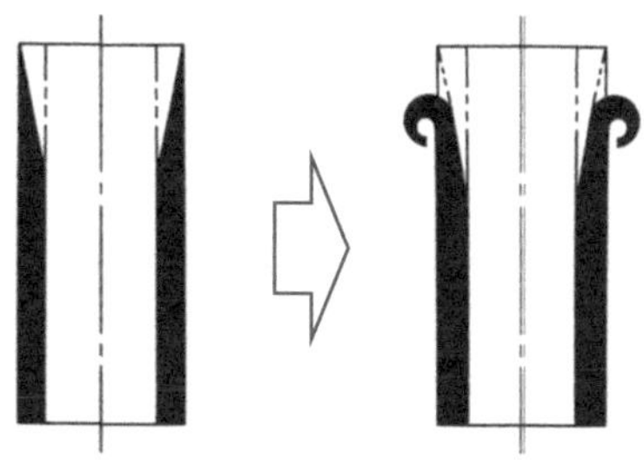

the tube end. Therefore, it is advantageous for the curling of thin-walled pipes that are prone to buckling.

5.4.5 Methods for Improving Forming Limits [2]

To improve the forming limits in necking and flaring processes, it is necessary to suppress defective deformations. Based on the factors influencing the forming limits, the methods for improving forming limits in press forming can be summarized as follows:

(a) **External factors**

1. Control of the frictional force between the tool and the tube [47, 48]
2. Buckling resistance improvement at the cylindrical section by restraint tools [52]

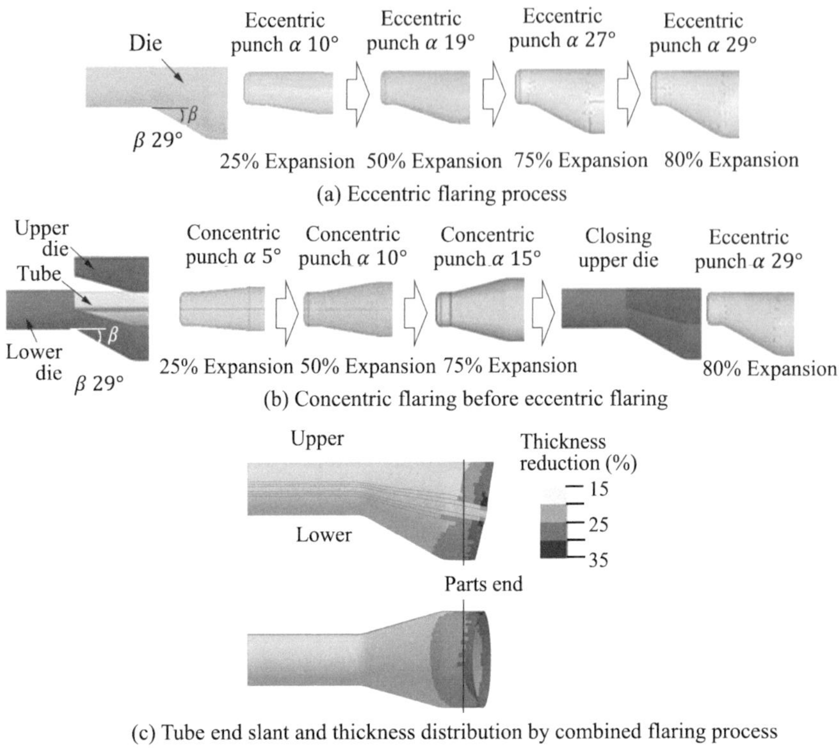

Fig. 5.48 Improvement of tube end slant and thinning by a combined flaring process

3. Wrinkle suppression in the forming section using restraint tools [52] and flexible tools (such as rubber) [49]
4. Mitigation of localized bending deformation at the entrance using arc-shaped entry sections for necking dies and flaring punches [44]
5. Improvement of tool shape and tool dimensions [37, 54]
6. Reduction of load burden on the cylindrical section through multi-step processes [55]

Alternatively, it is also possible to optimize the tool shapes for each process in multi-step processing that combines methods 5 and 6 above. For example, in addressing defective deformations in eccentric flaring, it has been reported that improvement can be achieved by initially performing partial flaring with concentric punches instead of eccentric punches, and then eccentrically expand in subsequent stages, as shown in Fig. 5.48 [52].

(b) **Material factors**

1. Localized annealing [45, 47, 56] and localized heating [57, 58] to control flow stress and to improve ductility.

2. Control of thickness (increase in the cylindrical section and reduction in the forming section) to enhance load carrying capacity of the cylindrical section.
3. Selection of tube materials with appropriate properties (n-value, r-value, C-value, etc.)
4. Low-temperature processing to improve ductility and buckling resistance [59] (for flaring of copper and aluminum).
5. Homogenization of welded zone (especially in flaring) [60]

In addition to these external and material factors, methods other than press forming can also be effective. Rotary forming methods can achieve higher forming limits than press forming when appropriate processing conditions are selected. As cold forming methods, (1) spinning, (2) rotary end-forming using roller tools [4, 61], and as hot forming methods, (3) rotary flaring utilizing frictional heat between the tool and the tube [62] are available. Method (3) is based on the concept of localized heating and involves compatible combinations of tube materials and tool materials. Generally, steel pipes with good self-lubrication properties are suitable. The method using roller tools has higher forming limits compared to methods using a press. Another effective method is orbital rotary forming, which is classified as rotary forming, where the tool relatively rotates and contacts the tube in a tilted orbit, resembling a planet movement. It has the following two characteristics: (1) suppression of the unconformity to the tool contour (curling, waviness deformation) by press forming, and (2) offering a versatile tool (semi-dieless processing) by creating various tool surfaces (envelope surfaces) with a single tool. These characteristics are also common to spinning and compete with these processing techniques.

5.5 Accuracy of Processing

5.5.1 Accuracy in Necking [9, 36, 46, 56]

The key aspects of accuracy to consider during the necking are the thickness distribution and the height of the formed product. Generally, the tapered portion with a high reduction ratio tends to experience thickening, and the thickness does not remain constant along the meridional direction. Additionally, due to elongation and contraction along the axial direction, the height of the formed product differs from the height of the initial tube. To evaluate the thickness distribution, attention should be given to the thickness strain ε_t, and for the evaluation of the formed product height, attention should be given to the meridional strain ε_z.

The die half-angle φ, friction coefficient μ, and material anisotropy influence both the thickness distribution and the formed product height. The influence of the die half-angle φ is shown in Fig. 5.49. A smaller φ leads to a wider region where ε_φ becomes negative at the die contact boundary, resulting in a lower formed product height. The

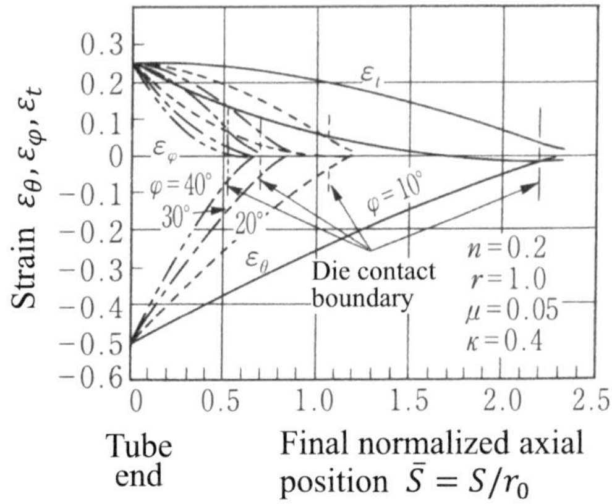

Fig. 5.49 Effect of die half-angle φ on strain distribution of conical necking by press [36]

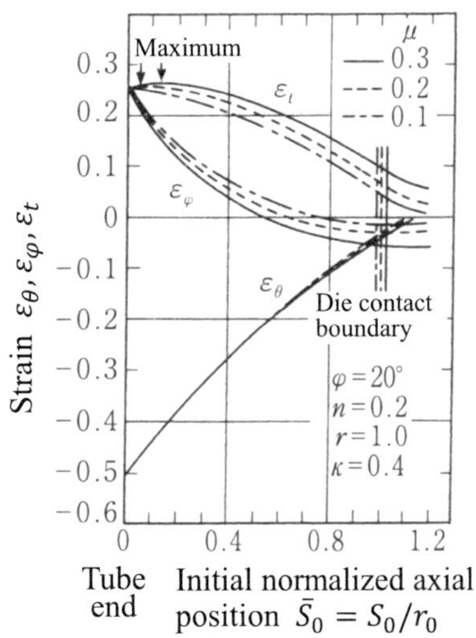

Fig. 5.50 Effect of coefficient of friction μ on strain distribution of conical necking by press [36]

influence of the friction coefficient μ is illustrated in Fig. 5.50. As μ increases, the maximum point of ε_t shifts towards the inner part of the formed section, and the overall level of ε_t increases, leading to overall thickening. Additionally, a larger μ widens the negative region of ε_φ near the die contact boundary, resulting in a lower formed product height.

The influence of perpendicular anisotropy (r-value) and in-plane anisotropy (r_φ-value, r_θ-value) is shown in Fig. 5.51a–c. A higher r-value contributes to a uniform thickness distribution ε_t and makes it more susceptible to tensile deformation along the meridional direction (Fig. 5.51a). A decrease in r_φ-value increases ε_t at the die contact boundary and enlarges the negative region of ε_φ (Fig. 5.51b). This phenomenon is more pronounced when the r_φ-value is smaller than 1.0. On the other hand, a decrease in r_θ-value increases ε_t at the top side of the formed section and

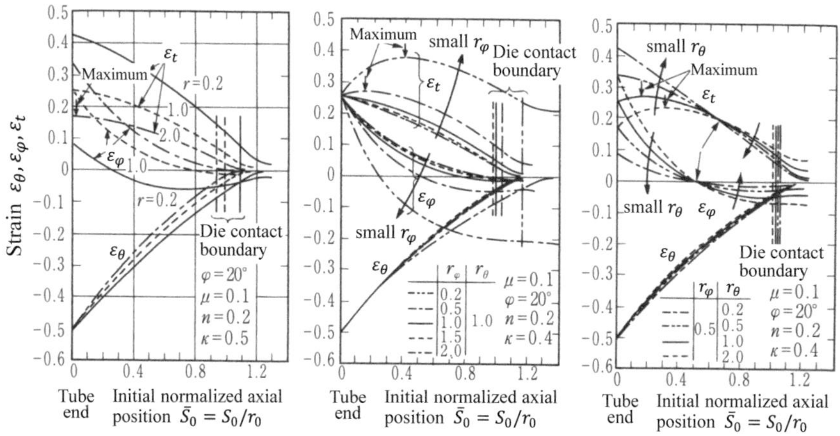

(a) Effect of r-value

(b) Effect of r_φ-value

(c) Effect of r_θ-value

Fig. 5.51 Effect of anisotropy on strain distribution of conical necking by press [36]

decreases ε_φ (Fig. 5.51c). The opposite trend occurs near the die contact boundary. Additionally, the n-value has little influence on the strain distribution. From these observations, to achieve a uniform thickness distribution in the formed section, larger r-value and r_θ-value, as well as smaller r_φ-value, are preferred. To enhance strength by increasing thickness at the top of the formed section, materials with smaller r-value and r_θ-value are favorable.

Furthermore, by adjusting the initial flow stress distribution of the initial tube to match the final degree of deformation, as shown in Fig. 5.52, the thickness distribution in the formed section can be made more uniform [56]. Specifically, in cases of low deformation, heating or annealing should be applied to the inner part rather than the top end, while in cases of high deformation, heating or annealing near the top end is sufficient. Additionally, there are various methods for thickness variation by pre-forming to control the post-formed thickness distribution [21].

Comparing the height of the product after necking and the initial height of the tube, it is observed that the product height can be higher than the initial tube height [9, 36, 46]. This phenomenon is due to the positive ε_φ strain distribution (Fig. 5.51a), which results in elongation along the meridional direction. According to elementary analysis, the relationship between the critical necking ratio and the die half-angle φ for the product height H to be equal to the tube height H_0 is expressed by Eq. (5.20) [9].

$$\varphi = \cos^{-1}\left[(2-\kappa_H)/4 \cdot \left\{1 + (1-\kappa_H)^{-1/(1+r)}\right\}\right] \tag{5.20}$$

For isotropic materials ($r = 1$), the product height never exceeds the tube height for $\varphi > 25.5°$ (Fig. 5.53). Therefore, from the perspective of material yield improvement and weight reduction, it is crucial to understand the relationship between these two factors during the design stage.

Fig. 5.52 Flow stress distribution for uniform thickness in conical necking [56]

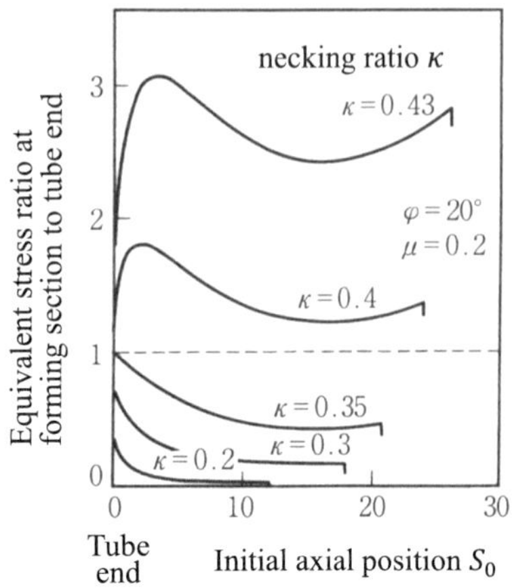

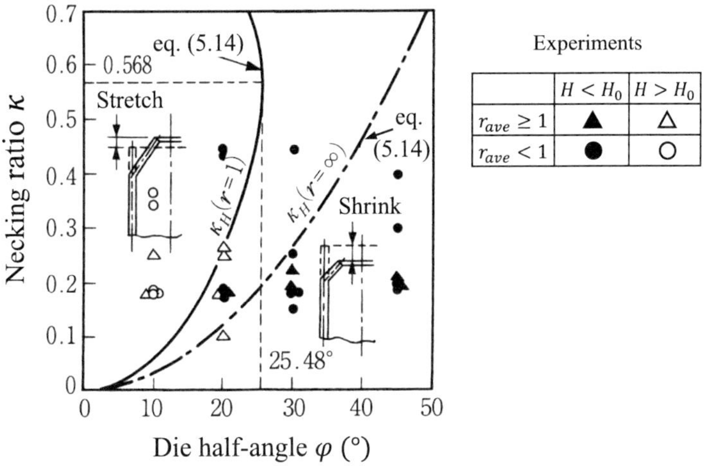

Fig. 5.53 Diagram of critical necking ratio κ_H for achieving same formed height H as initial height H_0 in conical necking by press [9]

5.5.2 *Accuracy of Flaring*

In flaring process using a press, achieving improved dimensional accuracy of the outer contour becomes crucial due to the possibility of tube lift-off from the punch. In flaring processes utilizing tools such as punches, the material tends to have difficulty conforming to sudden changes in tool contour. For instance, as shown in Fig. 5.18, in step flaring processes, the tube undergoes significant deformation, exhibiting waviness without conforming to the tool in the parallel section after expansion. Additionally, in the mouth edge region, 2–5 times t of the flared section deforms outwards [21]. Therefore, similar to the waviness deformation observed in flaring processes, step flaring processes also require attention to achieve precise shape accuracy. This phenomenon becomes more pronounced with larger punch half-angles φ. Hence, to achieve shape and dimensional accuracy, it is recommended to perform ironing on the expanded cylindrical section after step forming, as shown in Fig. 5.19 [21]. Understanding material flow and deformation behavior during the process is essential to improve the thickness accuracy of the formed section. In flaring processes using a press, the meridional strain in the formed section becomes compressive throughout the entire section, resulting in thickness reduction at the tip and thickening towards the root. As the degree of forming increases, this thickening region expands towards the tip [63]. To achieve uniform thickness in the formed section and enhance the aforementioned shape accuracy, the use of dies on the inner and outer of the tube is necessary, and applying ironing as depicted in Fig. 5.19 further improvement is achieved. To enhance surface accuracy, rotary processing using roller tools is effective, as it generates burnished surfaces, leading to highly precise finished surfaces [5].

5.5.3 *Accuracy of Curling and Reverse Forming*

In curling and reverse forming using a press, (1) thickness variation of the tube, and misalignments of tools such as (2) poor coaxiality between the tool axis and the tube and (3) poor parallelism of the die set are crucial and can affect significantly the shape of the curl. Particularly when using tools with a large r_d, the shape defects in the curl become prominent. Figure 5.54 illustrates the shape defects in curling and reverse forming. In most cases, these defects occur due to the deformed material lifting off from the tool and undergoing free deformation.

Therefore, to prevent shape defects, measures such as (1) incorporating guide sections in the entrance area, (2) avoiding excessively large r_d (in the case of using

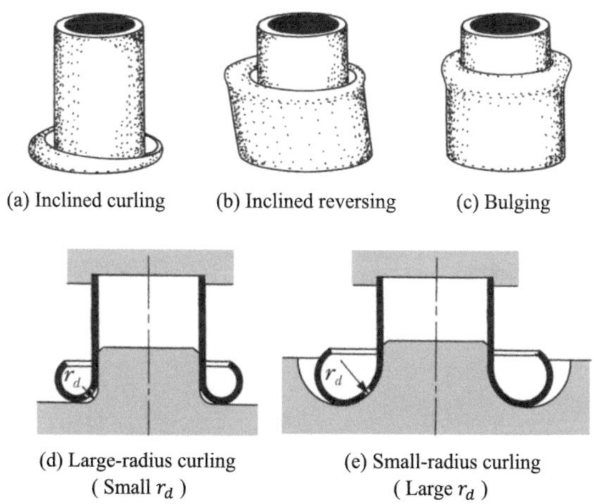

Fig. 5.54 Shape defects in curling and reverse forming

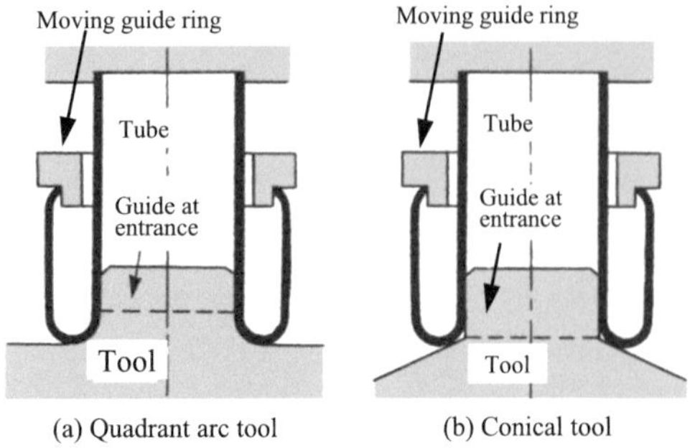

Fig. 5.55 Countermeasures for shape defects in reverse forming

arc-shaped tools), and (3) utilizing moving guide rings are often implemented, as shown in Fig. 5.55. On the other hand, in rotary forming, there is an advantage to achieve curling in a symmetric shape even if there is some misalignment between the tool axis and the tube axis.

5.6 Process Design and Die Design

5.6.1 Punch and Die Design for Flanging by Press

(1) **Process design**

In a single operation of necking (tube end reduction) as shown in Fig. 5.56, the achievable reduction ratio ($(d_0–d_1)/d_0$, where d_0 is the initial outer diameter and d_1 is the formed outer diameter) is approximately 20–28%, depending on the forming speed. If a higher reduction ratio is required, the process can be divided into multiple stages. The optimal die half-angle is around 20–30°, and the suitable entrance section shape is an R-shaped configuration. Additionally, defective deformations, as shown in Fig. 5.57, occur at the tube end. Therefore, the length of the blank tube should

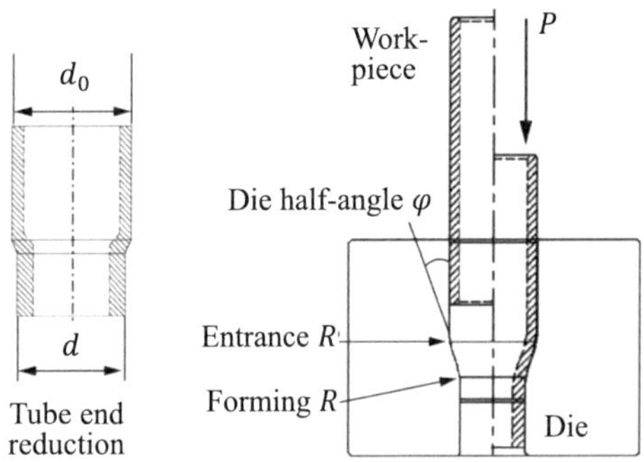

Fig. 5.56 Tube end reduction by die [64]

Fig. 5.57 Tube end reducing defects [64]

be estimated accordingly, and after forming, the defective deformation section is cut off [64].

(2) **Die design**

In taper forming, the influence of bending deformation at the die entrance becomes significant when the die half-angle φ exceeds 30°. Therefore, in conical dies, it is advisable to incorporate an arc-shaped introduction section to ensure a gradual increase of forming load. In die design, the following guidelines are used to determine the radius of the introduction section, rd. As shown in Fig. 5.58a, a bending deformation section with a curvature radius of ρ_f, as expressed in Eq. (5.1), is generated at the die entrance. When only bending occurs, it theoretically follows the following equation: $\rho_f = \sqrt{t_0 r_0 / 4(1 - \cos\varphi)}$

However, in practice, both bending and unbending occur, so Eq. (5.2) is used instead. To ensure contact between the tube and the introduction section, r_d should be larger than ρ_f. Equation (5.21) represents the radius of curvature ρ_C for a closing tube end with an arc die, and r_d is set smaller than the maximum value of ρ_C. In actual processing, before the closing is completed, buckling and wrinkling occur, reaching the forming limit. Therefore, r_d should be smaller than the radius of curvature ρ_B that corresponds to the maximum flanging ratio κ_{max}, as shown in Eq. (5.22). In other words, the relationship should be $\rho_f < r_d < \rho_B$. If improved forming limits are

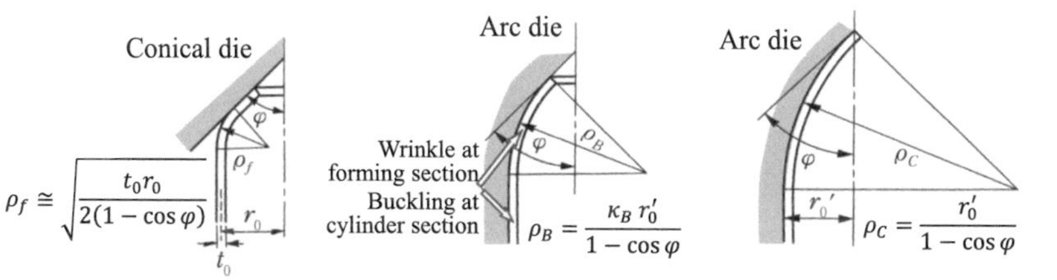

(a) Bending in free necking (b) Necking limit by arc-shaped die (c) Closing completed

r_d: radius of introduction section, $\rho_f < r_d < \rho_B < \rho_C$

Fig. 5.58 Guideline for introduction section radius r_d in necking process [45]

required, a value close to ρ_B should be adopted [45].

$$\rho_f = \sqrt{\frac{t_0 r_0}{2(1-\cos\varphi)}} \qquad \text{(5.2, restated)}$$

$$\rho_C = \frac{r_0'}{1-\cos\varphi} \qquad (5.21)$$

$$\rho_B = \frac{\kappa_{\max} r_0'}{1-\cos\varphi} \qquad (5.22)$$

When designing a curved (arc) die for curved surface forming as shown in Fig. 5.59, there are two basic approaches based on a conical (straight) die. One approach is to use the inscribed arc ab in the tube and conical die (Fig. 5.59a). The other is to use to the arc for which arc chord ab is to be conical die (Fig. 5.59b). Based on the comparison results of the forming limit shown in Fig. 5.60, design (b) is preferable. Design (a) is acceptable when the die half-angle φ is 40° or more. Additionally, by applying an arc-shaped die, as shown in Fig. 5.61, the distribution of die pressure becomes more uniform, which is advantageous in terms of die strength.

An example of the process design for a necking with die half-angle 45° is shown in Fig. 5.62. Two-stage necking with two different die half-angles is employed. A carbon steel tube for machine structures with a diameter of 54 mm and a thickness of 2 mm is used. In the first stage, the tapered section with a die half-angle of 30° is formed, followed by the second stage with a tapered section of 45°. The forming process of the second stage is shown in Fig. 5.63, where both outer and inner constraints are used to suppress buckling in the cylindrical and forming regions. After applying the outer constraint by the external die, the inner constraint punch is used to perform the 45° forming [64].

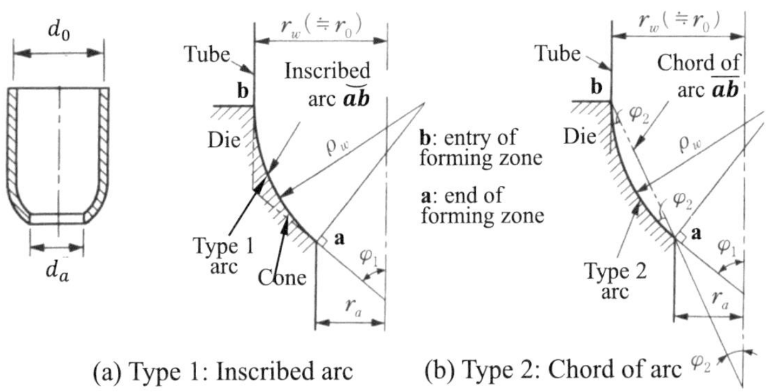

Fig. 5.59 Arc-shaped die designs based on conical die [37]

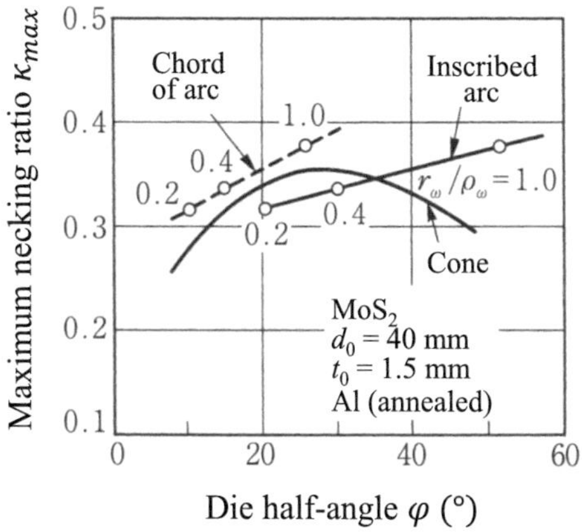

Fig. 5.60 Effect of tool shape on maximum necking ratio [37]

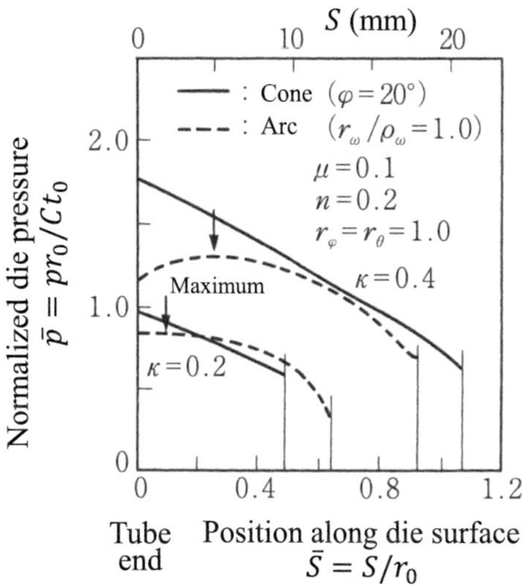

Fig. 5.61 Effect of die shape on distribution of die pressure in necking process [37]

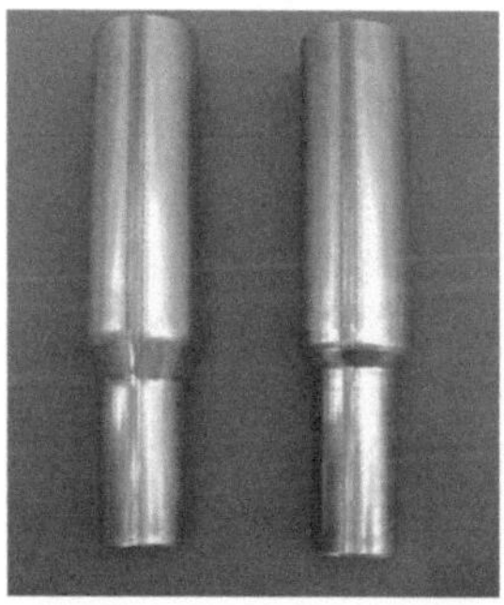

Fig. 5.62 Example of two-stage necking with die

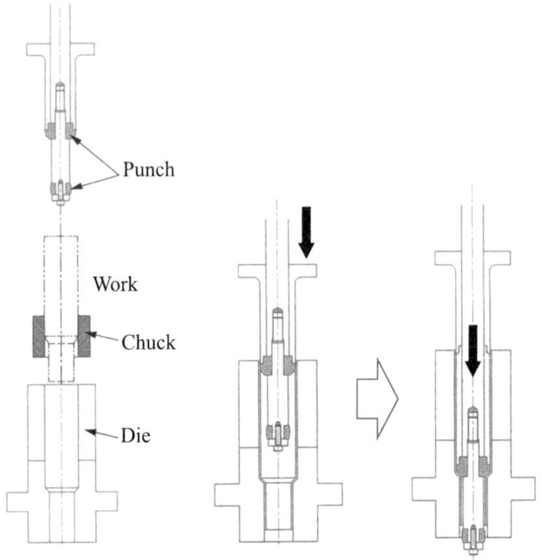

Fig. 5.63 Second stage of necking process with outer and inner constraints

5.6.2 Punch and Die Design for Flaring by Press

The tool surface pressure in the flaring is maximum at the tool entrance, similar to necking, but it is distributed almost uniformly. The level of surface pressure decreases as the degree of deformation increases, and the surface pressure itself is significantly smaller compared to necking. Therefore, die strength is not so much required [65]. Flaring using a punch needs to be performed with a smaller punch half-angle that prevents curling deformation. Curling deformation occurs when the minimum bending radius ρ_m required for continuous conical forming by unbending, given by Eq. (5.13) [65], is larger than the bending radius ρ_f, such that the curvature the tube begins to conform to the die [12, 13]. The bending radius ρ_f is expressed by Eq. (5.23) [7]. Here, Eq. (5.13) is given theoretically, and an experimental equation based on Eq. (5.13) is Eq. (5.24) [13].

The angle of starting curling deformation φ is approximately 45° according to Eq. (5.13) [12], and approximately 51° according to Eq. (5.24). From the standpoint of die design, it is possible to (1) design the punch half-angle φ smaller than φ_S (around 60° in experiments) [26, 27] and (2) provide an arc-shaped introduction section with a bending radius r_d greater than ρ_{f1} as in Fig. 5.64 to increase the curling starting angle from φ_S to $\varphi_S{}'$, among other considerations [2]. Option (2) also helps to reduce the tool surface pressure.

$$\rho_f = \sqrt{\frac{t_0 d_0}{8(1 - \cos\varphi)}} \tag{5.23}$$

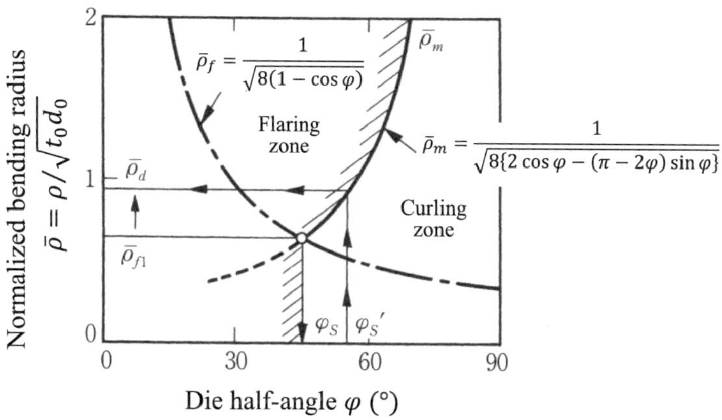

Fig. 5.64 Diagram for curling start angle φ_s in flaring process [2]

$$\rho_m = \sqrt{\frac{t_0 d_0}{8\{2\cos\varphi - (\pi - 2\varphi)\sin\varphi\}}} \quad (5.13, \text{restated})$$

$$\rho_m = \sqrt{\frac{0.28 t_0 d_0}{\sqrt{8\{2\cos\varphi - (\pi - 2\varphi)\sin\varphi\} + 0.36}}} \quad (5.24)$$

5.6.3 *Tool Materials and Coatings*

To improve processing limits and enhance product accuracy, reducing friction of the tool plays a significant role, making the selection of tool materials, surface treatment techniques, and coating technologies crucial factors. Table 5.3 provides examples of tool materials used in tube end-forming tools [66, 67]. The selection of tool materials not only improves forming limits through friction reduction with the workpiece but also significantly influences the tool life such as wear, galling, and stable quality of the product. Therefore, tool material is chosen based on the friction and forces applied on the tool during processing.

In particular, for necking dies and flaring punches subjected to high surface pressure, hard materials, including surface treatments, are selected. Tool materials can range from die steel and high-speed steel (requiring quenching surface treatment) to super-hard (cemented carbide) materials and ceramics. Generally, as the degree

Table 5.3 Examples of tool materials used in tube end-forming tools [66]

Item	Unit	Tool material					(Reference) Metal	
		Tool steel		Cemented carbide	Ceramics			
		Die steel	High speed steel		Silicon nitride	Zirconia	Carbon steel	Stainless steel
		SKD11	SKH3	WC+Co	Si_3N_4	ZrO_2	S45C	SUS304
Hardness	HV	653	800	1150	1850	1250	245	200
Density	g/cm^3	7.7	8.6	14.3	3.3	6.0	7.8	7.9
Bending strength (* Tensile strength)	MPa	–	1800*	3200	1400	1800	700*	520*
Young's modulus	GPa	210	210	550	310	210	210	200
Fracture toughness	MPa mm$^{1/2}$	–	–	12.6	7.5	6.0	–	–
Thermal expansion coefficient	$\times$ 10^6/K	12.0	11.2	5.7	3.4	10.5	10.7	17.1
Thermal conductivity	W/m K	29.3	21	100	25	3	44	16
Thermal shock resistance ΔT	°C	–	–	900	800	300	–	–

of deformation becomes more severe, die steel is used first, followed by high-speed steel, and then super-hard materials.

Ceramics have the advantage of high hardness and low affinity with the workpiece material (iron), making them less prone to adhesion. This makes ceramics particularly effective for high-pressure processing. However, caution must be taken as ceramics tend to retain heat during processing, reduce strength at high temperatures, and are susceptible to chipping and impact. Silicon nitride and zirconia are typical ceramic materials used. Additionally, for tool steels (such as die steels and high-speed steels) and super-hard materials, ceramic coatings can be applied to improve slipperiness, prevent adhesion, and enhance tool life. Table 5.4 shows representative types of coating films [67]. The production methods include chemical vapor deposition (CVD) and physical vapor deposition (PVD), which are both vapor-phase synthesis techniques. CVD involves synthesizing the film through chemical reactions at high temperatures of 800–1000 °C under reduced pressure. PVD involves evaporating

Table 5.4 Representative tool coatings for tube end-forming

Material		Membrane hardness [HV]	Production process	Features
TiC	Titanium carbide	3000	PVD/CVD	Usage according to HV
TiN	Titanium nitride	2000		Easy to production
TiCN	Titanium carbonitride	2500		Multilayer available
DLC	Diamond Like Carbon	1000–2000	PCVD/PVD	Hardness controllable
				Available below 300 °C

and ionizing the metal film in a vacuum chamber and depositing it onto the tool. The processing temperature is below 500 °C. Each method has its characteristics. CVD relies on high-temperature chemical reactions and allows for thicker films with good adhesion and the ability to coat the inner surface of the mold. However, the high-temperature process requires attention to the thermal expansion and thermal stress effects on the tool. PVD has the advantage of minimal thermal impact due to the lower processing temperature. However, some points to note include thinner film thickness in deep grooves or inside the tool and slightly inferior adhesion of the coating. Recently, there have been advancements in low-temperature CVD techniques (plasma CVD) and the development and practical use of methods to improve the adhesion of PVD coatings. Depending on the desired application of the tool, careful selection of tool material, appropriate processing methods, and choice of coating type are necessary. The table also includes recent references to diamond-like carbon (DLC) treatment.

References

1. Nishimura, H., Endo, J., Manabe, K.: J. JSTP **19-214**, 918–925 (1978)
2. Manabe, K.: J. JSTP **30-339**, 481–488 (1989)
3. Kuboki, T., Kajikawa, S.: Proceedings of the 144th JSTP Tube Forming Committee Research Meeting, pp. 15–23 (2017)
4. Maruo, T.: Nikko Mater. **2-6**, 36 (1984)
5. Kanayama, K., Tazaki, Y., Fukuda, M.: J. JSTP **29-333**, 1049–1056 (1988)
6. Kuboki, T., Abe, M., Yamada, Y., Murata, M.: CIRP Ann. Manuf. Technol. 269–272
7. Manabe, K., Nishimura, H.: J. JSTP **23-258**, 650–657 (1982)
8. Kitazawa, K., Kobayashi, M.: J. JSTP **28-316**, 481–487 (1987)
9. Manabe, K., Nishimura, H.: J. JSTP **23-256**, 451–457 (1982)
10. Manabe, K.: Doctoral Thesis (Tokyo Metropolitan University) (1985)
11. Manabe, K., Nishimura, H.: J. JSTP **24-264**, 47–52 (1983)
12. Kitazawa, K., Kobayashi, M., Yamashita, S.: J. JSTP **29-331**, 845–850 (1988)
13. Kitazawa, K., Kobayashi, M., Yamashita, S.: J. JSTP **29-33**, 1043–1048 (1988)
14. Kitazawa, K., Kobayashi, M., Yamashita, S.: Proceedings of the 1987 Spring Meeting, JSTP, pp. 513–514 (1987)
15. Al-Hassani, S.T.S., Johnson, W., Lowe, W.T.: J. Mech. Eng. Sci. **14-6**, 370 (1972)
16. Al-Hassani, S.T.S.: Proceedings of 15th International Machine Tool Design & Research Conference, p. 571 (1974)

17. Al-Qureshi, H.A., et al.: Design Engineering Conference on ASME paper 76-DE-9 (1976)
18. Al-Qureshi, H.A., Morais, G.A.: Design Engineering Conference on ASME paper 77-DE-35 (1977)
19. Kinkead, A.N.: J. Strain Anal. **18-3**, 177–188 (1983)
20. Manabe, K.: Handbook of Technology of Plasticity. Corona Publishing, pp. 755–758 (2006)
21. Nakamura, M.: Pipe Forming Techniques. Nikkan Kogyo Shimbun (1982)
22. Hamanishi, Y., Kato, K.: Proceedings of the 205th Symposium on the Technology of Plasticity, JSTP, pp. 37–43 (2001)
23. Nakamura, M., Iigo, S., Niikura, T., Nemoto, S.: Press Working. Nikkan Kogyo Shimbun **17-7**, 21–39 (1979)
24. Engel, H. E.: Patent-Anmeldeschrift P3422040
25. Гобунов, М.Н. et. al.: Кузнеу- Штамл. Лроизв **2**, 16 (1975)
26. Ichinose, K., Masuda, T.: J. Jpn. Inst. Light Metals **29-11**, 483–490 (1979)
27. Kitazawa, K., Kobayashi, M.: J. JSTP **28-323**, 1267–1274 (1987)
28. Ichinose, K., Masuda, T.: J. JSTP **26-288**, 25–31 (1985)
29. Guist, L.R., Marble, D.P.: NASA Technical Note, TND-3622, pp. 9–10 (1966)
30. Kato, K., Endo, K.: J. JSTP **46-530**, 211–215 (2005)
31. Japan Aluminum Can Recycling Association: Demand for Japanese Beverage Aluminum Cans (2015)
32. Koide, M., Tsuruta, A.: R&D Kobe Steel Eng. Rep. **55-2**, 75–80 (2005)
33. Daiwa Can Company: Japan Patent No. 5937337
34. Ito, T., Tsujimoto, K., Ogoe, T.: J. Jpn. Inst. Light Metals **52-2**, 82–87 (2002)
35. Manabe, K., Nishimura, H.: J. JSTP **23-255**, 335–342 (1982)
36. Manabe, K., Nishimura, H.: J. JSTP **23-260**, 878–885 (1982)
37. Manabe, K., Nishimura, H.: J. Mech. Working Tech. **10**, 287–298 (1984)
38. Miyagawa, M., Nishimura, H., Shimasoh, T.: The Machinist **3**, 6–11 (1969)
39. Kiuchi, M., et al.: J. Inst. Ind. Sci. **32-12**, 5 (1981)
40. Kuboki, T.: J. JSTP **55-646**, 989–994 (2014)
41. Manabe, K., Nishimura, H.: Proceedings of the 31st Joint Conference on the Technology of Plasticity, JSTP (1980)
42. Kitazawa, K., Seino, J.: Proceedings of the 1991 Spring Meeting, JSTP, pp. 639–640 (1991)
43. Schmid, W., et al.: 12th Beinn Congress International Deep Drawing Research Group, vol. 2, p. 31 (1982)
44. Kitazawa, K., Kobayashi, M., Tsujide, M.: Proceedings of the 39th Joint Conference on the Technology of Plasticity, JSTP, pp. 509–512 (1988)
45. Manabe, K., Nishimura, H.: Trans. Jpn. Soc. Mech. Eng. **51-463**, 641–647 (1985)
46. Manabe, K., et al.: Amada Tech. J. **19-93**, 19 (1986)
47. Miyagawa, M., Park, T.J.: J. JSTP **4-26**, 163–170 (1963)
48. Miyagawa, M., Park, T.J.: J. JSTP **7-62**, 142–144 (1966)
49. Mori, S., Manabe, K., Nishimura, H., Tamura, K.: Proceedings of the 1988 Spring Meeting, JSTP, pp. 479–482 (1988)
50. Kuboki, T., Kominami, A.: Metal Forming, pp. 547–550 (2012)
51. Manabe, K., Nishimura, H.: Proceedings of the 1982 Spring Meeting, JSTP, pp. 219–224 (1982)
52. Tamura, S., Iguchi, T., Mizumura, M., Sakagami, T.: Proceedings of the 2016 Spring Meeting, JSTP, pp. 155–156 (2016)
53. Аверкиев, ЮА: J. JSTP **1-4**, 349–356 (1960)
54. Miyagawa, M., Park, T.J.: J. JSTP **3-17**, 397–405 (1962)
55. Furuya, Y., Ochiai, I., Kitayama, Y.: J. JSTP **7-61**, 73–82 (1966)
56. Manabe, K.: Press Working, vol. 86, pp. 25–29. Nikkan Kogyo Shimbun (1987)
57. Takahashi, K.: Proceedings of the Tube Forming Committee Workshop, JSTP, pp. 84-4-4 (1984)
58. Sumitomo Metal Industries, Ltd.: Catalogue
59. Kobayashi, M., Kitazawa, K., Asao, H., Nakagawa, T., Kato, K., Tago, K.: Proceedings of the 1982 Spring Meeting, JSTP, pp. 647–650 (1982)
60. Ataka, M., Takazawa, A.: J. JSTP **30–339**, 497–504 (1989)

61. Maruo, T.: Proceedings of the 36th Joint Conference on the Technology of Plasticity, JSTP, pp. 225–228 (1985)
62. Kobayashi, M., Kitazawa, K., Tago, K.: Proceedings of the 1982 Spring Meeting, JSTP, pp. 643–646 (1982)
63. Manabe, K., Nishimura, H.: J. JSTP **24-266**, 276–282 (1983)
64. Ando, H., Morikawa, A.: J. JSTP **53-614**, 208–212 (2012)
65. Manabe, K., Nishimura, H.: J. Jpn. Inst. Light Metals **34-8**, 439–445 (1984)
66. Japan Tungsten Co., Ltd (2019) Product catalogue
67. Yamagata, K.: J. Mater. Process Technol. Center **52-10**, 12–17 (2011)

Chapter 6
Spinning, Ironing, and Rotary Forming

6.1 Spinning, Ironing, and Rotary Forming as Incremental Forming

In spinning, ironing, and rotary forming, the tubular material (workpiece) is plastically formed incrementally in localized regions to achieve the desired shape. Consequently, compared to stamping, these processes offer the following advantages: (1) reduced forming load and (2) improved formability. However, a drawback is (3) the slower forming speed. Another characteristic of incremental forming is (4) the elimination (or reduction) of dies, which has become feasible with the development of corresponding technologies since the 1980s. This advancement allows for the omission of complete dies that have the same contour as the product shape, and enables (5) profile forming. It is worth noting that rotary forming is defined as "a general term for the process of rotating bar, sheet, or tube blanks and creating the product shape through localized contact with tools" [1]. On the other hand, forming by rotating rolls around a fixed tube has long been referred to as "rolling" [2]. In this case, from the perspective of the rolls, the tube rotates relative to them, making "rolling" a form of rotary forming in this relative sense. Therefore, in this chapter, the term "rotary forming" includes forming processes where tools (rolls, dies) rotate around the tubular material. While spinning and rotary forming are incremental forming processes with the deformation styles of sequential bulging, sequential drawing, and sequential ironing, swaging is an incremental forming process with the deformation style of sequential forging. For tubular materials, there has been less technological development on sequential forging compared to the advancements seen in sequential forging of sheet materials regarding item (4). Existing swaging processes have enabled some profile forming, such as achieving rectangular cross-sectional shapes or introducing internal splines. However, these processes rely on mandrels or dies for shape transfer. On the other hand, using highly precise mandrels allows for obtaining products with high-precision inner surfaces, which is crucial in the medical field, but

Y. Kuriyama et al., *Tube Forming*, https://doi.org/10.1007/978-981-95-3421-0_6

it introduces a trade-off relationship between item (4) accompanied by roughness of the free surface.

6.2 Spinning

In the late 1990s, a technological innovation related to spinning of tubular materials began, specifically the introduction of non-axisymmetric spinning. Similar to sheet materials, tubular materials can undergo flow forming/shear spinning (rotary ironing) and simple spinning/draw spinning. The innovation in non-axisymmetric spinning is primarily focused on draw spinning. Tubular spinning can be performed using dedicated CNC spinning machines, as well as CNC spinning machines or lathe machines for sheet materials. Although the roller shapes used in flow forming and simple spinning differ, the roller shape, material, and lubrication conditions are essentially the same as those used in sheet metal spinning, allowing for the sharing of databases. It should be noted that expanding, flaring, and flanging, which involve stretch deformation, can also be accomplished through spinning. However, most of these operations are performed by rolling [2], which will be discussed in Sect. 6.4 on rotary forming.

6.2.1 *Rotary Ironing*

By performing rotary ironing (also known as tube spinning or flow forming), as shown in Fig. 6.1, it is possible to elongate and thin tubes. There are two methods of rotary ironing: forward rotary ironing, where the roller movement direction coincides with the metal flow direction as shown in Fig. 6.1a, and backward rotary ironing, where the roller movement direction is opposite to the metal flow direction as shown in Fig. 6.1b [3, 4].

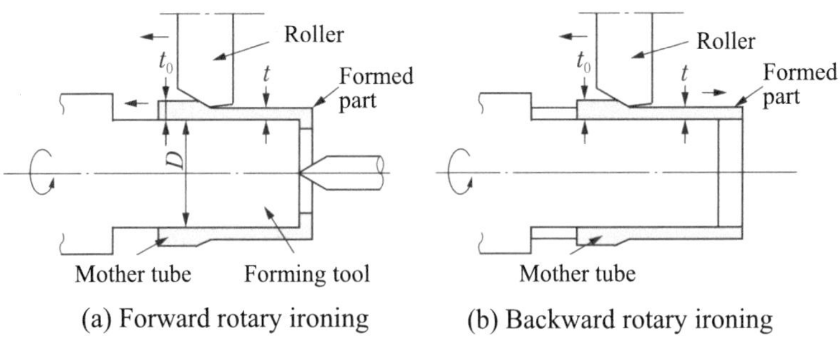

Fig. 6.1 Rotary ironing of tubes [4]

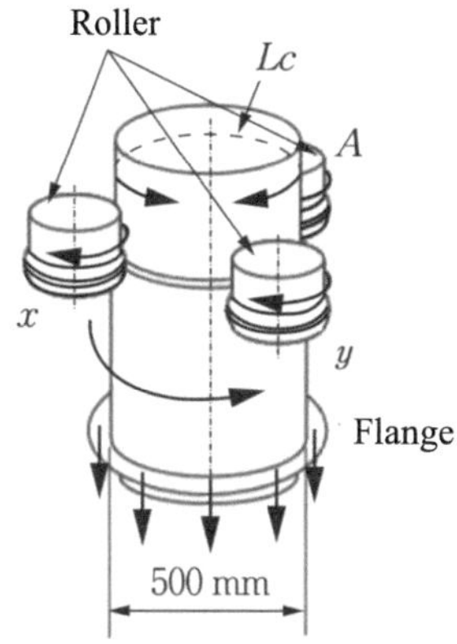

Fig. 6.2 Ironing spinning for thin-walled large-diameter pipe [6]

In the case of forward rotary ironing, the clamping method for the tube becomes complex. On the other hand, in backward rotary ironing, the clamping method becomes simpler. However, if there is eccentricity in the forming tool (mandrel), it can cause uneven thickness and result in distortion. Therefore, careful management to minimize eccentricity is necessary [3, 4]. For accuracy, it is recommended to maintain the roundness and runout of the forming tool within 0.01 mm and 0.03 mm, respectively. In the case of higher precision, these values should be within 3 μm and 0.01 mm, respectively [5].

A forming machine has been developed, as shown in Fig. 6.2, which performs ironing spinning with three rolls while applying tension to the flange section. Successful ironing spinning has been achieved from a thin-walled large-diameter welded SUS 304 pipe with diameter of 500 mm and thickness of 0.5 mm to a diameter of 500 mm and a thickness of 0.2 mm [6, 7].

6.2.2 Draw Spinning

The middle section of a tube can be narrowed (necked) by draw spinning, using multiple passes while rotating the tube, as shown in Fig. 6.3 [8]. In this process, reducing the feed amount in the radial direction and sequentially feeding in the axial direction of the tube are crucial points. An example of the application of draw spinning for the production of a muffler using an STKM 38 mother tube with an outer diameter of 76 mm, a thickness of 2.8 mm, and a length of 200 mm has been reported. The tube was rotated at 590 rpm, and a pass schedule of 6 reciprocating passes with a roll feed of 0.3 mm/rev resulted in necking with an outer diameter of 50 mm and length of 62 mm [8].

Draw spinning for necking tube ends can be performed either by rotating the tube or by keeping the tube stationary and revolving the rolls. In the latter case, either 2 rolls or 3 rolls, as shown in Fig. 6.4, are used for spinning around the fixed

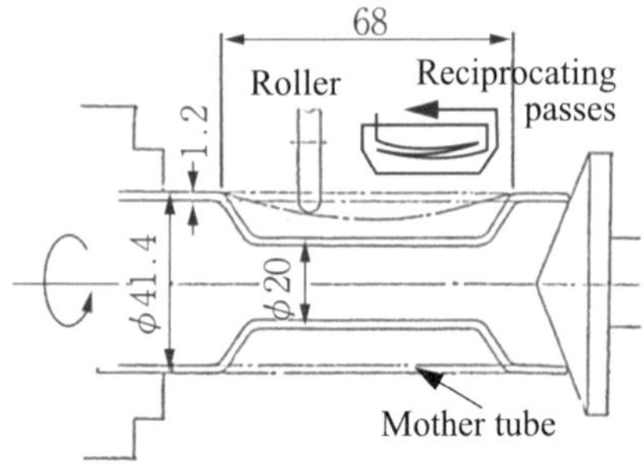

Fig. 6.3 Draw spinning of middle section of tube [8]

pipe (coaxial draw spinning). Improper pass schedules can lead to wrinkling at the mouth edges [9]. In necking by a press, the taper section monotonically thickens towards the end, while in spinning, although thickening occurs at the end, the central taper section is thinner than the raw tube section [10]. This thinning phenomenon is specific to spinning and occurs even in the case of rotating the tube for spinning [11]. The variation in thickness at the pipe ends depends on the roll pass schedule, and it is known that using a pass schedule that rolls back from the pipe end can result in thicker edge portions [11]. Figure 6.5 illustrates an example of necking using 2 rolls with a fixed tube, where an approximately 80% necking ratio was achieved for a roll-formed stainless steel pipe (SUS 409, TIG welded pipe) with an outer diameter of approximately 101 mm.

Tube end closing, also known as doming, is performed by rotating the tube. As shown in Fig. 6.6, the tube end is locally heated, and the roller is moved along an arc-shaped path to perform the closing process. Burner heating is also applied during the forming process. In the case of high-pressure gas cylinders, a temperature of over 1000 °C is required for melting the joint in the central region [9]. Determining the path schedule for spinning is crucial. A method has been developed to automatically determine the optimal path schedule based on image information of the forming

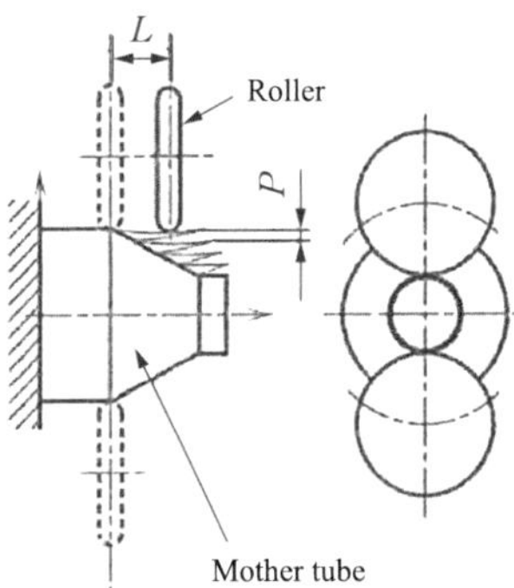

(a) Orbital path of the roller

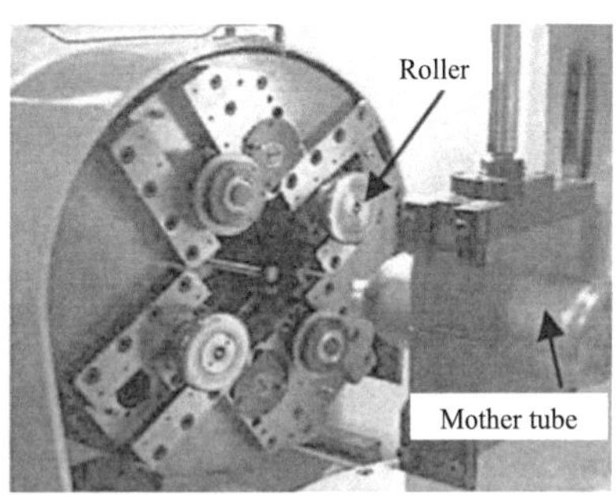

(b) Draw spinning machine

Fig. 6.4 Draw spinning for necking tube ends (fixed tube and revolving rolls) [9].

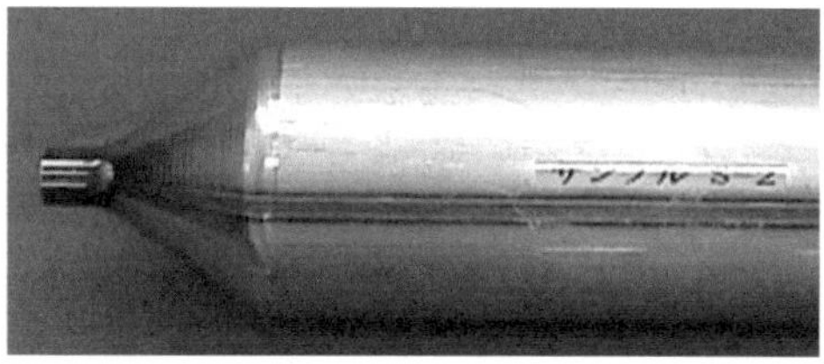

Fig. 6.5 Necking formed tube end by draw spinning[1]

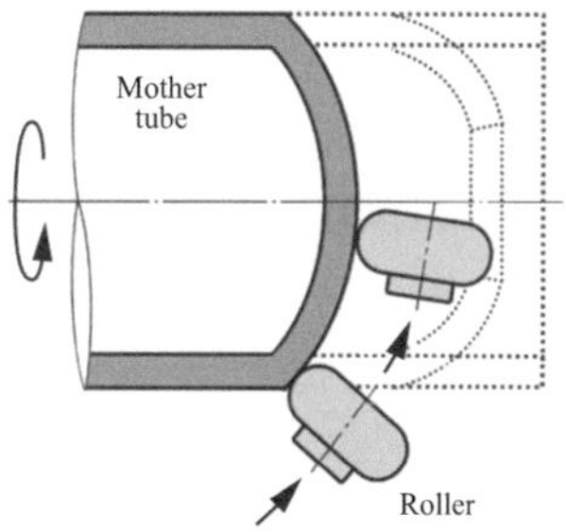

Fig. 6.6 Tube end closing (doming)

shape using fuzzy inference, and it has been successfully applied to the doming of a magnesium alloy tube end in conjunction with burner heating, demonstrating its effectiveness [12].

6.2.3 Eccentric and Inclined Spinning

In the case of conventional coaxial spinning used for automotive catalytic converter cases, the domed end and the straight tube section are coaxial, which can pose spatial constraints when assembling them in vehicles. In order to address this issue, eccentric and inclined spinning [13–15] has been developed.

This forming technique, as shown in Fig. 6.7, is performed using a 4-axis CNC spinning machine where three rollers revolve around the tube while, at each pass, the tube's central axis is eccentrically offset or inclined with respect to the revolving axis (central axis of the terminal forming section) [15]. The pass schedule for eccentric spinning is shown in Fig. 6.8a, and the path schedule for inclined spinning is shown in Fig. 6.8b [15]. An example of a formed shape is shown in Fig. 6.9 [15]. It should be noted that, similar to coaxial spinning, the tapered section in eccentric and inclined spinning also experiences thinning [15].

[1] Photo courtesy of SANGO Co., Ltd.

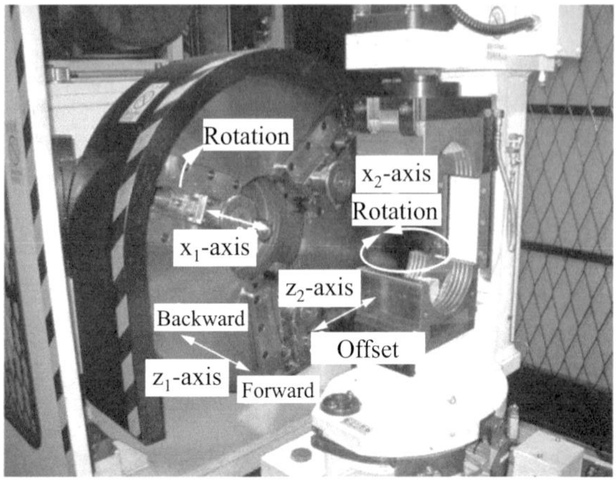

Fig. 6.7 4-axis CNC spinning machine for eccentric and inclined spinning of tube end [15]

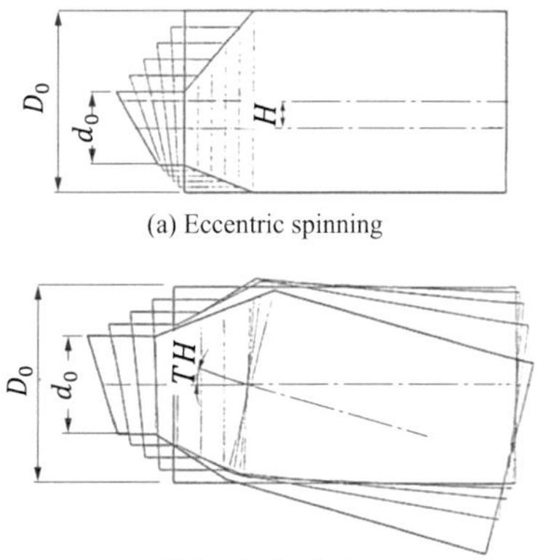

Fig. 6.8 Path schedule of eccentric and inclined spinning [15]

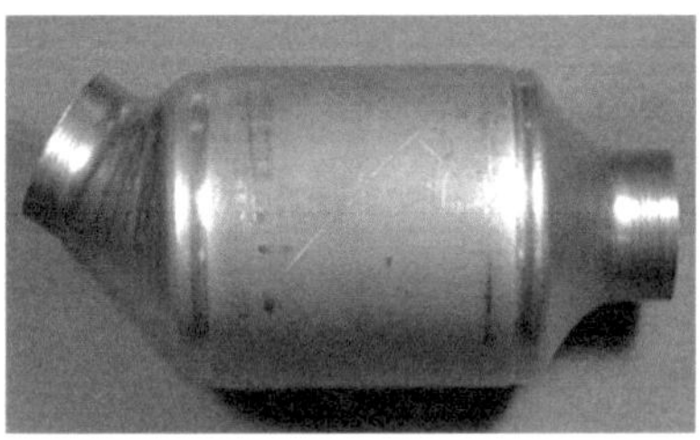

Fig. 6.9 Example formed part by eccentric and inclined spinning (exhaust catalyst case) [15]

6.2.4 Synchronized Spinning

In the spinning process where the tube is rotated, by synchronizing the rotation of the tube with the motion of the rollers, as shown in Fig. 6.10, it becomes possible to

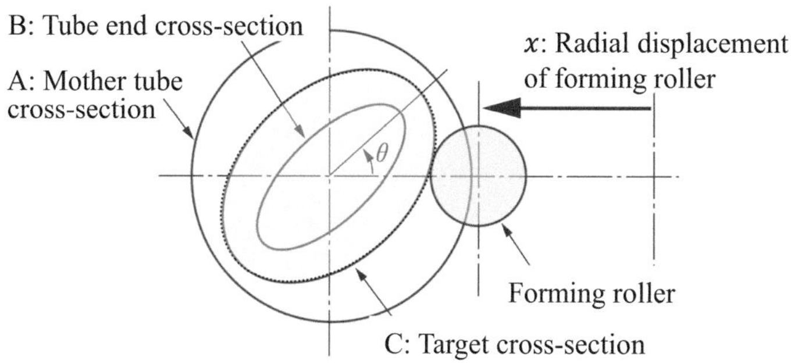

Fig. 6.10 Mechanism of CNC synchronized spinning [16]

shape the tube into complex cross-sectional shapes [16, 17]. Similar to incremental forming of sheet, the workpiece is formed into the shape of the tool enveloping surface which is formed by the relative motion of rollers to the tube.

A 5-axis, 2-roller CNC forming machine, as shown in Fig. 6.11, has been developed for this purpose. Examples of formed profiles are shown in Fig. 6.12. This technique allows for the formation of complex cross-sectional shapes not only at the tube ends but also in the central portion of the tube. Furthermore, the forming of curved tubes, as shown in Fig. 6.13, has also become possible [18].

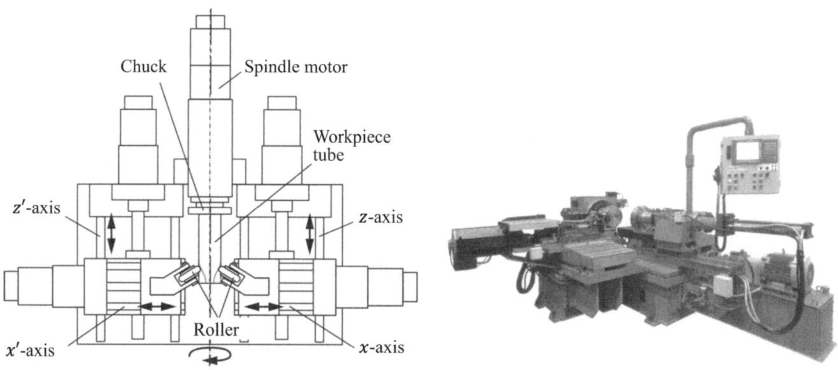

Fig. 6.11 2-roller CNC synchronized spinning machine for tube forming[2]

[2] Photo courtesy of Daito Spinning Corporation

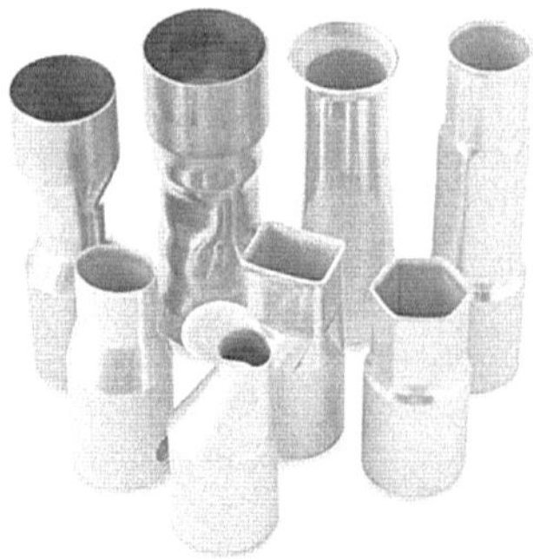

Fig. 6.12 Examples of formed profiles by CNC synchronized spinning (see Footnote 2)

Fig. 6.13 Example of curved tube formed by CNC synchronized spinning [18]

6.3 Rotary Swaging

Regarding rotary swaging, it has been discussed in technical books [19, 20] and papers [21] published in the United States in the 1960s, and there have been no significant developments in the fundamental aspects of the process up to the present. Due to the highly complex deformation phenomenon, many aspects of the know-how that determine the success of the forming process remain unclear, and simplification of tooling is still largely unexplored. Examples of rotary swaging for tube forming are shown in Fig. 6.14. Since it involves sequential compression forming (sequential forging) in the radial direction of the tube, it has been effective in various applications such as necking the tube ends and the center portion of tubes, crimping with wires, forming the threading guide of tube top for drawing, forming inner splines or rectangular cross-sections, and forming hard-to-work materials.

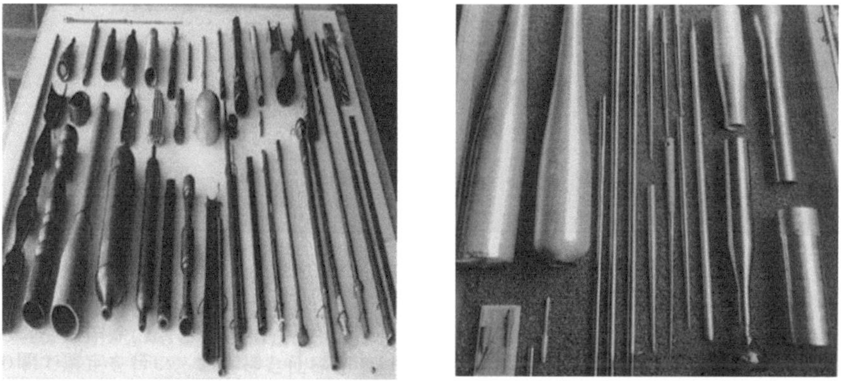

Fig. 6.14 Examples of formed profiles by rotary swaging[3]

6.3.1 Spindle Rotation Method

Figure 6.15 illustrates the rotary swaging process with a stationary (fixed) housing and rotational spindle. As the spindle rotates, the backer (hammer) in the state shown in Fig. 6.15a comes into contact with the rollers and pushes the die in the direction of the rotation center, compressing the tube using opposing dies. When the position of the backer moves between the rollers as shown in Fig. 6.15b, the dies open due to centrifugal force, and at that moment, the tube is slightly inserted into the dies. While the tube rotates in conjunction with the spindle, its rotation is slower than that of the spindle. However, due to this rotational difference, the compression location on the tube sequentially moves in the circumferential direction.

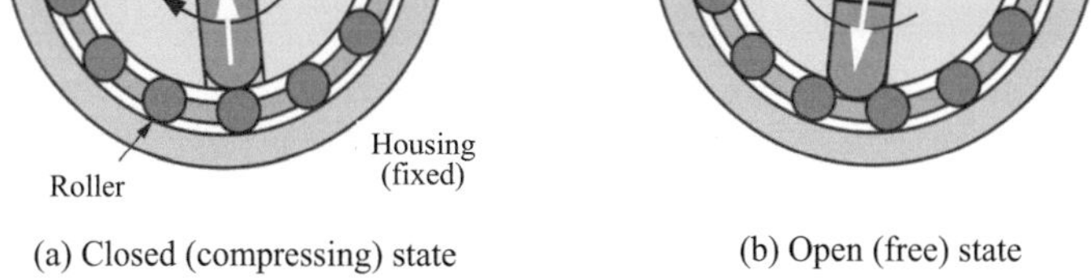

Fig. 6.15 Mechanism of spindle rotation type rotary swaging of tube

[3] Photo courtesy of Yoshida Kinen Co., Ltd.

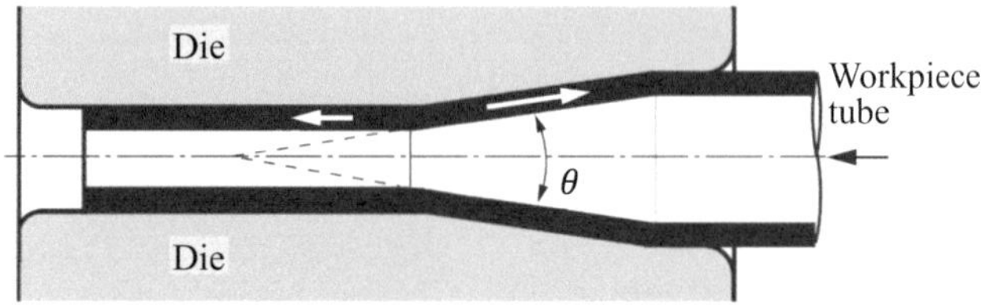

Fig. 6.16 Axial metal flow in rotary swaging of tube

The spindle rotation speed is typically around 300–500 rpm, and the number of impacts often reaches about 1,000–5,000 per minute. In this sequential forming process, the tube experiences compressive stress in the radial direction while there is no tensile stress acting in the axial direction. Therefore, the deformation occurs under high hydrostatic pressure. As a result, the material exhibits high formability and is suitable for forming hard-to-work materials.

In examining the process of sequential compression forging within the cross-section along the tube axis, as shown in Fig. 6.16, metal flow occurs in the opposite direction to the insertion of the tube in the taper section on the inlet side, divided by the watershed of metal flow. Hence, if the taper angle θ exceeds 8 degrees, an insertion device may be necessary [20]. In addition to the two dies shown in Fig. 6.15, there are also four-die and three-die configurations. The use of four dies is more common, but in the case of extremely small-diameter tubes, two dies are used [22].

6.3.2 *Stationary Spindle Method*

The rotary swaging process with a rotary housing and a stationary (fixed) spindle configuration is shown in Fig. 6.17. This method often utilizes four dies, and because of the stationary spindle, the tube remains stationary while undergoing biaxial sequential compression by the dies. As a result, it enables the forming of non-circular cross-sections, such as rectangular profiles.

Fig. 6.17 Mechanism of stationary spindle type rotary swaging of tube

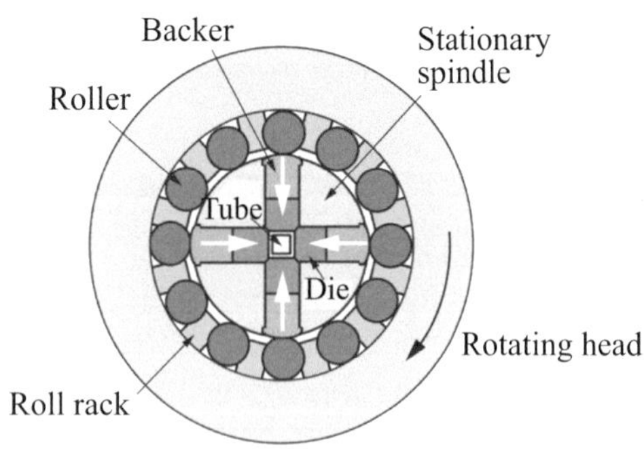

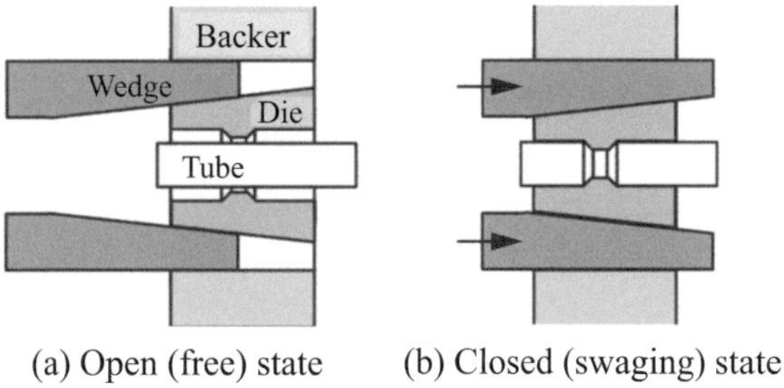

Fig. 6.18 Die closing type rotary swaging to form middle of tube

6.3.3 Die Closing Method

When forming the middle section of a tube using rotary swaging, the opening of the die is controlled by using wedges (shims), as shown in Fig. 6.18 [19, 20]. First, the insertion depths of wedges are kept small to enlarge the die opening and then the tube is inserted. Next, as shown in Fig. 6.18b, the wedges are pushed in to reduce the die opening and perform rotary swaging on the middle section of the tube. Afterward, the wedges are retracted to open the die and remove the tube after forming. Controlling the depth of wedge insertion during rotary swaging allows for the formation of a product with periodic bumps on its contour.

6.3.4 Rotary Housing with Low-Speed Spindle Rotation Method

In the case of forming large-diameter tubes using the rotary spindle method mentioned in Sect. 6.3.1, the circumferential speed of the tube increases, creating a hazardous condition during operation [19, 20]. Therefore, for large-diameter tubes, the rotary housing with a low-speed spindle rotation configuration (approximately 20 rpm [23]) is employed to reduce the circumferential speed of the tube during rotary swaging.

6.3.5 Mandrel Swaging

By using a mandrel as shown in Fig. 6.19, it is possible to form non-circular shapes such as splines on the inner surface and tubes with a rectangular inner shape through rotary swaging (Fig. 6.20) [19, 20].

When rotary swaging is performed without using a mandrel, surface unevenness becomes evident on the inner surface of the tube. On the other hand, when rotary swaging is performed with a mandrel, the surface morphology of the mandrel is well transferred on the workpiece, so if the mandrel surface is smooth, the inner

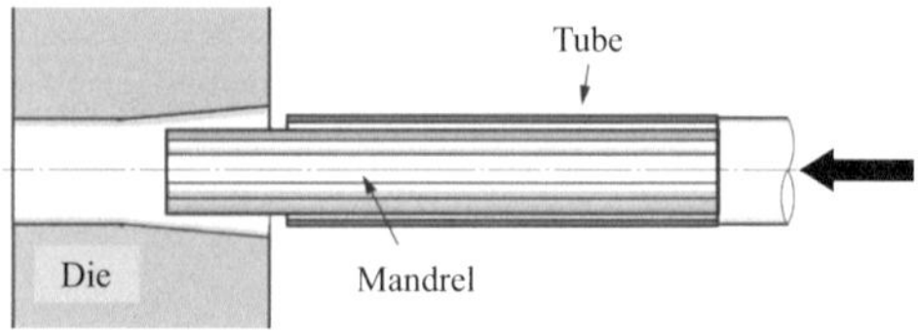

Fig. 6.19 Rotary swaging using a mandrel by rotary swaging using a mandrel

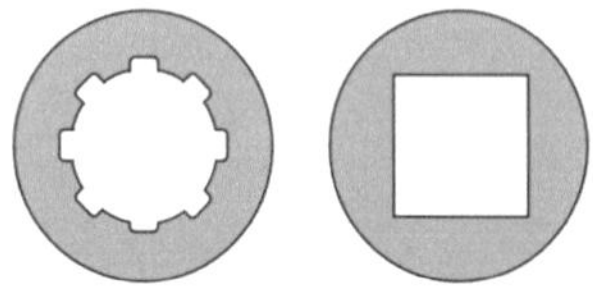

Fig. 6.20 Examples of cross-sections

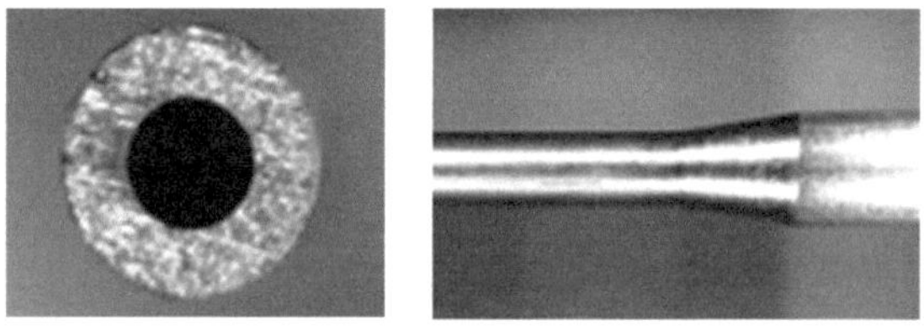

(a) Cross-section (b) Appearance of formed part

Fig. 6.21 High-precision swaging of stainless steel microbore tubing [22]

surface of the tube becomes smooth and eliminates surface unevenness. An example is shown in Fig. 6.21, where a mandrel made of a smooth-surfaced piano wire with a diameter of 0.09 mm is used to perform rotary swaging on a stainless steel tube (SUS 304) using two dies, achieving a surface roughness of 0.4 μm on the inner surface (allowing subsequent internal polishing). In this example, a mother tube with inner surface roughness Ra 0.8 μm is formed into a tube of minimum inner diameter 0.09 mm with tolerance ± 0.01 mm and inner surface roughness Ra 0.4 μm [22]. In the medical field, extremely small-diameter tubes with extremely low surface roughness are required for collecting and analyzing cells without damaging them [22]. In addition to piano wire, high-speed steel wire is sometimes used as a mandrel material.

6.3.6 Die Hole Clearance

In rotary swaging, the clearance (escape amount) applied to the left and right sides of the die hole, as shown in Fig. 6.22a, becomes extremely important. As a die hole shape including this clearance, an approximate elliptical hole created by two arcs has long been recommended. In this case, the clearance is obtained by grinding a round hole with a diameter *D*, divided by a 120° position, using a wedge (shim) of thickness s, as shown in Fig. 6.22b. This *D* is determined by the following equation

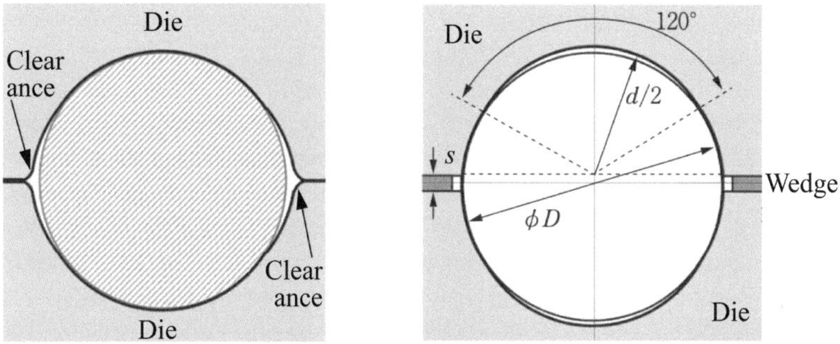

(a) Clearance at left and right

(b) Roll profile design comprising two arcs

Fig. 6.22 Roll profile design for appropriate clearance

based on geometric relationships [20]:

$$D = \sqrt{d^2 + sd + s^2} \tag{6.1}$$

Here, d is the outer diameter of the tube at the die taper inlet (initial outer diameter), and at the forming section, it is the outer diameter of the tube at that location. The recommended value of s varies depending on the material of the tube, the thickness-to- diameter ratio, and the use of a mandrel [20]. As a general guideline, an empirical value of $D = 1.02–1.15\ d$ is recommended [24].

6.4 Rotary Forming

Rotary forming (also known as rolling) for performing tube necking, flaring, flanging, and curling of tubes has been extensively discussed in a technical reference book published in the United States in the 1960s [2]. While these end-forming operations can also be achieved through spinning, rotary forming is generally superior in terms of productivity as it can be completed in a single pass. A key advantage compared to press forming lies in the significant reduction of forming loads and the consequent suppression of buckling, resulting from the avoidance of axisymmetric contact between the tube end and the die tool. Additionally, benefits include improved ductility resulting from a sequential strain distribution when using grooved or roller-equipped die tools, as well as the ease of achieving high rotational symmetry in the formed parts.

Closing by necking, achieved through localized heating utilizing frictional heat generated by rotation [25], has long been recognized as another beneficial technique. Although these forming processes were considered incremental forming, due to the one-to-one correspondence between formed tube end shapes and the contour of die tools (contoured dies), the elimination of die did not progress significantly. However, with the emergence of orbital die forming [26] in the 1980s, it became possible to use tool enveloping surfaces instead of complete contoured dies for shaping. Additionally, CNC forming machines capable of generating tool enveloping surfaces were developed [27], advancing the omission of traditional dies. Moreover, new possibilities such as the suppression of shape defects like mouth edge wrinkles, curling, and waviness [26], as well as metal flow control [27], have also been revealed. From the concept of imparting shape using tool enveloping surfaces, CNC incremental forming of bulge deformation type for sheet metal was developed. Beading, which involves the formation of bead shapes in the middle or near the tube ends, is often performed through rotary forming (rolling). While the cross-sectional shape of the bead formed in this rotary process can be considered as a contoured die due to the transfer of roller shape, it is possible to perform forming with a pair of rollers regardless of the tube diameter change, suggesting a state of die elimination for this particular aspect.

6.4.1 Rotary Flaring

Rotary flaring of tube ends can be achieved by either fixing the tube and allowing the tool to revolve around it (rolling) or by rotating the tube itself [28]. Both methods involve the same relative motion between the tube and the tool. The former method is illustrated in Fig. 6.23. As shown in Fig. 6.23a, a die (tube fixture) with the same contour as the flaring section and a tool (free rotation) are used. By introducing eccentricity between the tube axis and the revolving tool and moving the tool in the axis direction, incremental forming is achieved [2]. Consequently, the forming load is significantly reduced and the tool contact area can be effectively strained compared to press flaring. Moving the tool in the radial direction, as shown in Fig. 6.23b, further

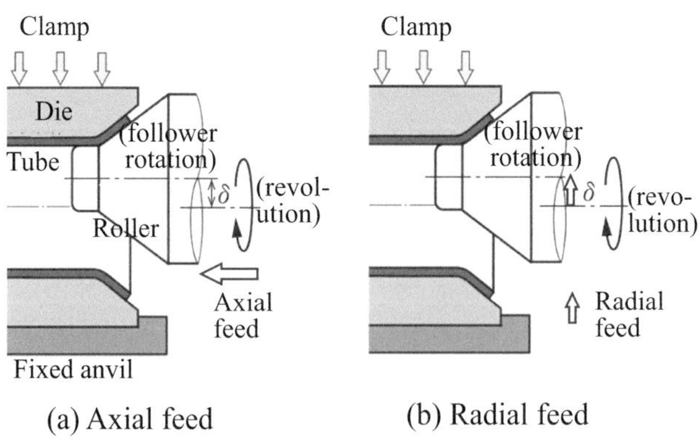

Fig. 6.23 Rotary flaring of tube ends (fixed tube and revolutionary roller)

reduces the axial component of the forming force [2]. This method is effective for high work-hardening materials prone to buckling in press flaring.

It also enables flange forming [2], although the flange expansion ratio is small. For flanges with a large expansion ratio, one-pass rotary forming with a tilted roll along the desired flange surface for forming purposes, and two-pass rotary forming with conical rotary flaring followed by roll flattening of the conical section, are possible. In both cases, it is necessary to insert a mandrel into the tube to prevent inward buckling of the tube at the flange root. This root buckling occurs because the watershed of the meridional metal flow in the tool contact area is closer to the root side. Due to incremental forming, complete suppression of undesired deformation modes such as curling and waviness (bending wrinkles) that are problematic in press forming can be achieved by reducing the contact area ratio between the tube and the tool surface [29]. When the tube axis and the tool axis are coaxial, full contact is achieved between the tube end and the tool, similar to the case of press forming. However, by attaching rollers to the tool surface [30] to create a localized contact state, an incremental forming state is achieved, resulting in effects similar to rotary flaring forming described above.

6.4.2 *Rotary Necking*

Rotary necking at the tube end can be performed using either the method of fixing the tube and allowing the tool to revolve around it (rolling), similar to the case of rotary flaring, or the method of rotating the tube itself [28]. In both methods, the relative motion between the tube and the tool is the same. The latter method is illustrated in Fig. 6.24. By using a tool (free rotation) with the same contour as the necking section shape and causing an eccentricity between the tube axis and the tool axis while moving the tool in the tube axis direction, an incremental forming state (eccentric necking state) is achieved [31]. As a result, the forming load is significantly reduced compared to press necking, and the tool contact area can be effectively strained. Additionally, the lower forming load contributes the suppression effect of buckling at the root of the necking, leading to an improvement in the forming limit and achieving a necking ratio exceeding 0.6.

Fig. 6.24 Rotary necking of tube ends (eccentric necking)

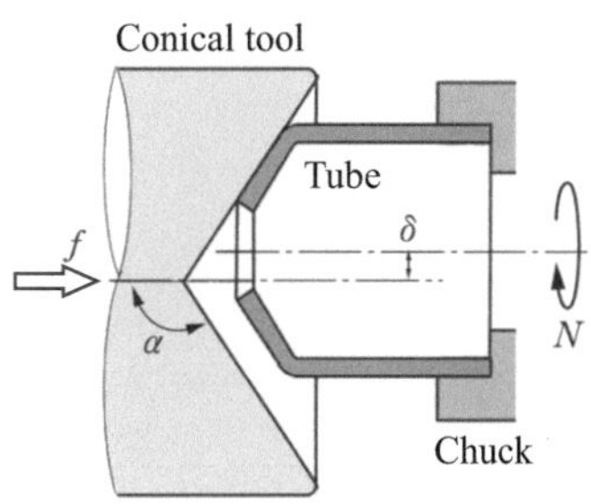

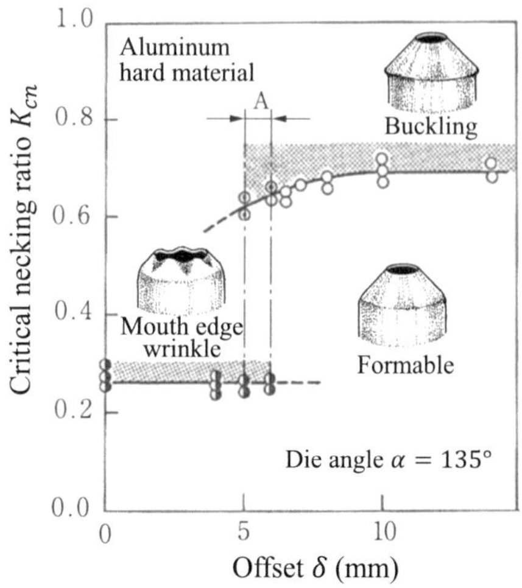

Fig. 6.25 Suppression of mouth edge wrinkles by eccentric rotary necking [31]

By setting the eccentricity to a certain value or higher, as shown in Fig. 6.25, the occurrence of mouth edge wrinkling can be completely suppressed [31]. Similarly, complete suppression of internal curling and bending defects (waviness) can be achieved by setting the eccentricity [31]. By reducing the contact area ratio between the tube end and the tool surface and increasing the degree of incremental deformation, the suppression of these irregular deformations can be achieved [31]. Similar effects can be obtained by moving the tool in the radial direction of the tube axis.

As shown in Fig. 6.26, by placing a mandrel inside the tube and moving a roller with a concave cross-section in the radial direction of the tube in a single pass, it becomes possible to form an internal flange that is close to vertical closing [32]. When the tube axis and the tool axis are coaxial, the tube end and the tool have full contact, similar to press forming.

However, as shown in Fig. 6.27, by providing oil grooves on an arc-shaped tool and performing rotary necking, a localized contact state is achieved, allowing for

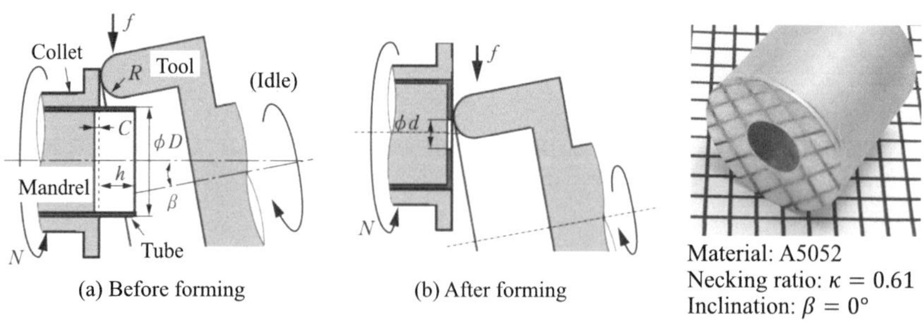

Fig. 6.26 Internal flange forming by single-pass rotary necking [32]

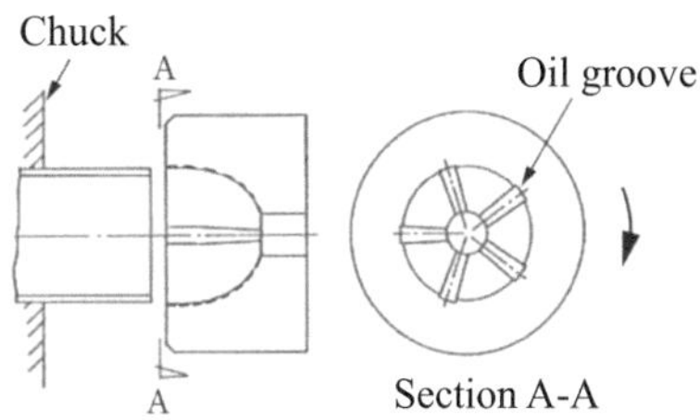

Fig. 6.27 Rotary necking using arc-shaped tool with oil grooves [33]

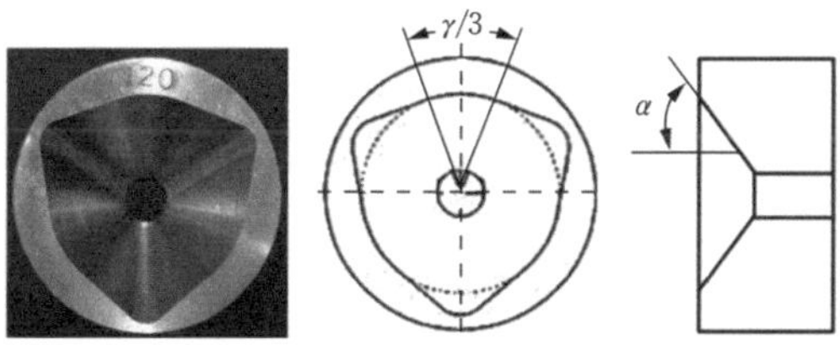

Fig. 6.28 Conical tool with relief surfaces for rotary necking [34]

the suppression of mouth edge wrinkling and the formation of an arc-shaped neck [33]. By providing reliefs to create a localized contact state, as shown in Fig. 6.28 [34, 35], an incremental forming state is achieved, resulting in similar effects to rotary necking (suppression of edge wrinkling and improvement in forming limit). It should be noted that when performing rotary necking while maintaining a full contact state, the necking section may undergo localized softening due to frictional heating, resulting in a hot forming condition where the forming load is reduced and the forming limit is improved [25]. However, in this case, the deformation state is different from incremental forming.

6.4.3 *Rotary Bead Forming*

Rotary bead forming, which applies a bead shape to the intermediate portion or near the end of a tube, can be performed in both outer bead forming and inner bead forming. Both outer and inner beads can be formed using either the method of fixing the tube and allowing the tool to revolve around it (rolling), as shown in Fig. 6.29, or the method of rotating the tube itself. The opposite side of the roller is restrained by a die or a support roll, and only the contact area of the roller is sequentially expanded or narrowed to form the bead. Without a supporting roll or die, the tube would undergo elastic deflection in the axial direction and flattening, making it difficult to achieve the desired shape. For small-diameter tubes, it becomes challenging to ensure the rigidity of the roller tool inserted inside the tube. Therefore, a tube diameter of approximately 19 mm (3/4 inch) or larger is considered a guideline for rotary bead forming [2]. When viewed in the axial cross-section, bead formation seems to fully transfer the contour of the roller shape to the tube. However, due to

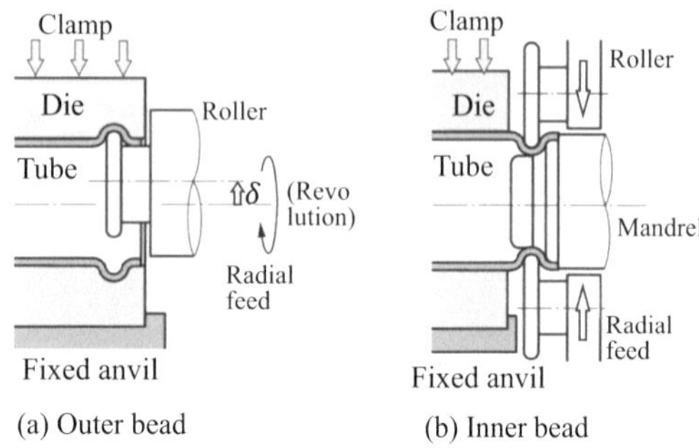

Fig. 6.29 Rotary bead forming of tube ends (fixed tube and revolutionary roller)

incremental forming, the shape transferability is good, allowing for the application of high-precision and highly circular beads.

6.4.4 *Orbital Rotary Forming*

This is one of the incremental forming techniques aimed at die-less forming. In this process, the tool enveloping surface created by the relative motion of the tool with respect to the tube is used instead of a die. This allows the formation of various shapes on the tube end using just one tool. When the tube is rotated and the inclined conical tool is moved axially along the tube axis, as shown in Fig. 6.30a, a conical surface with a half cone angle γ is obtained as the tool enveloping surface. Using this tool enveloping surface, the tube end is flared as shown in Fig. 6.30b [29]. Additionally, as shown in Fig. 6.30c, it is also possible to perform necking operations [26].

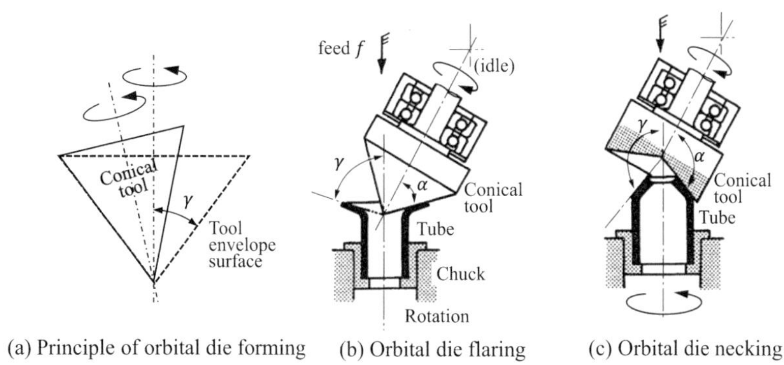

Fig. 6.30 Tube end-forming by orbital die forming

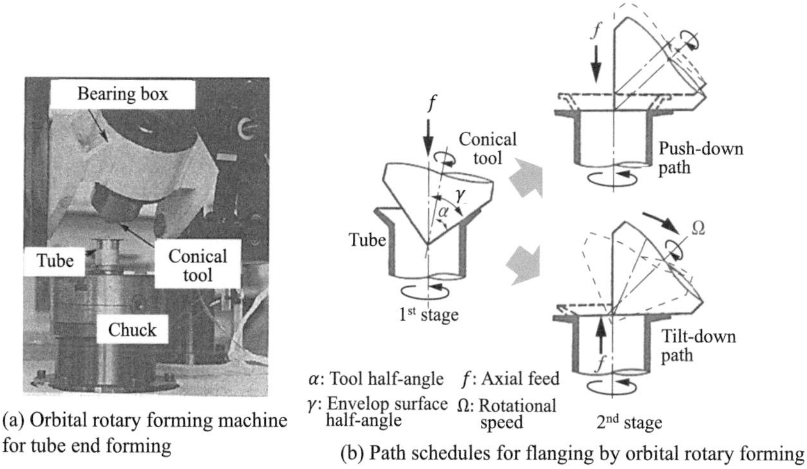

Fig. 6.31 Orbital rotary forming machine and path schedules [27]

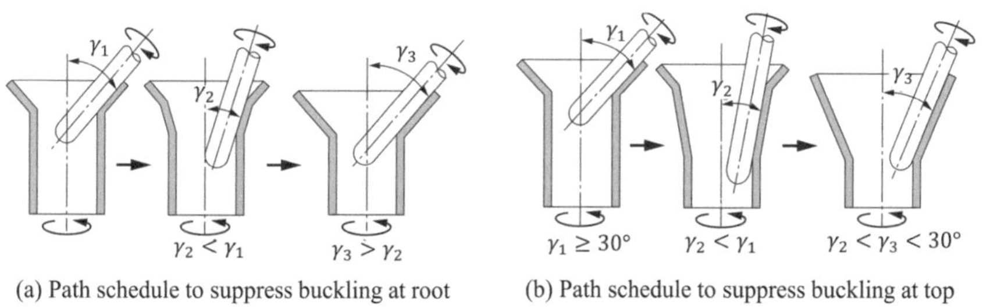

Fig. 6.32 Path schedule to suppress forming defects in orbital rotary forming using rod-shape tool [36]

In orbital rotary forming, by setting the contact area ratio between the tube end and the tool below a specific value, it is possible to completely suppress the shape defects such as curling, bending (waviness), and mouth edge wrinkles that occur during flaring or necking by a press. It also improves the forming limit [26, 29]. Dedicated CNC forming machines have been developed, as shown in Fig. 6.31, revealing the potential for metal flow control that varies the thickness distribution of the flange depending on the path schedule [27]. Two types of path schedule, push-down path and tilt-down path, are shown in the figure. When using a rod-shaped tool instead of a conical tool, as shown in Fig. 6.32, it becomes possible to suppress both buckling at the root and fracture at the top by varying the path schedule [36].

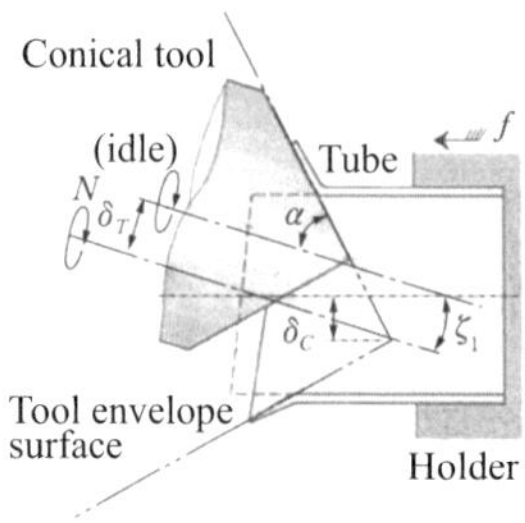

(a) Inclined flaring

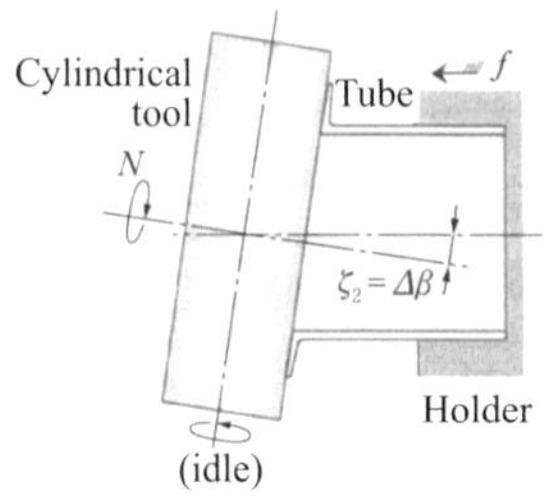

(b) Inclined flange forming

(c) Examples of inclined flange forming

Fig. 6.33 Inclined flaring and inclined flange forming of tube ends [37]

6.4.5 Inclined Flange Forming

This forming method is illustrated in Fig. 6.33. First, as shown in Fig. 6.33a, the tube end is flared with inclination to the tube axis using orbital rotary forming, by fixing the tube, and a conical roller is revolved around an axis inclined by ζ_1 with respect to the tube axis. Next, as shown in Fig. 6.33b, by using a roller to flatten the inclined flange surface, the inclined flange shown in Fig. 6.33c can be formed [37].

6.5 Incremental Flanging of Square Tube Ends

Using a CNC milling machine, similar to incremental forming of sheet metal, flaring or flanging of square tube ends is possible, as shown in Fig. 6.34 [38]. The mother tubes of these examples are aluminum A6063 square tubes with side length 30 mm and thickness 1.5 mm. Even with the use of general-purpose CNC machine tools, by applying the technique of incremental forming, various types of tube forming become possible, although it takes a longer processing time.

Fig. 6.34 Examples of inclined flaring and inclined flange of square tube ends [38]

References

1. The Japan Society for Technology of Plasticity (ed.): Technology of plasticity terminology dictionary, vol. 136. Corona Publishing, (1998)
2. Kervick, R.J., Springborn, R.K.: Cold bending and forming tube and other sections, pp. 185–205. ASTME, (1966)
3. ASM Handbook Committee.: Metals Handbook, **4**, 317–322, ASM, (1969)
4. The Japan Society for Technology of Plasticity (ed.): Spinning processing technology, vol. 74. Nikkan Kogyo Shimbun, (1984)
5. The Japan Society for Technology of Plasticity (ed.): Spinning processing technology, vol. 181. Nikkan Kogyo Shimbun, (1984)
6. Kuboki, T., Takahashi, K., Sanda, K., Moriya, S., Ishida, K.: Mater. Trans., **53**(5), 853–861 (2012)
7. Mita, K., Moriya, S. et al.: In: Proceedings of the 2010 Spring Meeting, JSTP, pp. 185–186 (2010)
8. The Japan Society for Technology of Plasticity (ed.): Technology of plasticity terminology dictionary, vol. 135, Corona Publishing, (1984)
9. Takada, Y., Takahashi, Y.: J. JSTP., **54**(628), 403–407 (2013)
10. Andou, H., Morikawa, A.: J. JSTP,. **53**(614), 208–212 (2012)
11. The Japan Society for Technology of Plasticity (ed.): Spinning processing technology, vol. 86. Nikkan Kogyo Shimbun, (1984)
12. Yoshihara, S., Mac Donald, B., Hasegawa, T., Kawahara, M., Yamamoto, H.: J. Mater. Process. Tech. **153–154**, 816–820 (2004)
13. Irie, T.: Japan Patent No. 2957153
14. Irie, T.: Japan Patent No. 2957154
15. Shindo, K., Ishigaki, K., Kato, K.: In: Proceedings of the 50th joint conference on the technology of plasticity, JSTP, pp. 173–174 (1999)
16. Arai, H., Fujimura, S., Okazaki, T.: In: Proceedings of the 56th joint conference on the technology of plasticity, JSTP, pp. 687–688 (2005)
17. Arai, H.: In: Proceedings of the 2014 Spring Meeting, JSTP, pp. 199–200 (2014)
18. Arai, H.: Proceedings of the 2016 Spring Meeting, JSTP, pp. 157–158 (2016)
19. Kervick, R.J., Springborn, R.K.: Cold bending and forming tube and other sections, pp. 205–220. ASTME, (1966)
20. ASM Handbook Committee.: Metals handbook, vol. 4, pp. 333–346. ASM, (1969)
21. Kegg, R.L.: Trans. ASME, J.Eng. Ind., 317–325 (1964)
22. Tada, M., Aizawa, J.: J. JSTP, **58**(683), 1094–1098 (2017)
23. The Japan Society for Technology of Plasticity (Ed.): Rotary forming, pp. 218–223, Corona Publishing, (1990)
24. Nakamura, M.: Pipe forming techniques, vol. 48. Nikkan Kogyo Shimbun, (1982)
25. Okamura, S., Tanaka, H.: J. Jpn. Soc. Precis. Eng., **43**(512), 898–904 (1977)
26. Kitazawa, K., Kobayashi, M., Maruno, H.: J. JSTP., **29**(330), 767–774 (1988)

27. Kitazawa, K., Kobayashi, M., Iida, M.: J. JSTP., **33**(380), 1080–1083 (1992)
28. Kitazawa, K.: J. JSTP., **30**(345), 1395–1402 (1989)
29. Kitazawa, K., Kobayashi, M., Maruno, H.: J. JSTP, **30**(344), 1259–1266 (1989)
30. Maruo, T.: Nikko Material, vol. 36. Japan Industrial Publishing, pp. 2–6 (1984)
31. Kitazawa, K., Kobayashi, M., Iida, M.: J. JSTP, **30**(339), 563–569 (1989)
32. Kitazawa, K.: In: Proceedings of the 2016 spring meeting, JSTP, pp. 159–160 (2016)
33. Nakamura, M., Maruyama, K., et al.: Pipe forming techniques, 2nd edn, pp. 118–119. Nikkan Kogyo Shimbun, (1998)
34. Murata, M., Kominami, A., Kuboki, T.: In: Proceedings of the 2008 spring meeting, JSTP, pp. 221–222 (2008)
35. Kuboki, T., Abe, M., Yamada, Y., Murata, M.: CIRP Annals–Manuf. Tech., **64**(1), 269–272 (2015)
36. Kitazawa., K., Seino., J.: Trans. Jpn. Soc. Mech. Engineers. C, **57**(539), 2453–2459 (1991)
37. Kitazawa, K., Nishida, M.: In: Proceedings of the 50th joint conference on the technology of plasticity, JSTP, pp. 459–460 (1999)
38. Matsubara, S.: In: Proceedings of the 50th joint conference on the technology of plasticity, JSTP, pp. 761–762 (1994)

Chapter 7
Cutting, Contouring, Piercing, and Burring

7.1 Introduction

In order to use tube materials as components, before forming the tube it is necessary to first cut the tube to the required dimensions. Additionally, contouring the tube end and creating holes of various shapes on the tube wall, as well as performing burring operations, are also necessary. There are various methods available for cutting tube materials, and each method is selected according to the characteristics of the respective components [1]. Likewise, there are various methods for contouring the tube ends and creating holes in the pipe wall, and these methods are also selected based on the characteristics of each component [2]. Burring operations on tube materials can be performed using rigid tool extraction, cross-pin, or incremental burring methods, depending on the dimensions and features of the components.

Tube materials, as a raw material for fabrication, have distinct characteristics compared to sheet materials. These characteristics include having inherent rigidity against deformation, having a closed cross-section that limits the insertion of tools inside, and the direction of tool movement cannot be limited to the thickness direction. When performing cutting, contouring, piercing, and burring operations on tube materials, it is essential to fully understand these characteristics.

7.2 Tube Cutting Methods

There are various methods for cutting tubes, as shown in Table 7.1. Here, we will discuss representative cutting methods using plastic deformation. For detailed information, it is recommended to refer to the textbook "Shearing Processes" published by JSTP.

Y. Kuriyama et al., *Tube Forming*, https://doi.org/10.1007/978-981-95-3421-0_7

Table 7.1 Various methods for cutting tubes

Mechanism		Method	Diagonal	Profile tube	Thin tube	Long tube	Short tube
Plastic deformation	Press	Roll pressing	–	–	–	Excellent	Average
		Roll pressing with tension	–	–	–	Excellent	–
		Roll pressing after turning	–	–	–	Excellent	Excellent
	Shear	Parting-off	–	Poor	Poor	Excellent	–
		Double parting-off	–	Poor	Poor	Excellent	–
		Shear with mandrel	–	Good	Average	Poor	Excellent
		Double shear with mandrel	–	Good	Good	Good	Excellent
Removal	Cutting	Tool cutting	Excellent	Excellent	Poor	Excellent	Good
		Band saw cutting	Excellent	Excellent	Poor	Excellent	Good
		Circular saw cutting	Excellent	Excellent	Average	Excellent	Excellent
		Abrasive disc cutting	Excellent	Excellent	Excellent	Excellent	Excellent
	Fusion	Laser cutting	Excellent	Excellent	Average	Excellent	Excellent
		Plasma cutting	Excellent	Excellent	Average	Excellent	Excellent
		Wire discharge cutting	Excellent	Excellent	Average	Excellent	Excellent

(continued)

Table 7.1 (continued)

Mechanism		Method	Sectional accuracy	Burrs	Chip disposal	Cutting speed	Equipment cost
Plastic deformation	Press	Roll press	Average	Average	Excellent	Good	Average
		Roll press with tension	Good	Good	Excellent	Good	Average
		Roll press after turning	Good	Average	Poor	Poor	Average
	Shear	Parting-off	Poor	Good	Good	Excellent	Good
		Double parting-off	Average	Good	Good	Good	Average
		Shear with mandrel	Average	Good	Excellent	Good	Average
		Double shear with mandrel	Good	Good	Excellent	Good	Average
Removal	Cutting	Tool cutting	Excellent	Good	Poor	Average	Average
		Band saw cutting	Excellent	Poor	Poor	Good*	Good
		Circular saw cutting	Excellent	Poor	Poor	Good*	Excellent
		Abrasive disc cutting	Excellent	Average	Average	Average*	Excellent
	Fusion	Laser cutting	Good	Good	Average	Good	Poor
		Plasma cutting	Average	Average	Average	Average	Average
		Wire discharge cutting	Excellent	Excellent	Average	Poor	Poor

* Multiple simultaneous cutting is possible

7.2.1 Roll Press Cutting

The roll press cut, as shown in Fig. 7.1, is a method of cutting by radially pushing rotating wedge-shaped rolls into the tube while revolving them around the tube. It is widely used for tube cutting due to its advantages of producing no cutting chips and allowing cutting without the use of a mandrel, although it cannot be used for cutting non-circular tubes or diagonal cuts. There are two methods: one involves rotating the rolls, and the other involves rotating the tube. The choice of method depends on the specific situation. However, when the tube is long or coiled, it is more common to rotate the rolls. The cut surface obtained by this roll press cut unavoidably has a taper due to the angle of the roll, and in addition, it exhibits a swell on the outer circumference as shown in Fig. 7.2, as well as a significant inward burr. Figure 7.2 shows an example of a cross-section of a tube for a small n-value material. In materials with a large n-value, the amount of swell is small, but under certain processing conditions, the diameter of the cut section may become slightly narrower. To prevent an inward burr, the tube should be subjected to tension while roll press cutting. By applying tension, tensile fracture occurs at the tip of the wedge before the burr occurs, preventing an inward burr. In this case, the cut surface has a fractured surface as shown in Fig. 7.3. Furthermore, to prevent a swell, it is recommended to pre-cut a groove, as shown in Fig. 7.4, before roll press cutting.

The pressing force of the rolls can be estimated by the theory of slip-line fields, where the shape of the rolls is considered as a collection of thin wedges, and the penetration force of these wedges into the semi-infinite body is calculated. However, in general, the pressing force is determined through actual measurements. Typically, rolls are arranged radially in sets of 3 or 4, and it is preferable to have larger roll diameters from a wear perspective, within the range where interference between

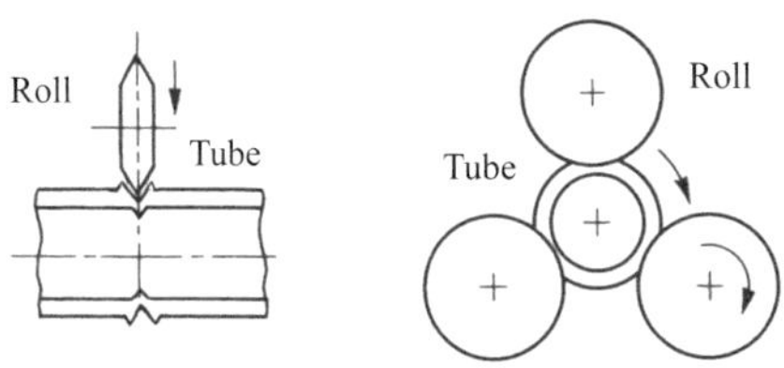

Fig. 7.1 Roll press cut

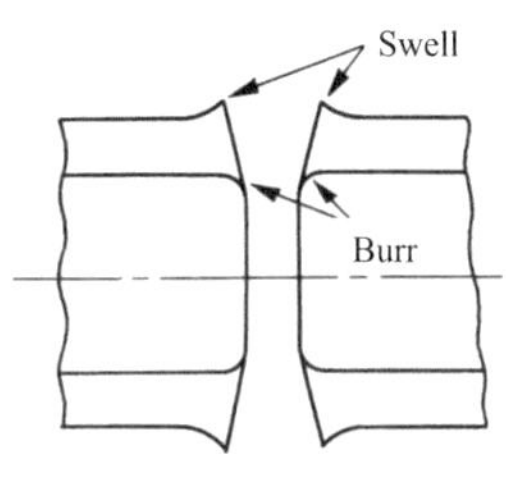

Fig. 7.2 Axial cross-section of roll press cut tube

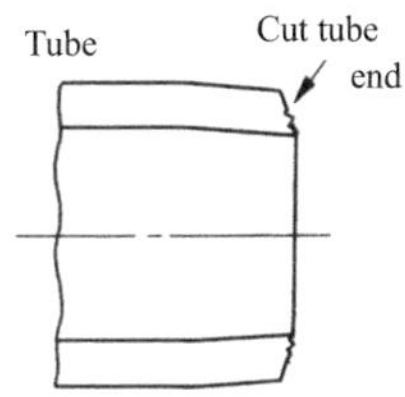

Fig. 7.3 Cross-section of tube cut by roll press under tension

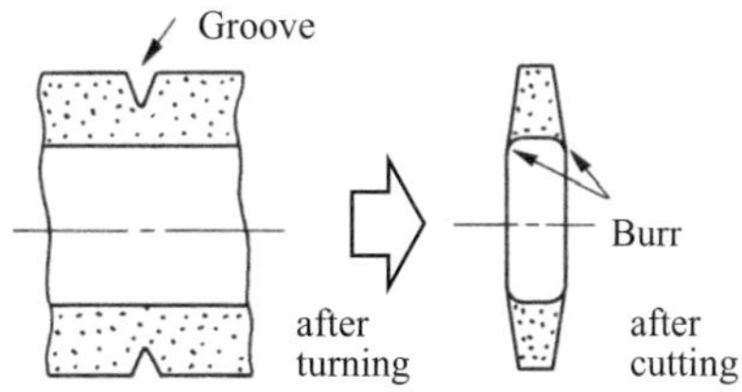

Fig. 7.4 Cross-section of tube (ring) cut by roll press after turning

the rolls can be avoided. The angle of the wedge at the tip of the rolls is typically selected within the range of 20–60°, but smaller angles are preferred from an accuracy standpoint, as long as they are within a range that does not pose strength-related issues.

7.2.2 *Parting-off Cutting*

Parting-off cutting is a method that utilizes the rigidity of a tube by pushing a blade-like punch from the outside to cut it. During this process, the tube's outer surface is constrained by a split die, as shown in Fig. 7.5. Although scrap is generated due to the thickness of the punch, this method allows for cutting with only vertical movement of an outer punch, resulting in the highest cutting speed. It is widely used for cutting of tubes. The equipment is relatively simple, allowing for cutting on-the-fly without stopping the moving tube, with examples of moving speeds exceeding 100 m/min.

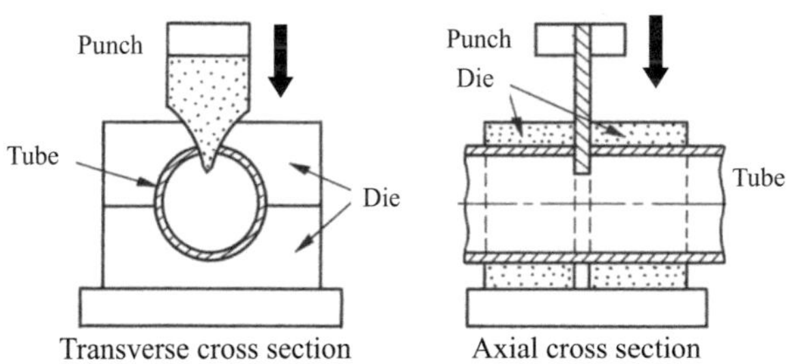

Fig. 7.5 Parting-off cutting

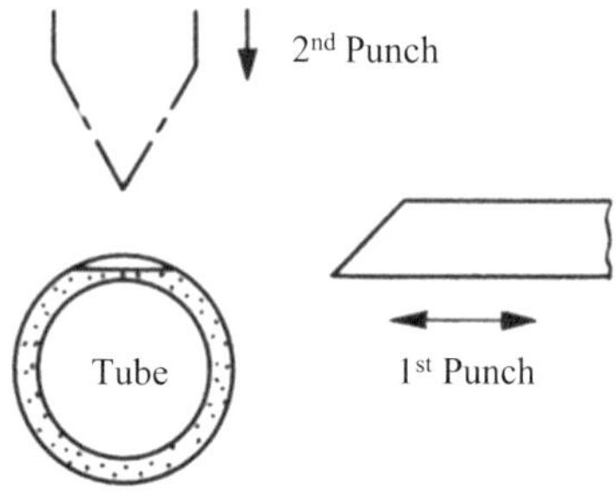

Fig. 7.6 Punches of double parting-off cut and transverse section after first cut

In this process, since there is no support inside the tube, it is prone to inward deformation. To prevent this, there is a method called "double parting-off cutting" where the upper part of the tube is thinned by cutting from the side with a punch before the main punch is pressed, as shown in Fig. 7.6. This method requires complex mechanisms and specialized machines, and the cutting speed is slightly slower, but it helps suppress deformation.

In this processing method, as shown in Fig. 7.7, the initial stage of the cutting process involves parting deformation through pressing and shearing deformation. However, it transitions to shearing by the scrap, and in some cases, at the final stage only in the central part shearing occurs by a punch and die. The shearing condition undergoes complex changes throughout the process, with a relatively small burr. The issue with this method is the occurrence of dents in the area where the punch contacts, as depicted in the figure. In many cases, it is necessary to correct these dents afterward, but they can be suppressed by applying the aforementioned double parting-off cutting method.

Regarding the processing force, although the shearing condition changes during the process, it can be considered as a load that punches out two surfaces of approximately 1/4 of the cross-sectional area of the tube at once, considering this process involves the natural shear angle. For the punch, it is preferable to have a thin thickness to minimize scraps and depressions. However, in the section where shearing by the scrap occurs, punch damage is more likely to occur, so the thickness needs to be determined considering its strength. The angle at the tip of the punch is preferably small to suppress depressions, but it is typically selected around 40–90°, taking into account the balance with strength. As for the die, using a cam mechanism that descends the die before the punch and provides strong support can help suppress depressions.

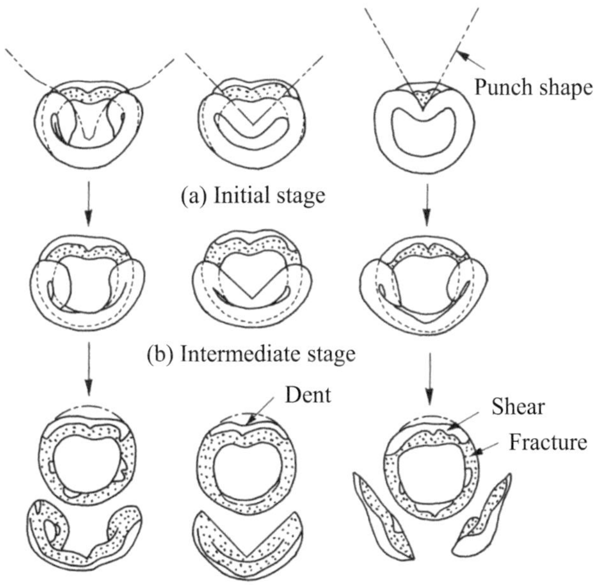

Fig. 7.7 Change of deformation by punch shape and processing stage in parting-off cutting

7.2.3 Shearing Using Mandrel

By inserting a mandrel into the tube and utilizing its end face as the inner cutting edge in conjunction with an outer punch, it becomes possible to eliminate crush deformation and scrap. There are two methods: the mandrel shearing method, as shown in Fig. 7.8, and the double mandrel shearing method, as shown in Fig. 7.9.

The mandrel shearing method shown in Fig. 7.8 has the advantage of not requiring special equipment using a floating die set mechanism. However, due to constraints such as the limited length of the movable mandrel and the need for overhanging the fixed die for knockout purposes, this method is only used for cutting very short tubes.

Fig. 7.8 Shearing with mandrel

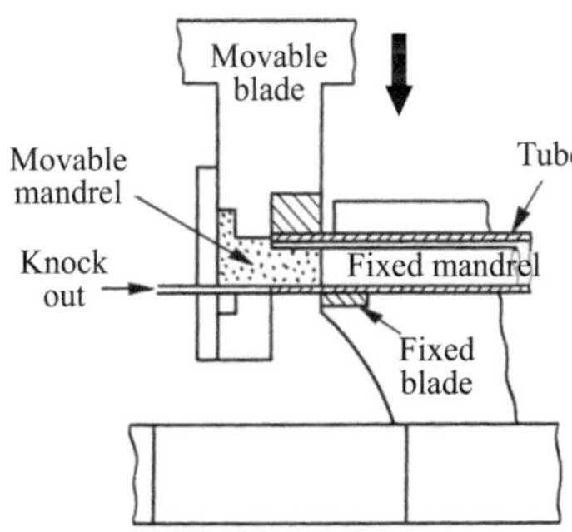

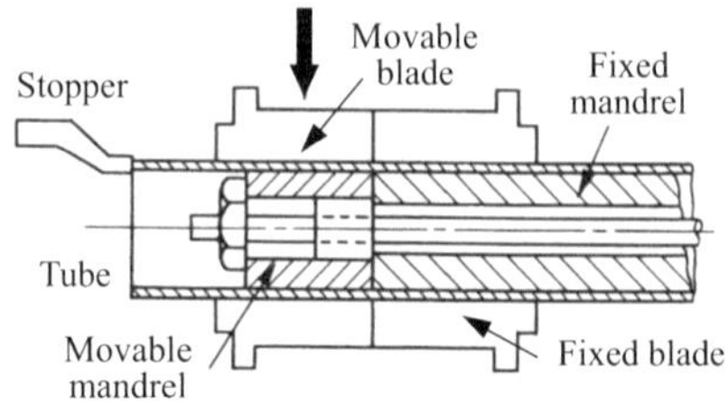

Fig. 7.9 Double shearing with mandrel

Additionally, this method requires some ingenuity for inserting the mandrel into the tube. The double mandrel shearing method shown in Fig. 7.9 resolves the limitations of the mandrel shearing method, which cannot handle long tube cutting and faces difficulties in placing the tube over the mandrel.

This is achieved by using two mandrels: one is a fixed mandrel, and the other is a movable mandrel that can shift independently. The mandrel is supported solely by a fixed mandrel holder. The movable mandrel is equipped with a spring mechanism for retraction. The movement of the movable blade is limited to slight vertical and horizontal motions, slightly greater than the thickness of the tube, to minimize mandrel displacement. With advancements in the accuracy of raw tube and mandrel shifting mechanisms, there are devices capable of cutting 6-m-long tubes with diameters ranging from 6.35 to 150 mm, thicknesses ranging from 0.152 to 6 mm, and cutting lengths ranging from 3 to 600 mm at high cutting speeds, following the speed of the parting cutting method. While the mandrel shearing method generally provides better cutting surfaces than the parting cutting method, there may be occurrences of secondary shearing at both ends of the cut surface when the tube thickness is substantial. The double mandrel shearing method achieves the best cutting surface among the shearing-based cutting methods. As for the processing force in the mandrel shearing method, it is sufficient to consider the load required to cut the tube section in one pass. However, it is important to consider spring retraction and cam operation, as is the case with any method. In the mandrel shearing method, it is crucial to ensure that the movable and fixed mandrels align properly in the initial state after the die set returns to its original position.

7.2.4 Cutting Examples

Figure 7.10 shows a cross-sectional photograph of a copper tube with a diameter of 9.53 mm and a thickness of 0.3 mm that was cut by roll cutting while applying tension. Although there is a taper, there is a minimal burr and few fracture surfaces. However, the vicinity of the roll cut end undergoes work-hardening. Therefore, when performing tube expansion in post-processing, it is necessary to chamfer, anneal, or expand the tube while avoiding the cut end [3]. Figure 7.11 shows a cross-sectional photograph of a copper tube with a diameter of 6.3 mm and a thickness of 0.6 mm

Fig. 7.10 Cross-section of copper tube using roll press cutting under tension

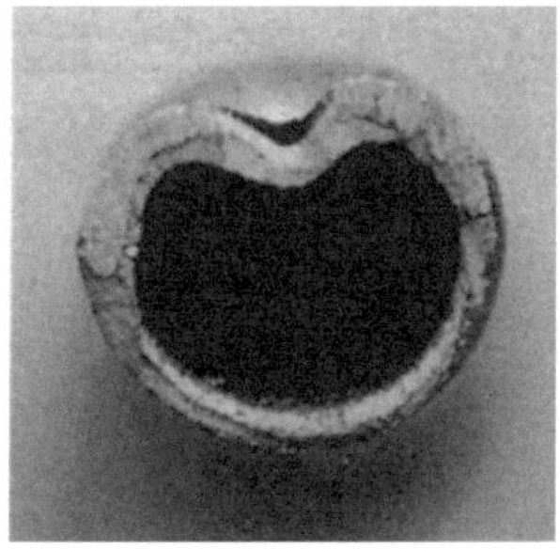

Fig. 7.11 Cross-section of small-diameter copper tube using parting-off cutting

that was cut by the parting-off cutting method. A dent, droop, and marks from contact with the punch side are visible.

7.3 Contouring at Tube Ends and Hole Piercing

There are various methods for contouring tube ends and piercing tubes, as shown in Table 7.2. Here, we will discuss the processing methods that utilize shearing.

Table 7.2 Various methods for contouring tube ends and piercing tubes

Mechanism		Method	Sectional accuracy	Burs	Shape variety	Thin tube	Long tube	Chip disposal	Cutting speed	Equipment cost
Plastic deformation	Shear	Penetration*	Poor	Poor	Average	Poor	Excellent	Average	Excellent	Good
		Punch and die	Good	Good	Good	Good	–**	Average	Good	Good
		Punch with 2 dies	Good	Good	Average	Average	–**	Good	Good	Good
	Plastic flow	Flowdrill*	Poor	Poor	Poor	Average	Excellent	Excellent	Good	Excellent
Removal	Cutting	Drilling*	Excellent	Poor	Poor	Poor	Excellent	Poor	Good	Good
		Milling	Excellent	Poor	Average	Poor	Excellent	Poor	Average	Average
	Fusion	Laser cutting	Good	Good	Excellent	Excellent	Excellent	Average	Good	Poor
		Plasma cutting	Average	Average	Good	Average	Excellent	Average	Good	Average
		Wire discharge	Excellent	Excellent	Excellent	Excellent	Poor	Average	Poor	Poor

* Only for hole, ** For middle of the tube

7.3.1 Contouring

For contour shearing of a tube end, similar to shearing with a mandrel, a punch and die are set in a floating die set, as shown in Fig. 7.12, and the shearing process is performed. The width of the contour cut in one pass is approximately 1/4 of the circumference. In this case, the burr occurs on the outer circumference side. To have the burr occur on the inner circumference side, the type of die shown in Fig. 7.13 is used, but in this case, the width of the cut in one pass is approximately 1/6 of the circumference due to the strength of the die. If the tube has the same notch shape at both ends, as shown in Fig. 7.14, another die is provided on the lower side of the tube, and both notches can be processed simultaneously by passing the same punch through. In the case of contour cutting, if the width of the cut in one pass is kept to about 1/4 or less, a cut surface with good accuracy similar to that of a general punching cut can be obtained. In the case of punching as shown in Fig. 7.14, the scrap size conforms to the dimensions of the die hole, so the lower hole that is punched out by the scrap becomes slightly larger than the upper hole made by the punch. Also, the lower hole surface mostly becomes a shear surface [4].

The processing force is the same as in the case of general punching. It is also almost the same even in the case of punching as shown in Fig. 7.14. Regarding components that are put inside the tube, sufficient consideration should be given to rigidity and strength. Special attention is required for finishing and heat treatment of sections with sharp changes in cross-section.

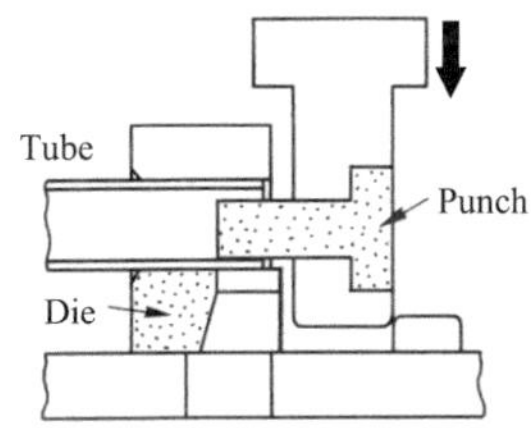

Fig. 7.12 Contouring tube ends (burr on outer circumference)

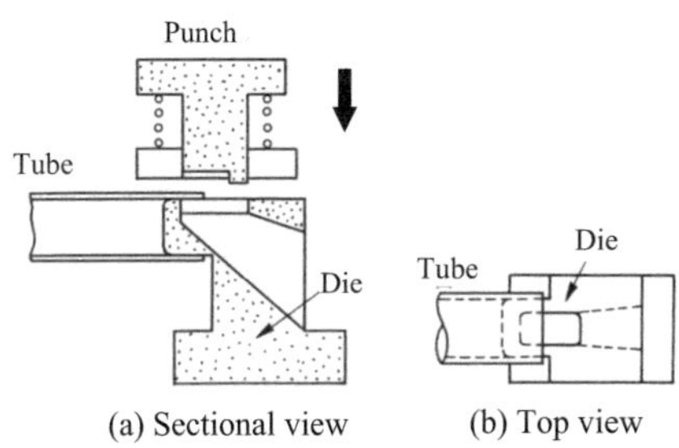

Fig. 7.13 Contouring tube ends (burr on inner circumference)

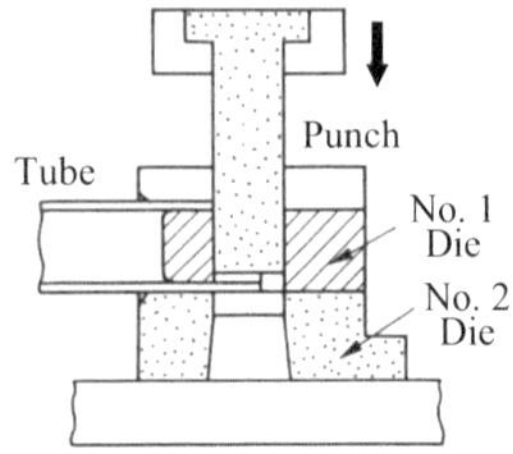

Fig. 7.14 Contouring tube end with punch and two dies

7.3.2 *Hole Piercing*

In the hole piercing process for tubes, unlike in sheet materials, there are cases where hole piercing is performed without using an inner die for the tube material (Fig. 7.15). In this case, since the tube's rigidity receives the punching force, it becomes a combination of punching and shearing. Moreover, in tube materials, there are cases where hole piercing is performed simultaneously on both opposing tube walls. However, in this case, as shown in Fig. 7.16, the scrap removed in the first half of the process remains on the punch tip, and the material being punched in the second half is not directly punched out by the edge of the punch but is punched out by the relatively soft scrap. Therefore, crack formation is delayed, resulting in a cut surface that is closer to finish blanking.

For holes near the tube end, they are punched out using a set of floating die sets as shown in Fig. 7.17a. If the burr is desired on the inner circumference side, a type of die shown in Fig. 7.17b is used. In this case, hole punching can also be performed at locations away from the tube end, but care must be taken to ensure effective removal of the scrap. In cases where holes pass through both sides of the tube wall, a die is also set on the lower side of the tube as shown in Fig. 7.18, and the same punch is used to punch through.

The processing force is the same as in the case of general punching. Similar to contouring, sufficient consideration should be given to the rigidity and strength

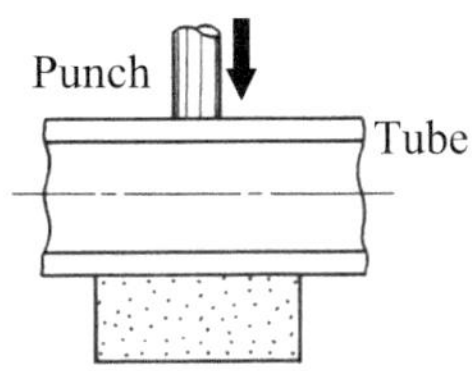

Fig. 7.15 Hole piercing without die

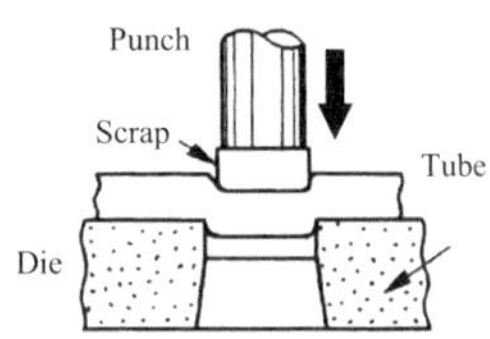

Fig. 7.16 Shearing by scrap in hole piercing

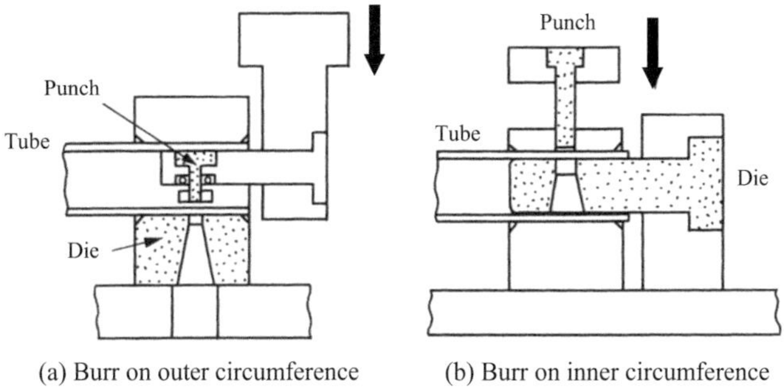

Fig. 7.17 Hole piercing near tube ends

Fig. 7.18 Hole piercing with punch and two dies

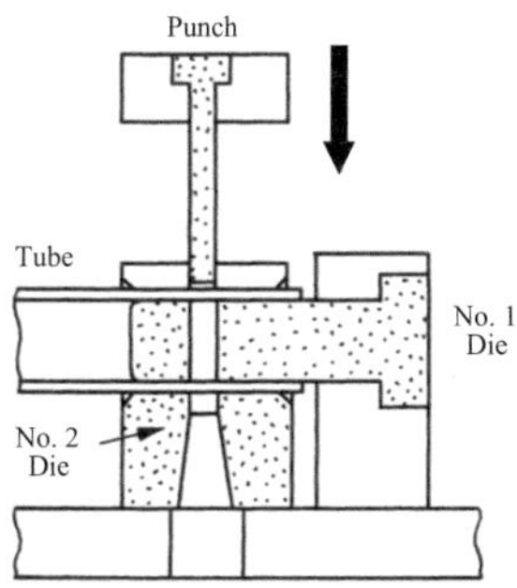

of components that are put inside the tube. One advantage of tube hole piercing compared to hole piercing in sheets is that it can easily generate high compressive forces within the plane of the tube wall due to the constraint on the outer circumference. This has been utilized in successfully piercing holes in fragile materials such as flake graphite cast iron tubes [5].

7.3.3 Examples of Contouring and Hole Piercing Processes

Figure 7.19 shows an example of deep-drawn mild steel sheet similar to a tubular shape with a diameter of 21.5 mm and a thickness of 1.55 mm. Subsequently, contouring is performed using the method illustrated in Fig. 7.14.

Furthermore, Fig. 7.20 presents examples of hole piercing, with burr on the inner circumference, of a copper tube with a diameter of 18 mm and a thickness of 1.8 mm. The diameter of the tube is reduced, and then two holes are pierced from opposite directions using the method shown in Fig. 7.17b.

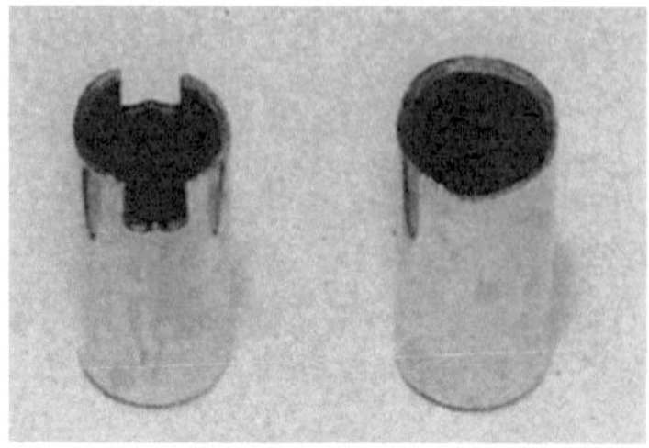

Fig. 7.19 Example of end contouring by punch and two dies

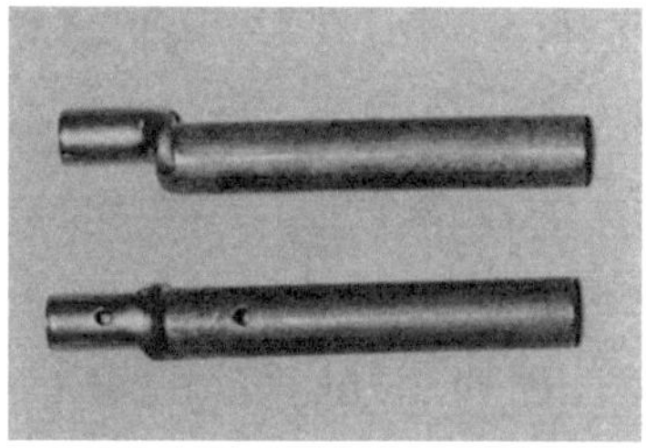

Fig. 7.20 Example of pierced holes on mother tube and reduced part

7.4 Burring Process

"Burr" refers to protrusions or rough edges, specifically small residual projections left during processing. Burring, on the other hand, refers to the formation of protrusions, which involves flaring the hole and raising the edge to a cylindrical shape. Burring of tubes is a process that involves raising the edge of a hole after performing hole piercing on the tube wall. Table 7.3 illustrates representative burring methods using

Table 7.3 Representative burring methods using rigid tools

	Rigid plug	Cross-pin	Incremental burring
Mechanism	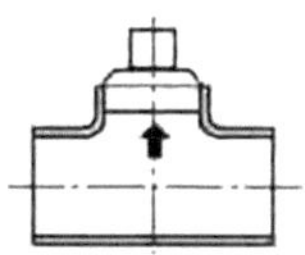	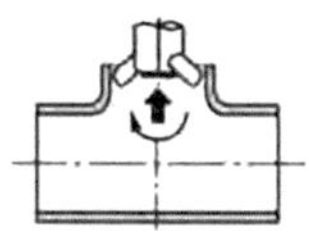	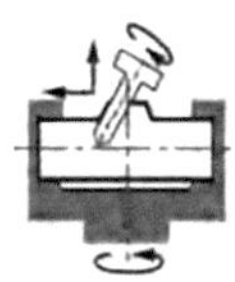
Number of stages	Good	Good	Average
Cost of die and tool	Poor	Good	Good
Ratio of branch to mother tube diameter	Poor	Good	Excellent
Circularity	Poor	Poor	Good
Cold forming	Poor	Average	Good
Processing force	Poor	Good	Excellent
Equipment cost	Good	Average	Poor

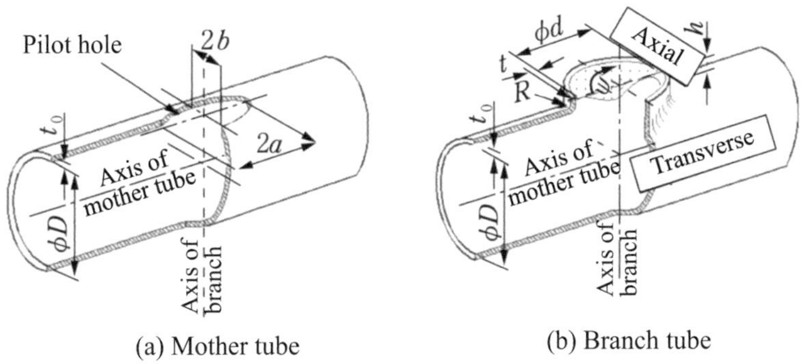

Fig. 7.21 Shapes of mother tube and branch formed by burring process

rigid tools and their characteristics. Hydraulic burring processes can be found in Chap. 4, "Hydroforming". Tube burring is used to fabricate branching shapes from tubes and is commonly applied to pipe fittings in piping systems as well as structural joints like bicycle joints and automobile exhaust manifolds. Additionally, internal tube burring is employed for purposes like creating threaded pilot holes. Tube burring follows a deformation mode similar to that of flange forming and hole expansion in sheets. However, since the workpiece is a tube with curvature, careful consideration is required to ensure that the branching height becomes uniform in the circumferential direction. To achieve this pilot hole is not circular but elliptical. Figure 7.21 shows the shapes of the mother tube and the branching tube formed by the burring process. The expansion ratio λ in burring is calculated by determining the three-dimensional elliptical shape on the side of the mother tube using the relationship between the base tube diameter *D*, branch tube diameter *d*, and the height of the raised edge *h*. It can be obtained by comparing the elliptical circumference L_0 calculated from the perimeter of the pilot hole with the circumference πd of the branching tube, as shown in Eq. (7.1).

$$\lambda = \frac{\pi d - L_0}{L_0} \tag{7.1}$$

7.4.1 Hydraulic Bulge Forming

Hydraulic bulge forming is a method where the outer circumference of a pipe material is restrained by a die, and the tube end is pushed in axial direction while hydraulic pressure is applied inside the tube, thereby deforming a portion of the tube into a branching shape. The branching top is then cut off to complete the fabrication process [6]. This method requires a die shape that corresponds to the branching tube shape, making it less suitable for high variety and low volume production due to die costs.

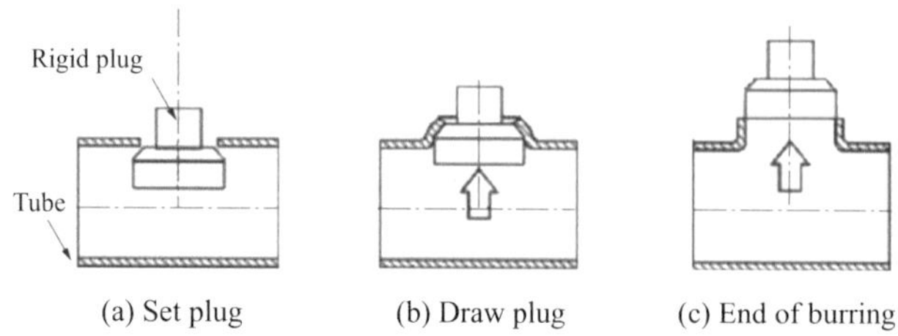

Fig. 7.22 Process of burring by rigid plug drawing

However, it has been implemented in applications such as automotive parts [7]. By optimizing the internal pressure and axial feed amount [8, 9], it is possible to increase the height of the branching section and suppress the reduction of its thickness, thus achieving a uniform thickness distribution in the branching tube.

7.4.2 *Rigid Plug Drawing Method*

The rigid plug drawing method involves piercing a pilot hole in the wall of the raw tube, inserting a rigid plug into the tube, and then vertically pulling the rigid plug in relation to the pilot hole to raise the edge and form the branching section [10]. Figure 7.22 illustrates the process of the rigid plug drawing method. The rigid plug is inserted into the raw tube from the axial direction, and a lifting jig is inserted from the outer side of the raw tube through the elliptical pilot hole, connecting them together inside the raw tube. The rigid plug is then pulled outward, aligning the center of the elliptical pilot hole with the center of the rigid plug.

This method is the most commonly used for one-piece formed branching sections. However, there are limitations on the achievable branching tube diameter ratio (branching diameter/mother tube diameter) due to the relationship between the mother tube diameter and the rigid plug diameter. For example, branching tubes with a branching diameter/mother tube diameter ratio of 1.0 cannot be formed. Additionally, there are challenges such as the need for partial heating of the mother tube material to facilitate deformation [11], which increases equipment requirements and forming steps, as well as the necessity of specific rigid plug sizes for different branching diameters, leading to higher tool costs.

7.4.3 *Cross-Pin Method*

The cross-pin method, also known as the T-drill method, is a forming technique developed by T-drill Industries Inc. It utilizes two pin tools that open up like a cone, allowing for versatility in branching section diameters ratio (branch diameter/mother

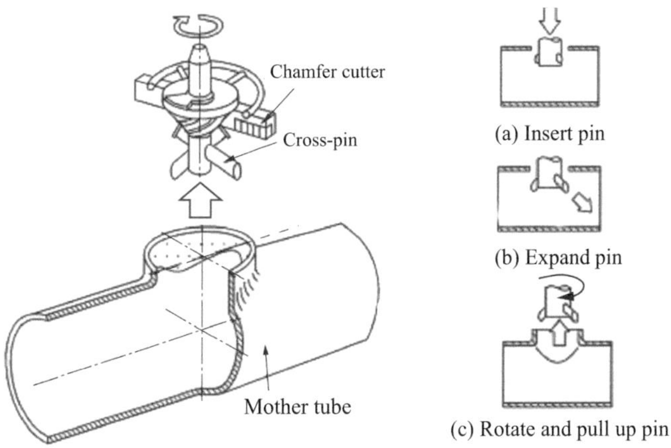

Fig. 7.23 Mechanism and process of burring by cross-pin

tube diameter ratio) using a single tool [12, 13]. However, if suitable lubricants are not used, there can be issues of adhesion and sticking between the pin tools and the workpiece contact area, especially with less ductile materials, requiring additional equipment and processing step for partial heating. Figure 7.23 provides an overview of the branching tube formation using the cross-pin method. The pin tools are inserted into the forming head pilot hole, and by expanding and rotating the cross-shaped pin tools while pulling them up, the edge of the pilot hole is raised to form the branching section. A chamfer cutter attached to the forming head allows for end surface processing of the branching section after the pin tools are withdrawn.

7.4.4 Incremental Burring Method

The incremental burring method is an approach that expands on the incremental forming technique developed for sheet materials and applies it to tube materials, utilizing rod-shaped tools for burring processes [14, 15]. As shown in Fig. 7.24, this method employs a driven rotating rod-shaped tool to raise the edge of the pilot hole and deform specific portions of the branching section during its movement [16]. The processing stages of the incremental burring method using rod-shaped tools are illustrated in Fig. 7.25. The forming limits of the three burring methods described so far are determined by the following factors.

To enhance the forming capabilities, it is desirable to achieve uniform reduction in thickness at the edge of the branching section. However, in both the rigid plug and cross-pin methods, the tool radius remains constant, making it impossible to vary the distribution of thickness reduction during the forming process, as shown in Figs. 7.26a and b respectively. Additionally, when the branch diameter/mother tube diameter ratio (d/D) is large, it is crucial to impart deformation specific to the location

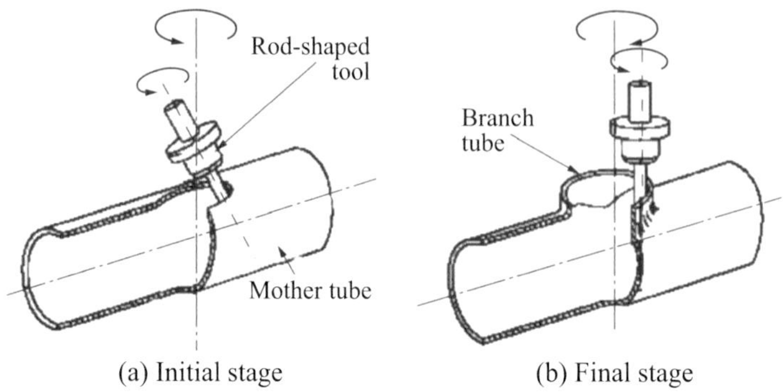

Fig. 7.24 Schematics of incremental burring using rod-shaped tool

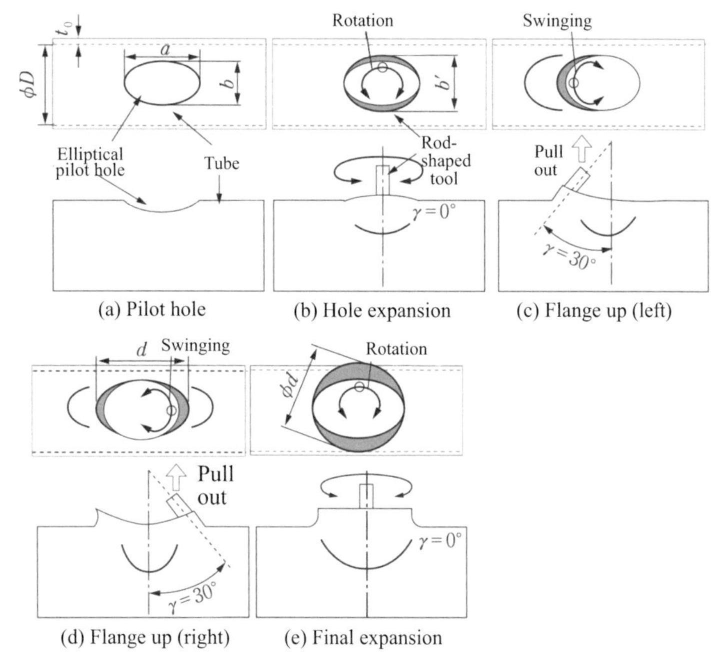

Fig. 7.25 Stages of incremental burring using rod-shaped tool

of the end of branching section. However, the rigid plug and cross-pin methods do not allow for deformation based on the position of the pilot hole edge.

In contrast, the incremental burring method with rod-shaped tools enables deformation tailored to each position of the end of the branching section during the forming process, as shown in Figure (c). By varying the distribution of thickness reduction during forming, it allows for the eventual achievement of a uniform thickness distribution [16].

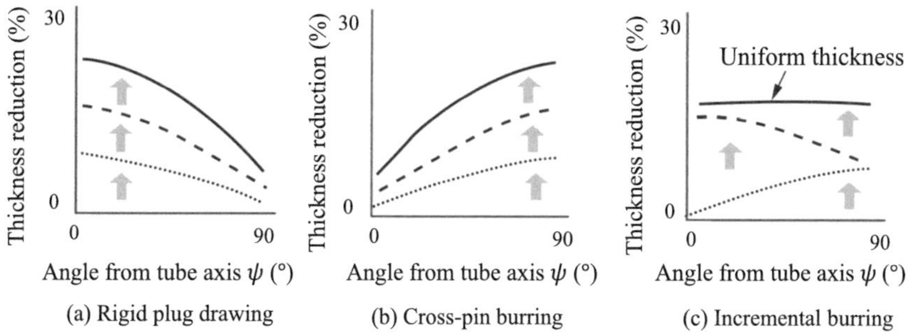

Fig. 7.26 Changes in thickness distribution at edge of branch for three burring methods

Specifically, in the first step, as shown in Fig. 7.25b, the part perpendicular to the tube axis is expanded. Then, as illustrated in Figures (c) and (d), while swinging the rod-shaped tool, the part in tube axial direction is raised to achieve the desired dimensions of the branching section. Finally, as depicted in Figure (e), the part perpendicular to tube axis is expanded to the predetermined branching section diameter by increasing the revolving radius of the rod-shaped tool, completing the formation of the branching section. The incremental burring method presents challenges such as the requirement for an orbital rotation mechanism and longer processing time. However, it offers the advantages of die-less forming and improved forming limits, which are characteristics of incremental forming.

7.4.5 *Examples of Burring Processes*

Figure 7.27 illustrates an example of forming a branching section with a branch diameter of 100 mm using the incremental burring method with rod-shaped tools on the side surface of an aluminum alloy (A5083-O) tube with a diameter of 100 mm and a thickness of 1 mm [17]. Furthermore, Fig. 7.28 presents examples of forming branching sections with branch diameter ratios (d/D) of 1.0, 0.9, and 0.8 on the side surface of an aluminum alloy (A5083-O) tube with a diameter of 400 mm and a thickness of 3 mm, using the same method [17]. In this way, the incremental burring method with rod-shaped tools enables the formation of branching sections with various branch diameters.

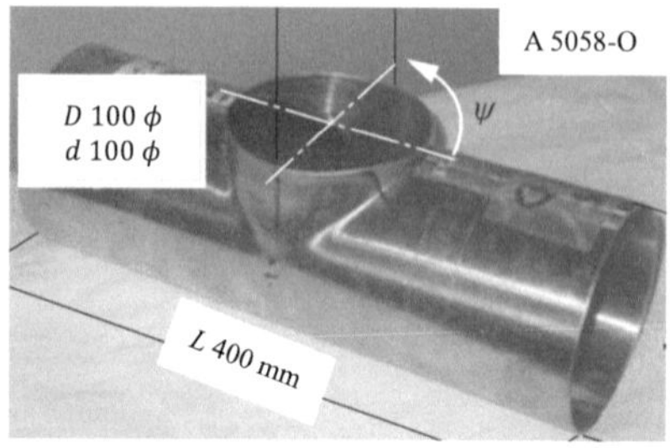

Fig. 7.27 Example of incremental burring

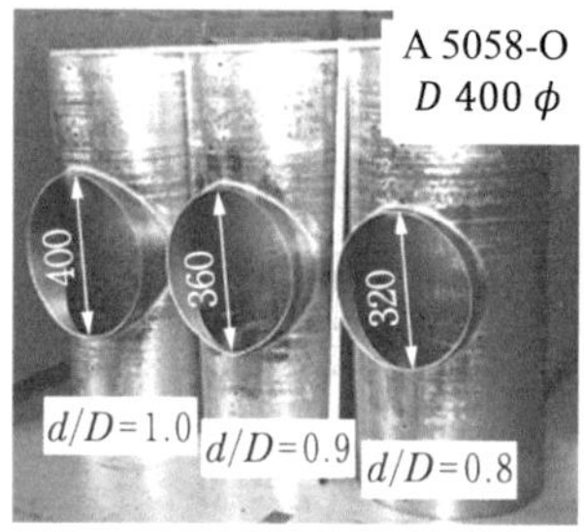

Fig. 7.28 Examples of incremental burring

References

1. The Japan Society for Technology of Plasticity (ed.): Press processing handbook, vol. 192. Maruzen Publishing, (1975)
2. Nakagawa, T.: J. JSTP, **23**(255), 307–311 (1982)
3. Shiramizu, H., Nihashi, I.: J. JSTP, **23**(255), 290–294 (1982)
4. Ochiai, I.: J. JSTP, **23**(255), 296–300 (1982)
5. Koike, T., Yamagata, H.: J. JSTP, **31**(351), 474–478 (1990)
6. Christensen, W.L.: Metal forming. **29**(10), 36–43 (1995)
7. Endo, J.: J. JSTP., **35**(400), 511–515 (1994)
8. Fuchizawa, S., Kitamura, K., Narasaki, M. et.al.: J. JSTP, **36**(408), 80–84 (1995)
9. Fuchizawa, S.: J. JSTP, **30**(339), 473–477 (1989)
10. Sumitomo Metals (ed.): Press working. **34**(7), 112 (1996)
11. Okada, K., Asao, H., Watanabe, S.: J. JSTP, **32**(364), 583–587 (1991)
12. Atkey, M.: Machinery and production engineering. **141**(3637), 22–23 (1983)
13. Watanabe, A.: The piping engineering, vol. 32, no. 14, pp. 138–141. Japan Industrial Publishing, (1990)
14. Kitazawa, K.: J. Jpn. Inst. Light. Metals, **49**(6), 233–237 (1999)
15. Kitazawa, K., et. al.: Proceedings of the 1998 spring meeting. JSTP,491–494 (1998)
16. Teramae, T., Manabe, K., Ueno, K., Nakamura, K., Takeda, H.: J. JSTP, **48**(553), 125–129 (2007)
17. Teramae, T., Manabe, K., Ueno, K., Nakamura, K., Takeda, H.: J. JSTP, **50**(581), 560–564 (2009)

Chapter 8
Joining

8.1 Overview

Among the joining methods for tubular materials, techniques that induce plastic deformation during the joining process and directly utilize it involve at least one of the following four joining mechanisms:

(1) Residual contact pressure at the interface between the joining component and the joined component (or between joining components).
(2) Structural interlocking between the joining component and the joined component (or between joining components).
(3) Plastic flow of the joining component into the groove provided in the joined component.
(4) Diffusion of metal atoms at the interface between the joining component and the joined component (or between joining components).

Here, the "joining component" refers to the component that undergoes plastic deformation during the joining process, while the "joined component" refers to the component that does not undergo plastic deformation (or undergoes only minimal plastic deformation). In the case of joining tubular components, the first joining method involves expanding or contracting the tubular component using a tool to bring it into contact with the joined component. The joining force is determined by the residual contact pressure and contact area on the interface, as well as the friction on the contact surface after removing the tool. In the second joining method, the joining component is plastically deformed to achieve structural interlocking with the joined component. When separating the components joined using this method, either one or both of the joining component and the joined component undergo plastic deformation. The third joining method takes into consideration the plastic flow in the thickness direction of the joining component during plastic deformation. The joining force is obtained through an anchoring effect, where a portion of the joining component enters the groove provided in the joined component. In the fourth joining

Y. Kuriyama et al., *Tube Forming*, https://doi.org/10.1007/978-981-95-3421-0_8

Table 8.1 Combinations of tubes and joining components and their applications

	Combination	Application	Examples
A	Circular tube and thin walled components	– Container – Shock absorber	Food and beverages cans (2), Pails and drums (2), Washing machine drum (2), Bumper (2)
B	Circular tube and thick walled components	– Collector – Rotating part ($D > L$) – Torsion spring	Boiler (1, 2), Rotating disc (1, 3), Camshaft (1, 2, 4)
C	Circular tube and Shaft components	– Rotating part ($D < L$)	Copier rollers (1, 2), Automotive parts (1, 3), Building structural components (2, 4)
D	Circular tube and Pipe joints	– Piping for fluids	Mechanical pipe joints (1, 2)

* The digits indicate the correspondence to the four types of joining mechanisms defined in Sect. 8.1 (1–4)

method, surface films such as oxide films on the joining surface are broken to expose fresh surfaces. The joining force is influenced by factors such as surface pressure, contact area, susceptibility of surface films to breakage, temperature, and relative sliding amount.

When classified by the combination of circular tubes and joining components, as well as their applications, it can be categorized as shown in Table 8.1. Circular tube materials can be joined to other circular tube materials, thin-walled components, thick-walled components, shaft components, and pipe joints. The applications include not only piping but also shock-absorbing parts, springs, lightweight structural components, and more. The specific joining mechanisms used for each type of joint can vary depending on the specific product, but for the joining of circular tubes and thin-walled components, the second joining mechanism is primarily used. For the joining of thick-walled components and shaft components, the first mechanism utilizing their rigidity, or the second and third mechanisms with grooves can be applied.

In the following sections, plastic joining methods will be explained according to their respective applications.

8.2 Plastic Joining Primarily Used in Piping

8.2.1 Roller Expansion Method

(1) **Overview**

The roller expansion method is a cold plastic forming technique using a tube expander shown in Fig. 8.1 [1]. In this method, joining components (tubes) are expanded from

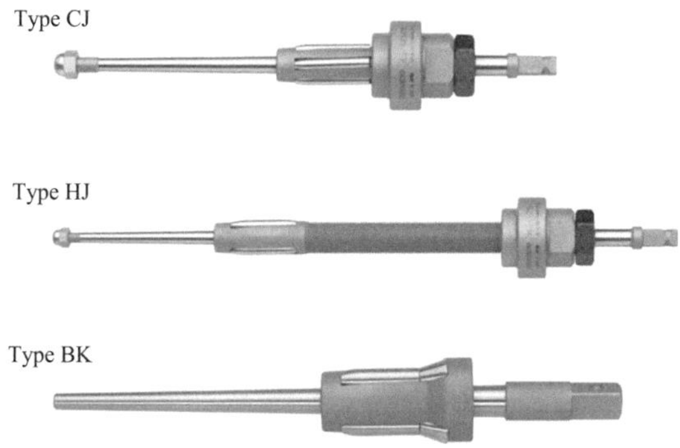

Fig. 8.1 Tube expanders

inside to bring them into close contact with the joined components (tube plates) as shown in Fig. 8.2.

As depicted in Fig. 8.3, a tube expander consists of a tapered mandrel and multiple rollers evenly arranged in the circumferential direction. The rollers are inclined at a certain angle relative to the mandrel. By inserting the tube expander into the tube and rotating the mandrel, the rollers rotate and make contact with the inner surface of the tube, causing the mandrel to advance along the axis of the tube. By the taper of the mandrel, rollers expand the tube radially. The tube undergoes polygonal deformation corresponding to the number of rollers, with its outer surface coming into contact with the holes in the tube plate. As the expansion progresses, the holes in the tube plate also widen, requiring a sudden increase in rotational force on the mandrel. Due to the thinness of the tube, it starts undergoing plastic deformation first. The radial deformation of the tube is constrained by the tube plate, causing the tube to elongate

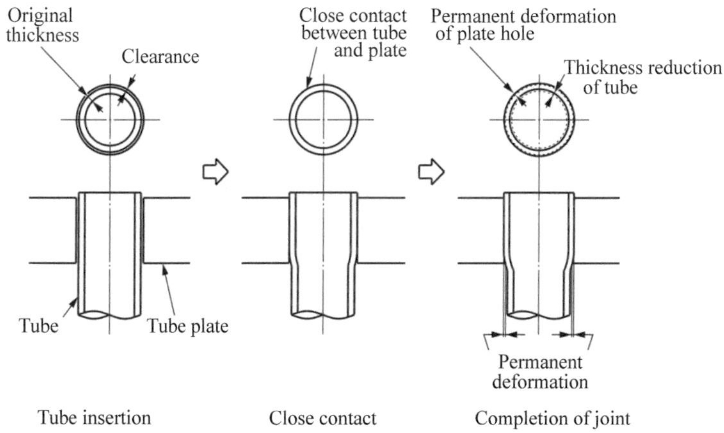

Fig. 8.2 Joining process using tube expander

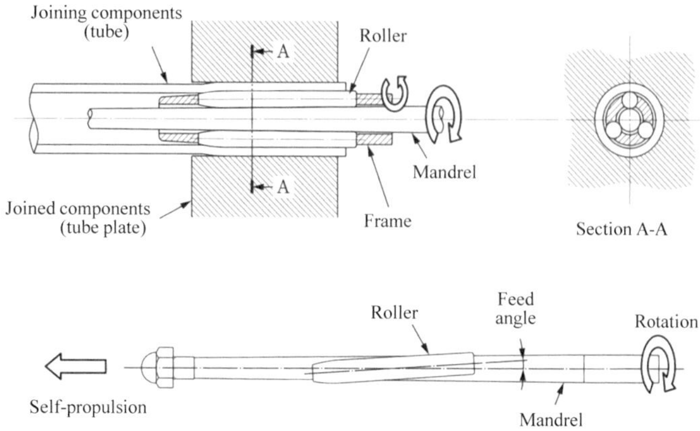

Fig. 8.3 Roller expansion using tube expander

in the axial (forward–backward) direction. As the tube plate holes undergo elastic deformation and become slightly larger than their original dimensions, removing the tube expander after expansion causes the tube plate holes to contract, exerting residual contact pressure that compresses the outer surface of the tube. As a result, the tube is securely fastened and airtightness is maintained.

(2) Mandrel and rollers

In the case of the roller self-propelled parallel tube expanding method shown in Fig. 8.3, it consists of three main components: a tapered mandrel at the center, 3 to 5 rollers arranged evenly around it, and a frame that holds the rollers. The taper of the mandrel is typically 1/50 to 1/25, with larger tapers resulting in higher processing speeds and the ability to shorten the overall length of the tool. The roller axes are slightly inclined with respect to the mandrel axis, creating a feed angle. As a result, simply rotating the mandrel automatically feeds it in the axial direction. A larger feed angle leads to faster tube expansion operations.

(3) Joining conditions

The most crucial point in roller tube expanding operations is to determine the optimal tube expansion ratio that achieves secure fixation with the appropriate residual contact pressure for a given combination of tube material and tube plate. Additionally, when joining multiple tube plate holes as depicted in Fig. 8.4 [1], it is necessary to ensure uniform joining conditions. If the tube material or tube plate shape and material are unfamiliar, it is advisable to first verify the optimal tube expansion ratio using models of actual dimensions and the same material. Methods for examining the characteristics of the expanded section include (1) measuring residual contact pressure, (2) measuring residual stress, and (3) performing tube drawing or extrusion tests.

To uniformly control the optimal tube expansion ratio for all tube materials, two methods are commonly used. The first method involves expanding the tube until it

Fig. 8.4 Example of joining work on multiple holes of tube plate [1]

reaches the desired dimensions, ensuring consistent dimensional changes caused by the processing.

The second method involves stopping the mandrel rotation when the torque reaches a predetermined value, thereby controlling the tube expansion ratio consistently. There are parameters used to express the optimal tube expansion ratio, such as the thickness reduction rate W_t and the inner diameter increase rate W_d, as shown in Eqs. (8.1) and (8.2).

$$W_t = \frac{t_1 - t_2}{t} \times 100\,(\%) \tag{8.1}$$

$$W_d = \frac{d_2 - d_1}{d_1} \times 100\,(\%) \tag{8.2}$$

Here, t is the thickness of the tube before processing, t_1 is the thickness of the tube when it contacts the tube plate hole, t_2 is the thickness of the tube after processing, d_1 is the inner diameter of the tube when it contacts the tube plate hole, and d_2 is the inner diameter of the tube after processing. As it is challenging to measure t_1 and d_1 directly, a simplified expression is often used, assuming no change in thickness until the tube contacts the tube plate hole, as shown in Eqs. (8.3) and (8.4).

$$W_t = \frac{d_2 - (d + C)}{2t} \times 100\,(\%) \tag{8.3}$$

$$W_d = \frac{d_2 - (d + C)}{(d + C)} \times 100\,(\%) \tag{8.4}$$

Here, d is the inner diameter of the tube before processing, and C is the gap between the outer diameter of the tube before processing and the diameter of the tube plate hole.

The relationship between the thickness reduction rate and residual contact pressure is shown in Fig. 8.5 [2]. There is an upper limit to the residual contact pressure, and if

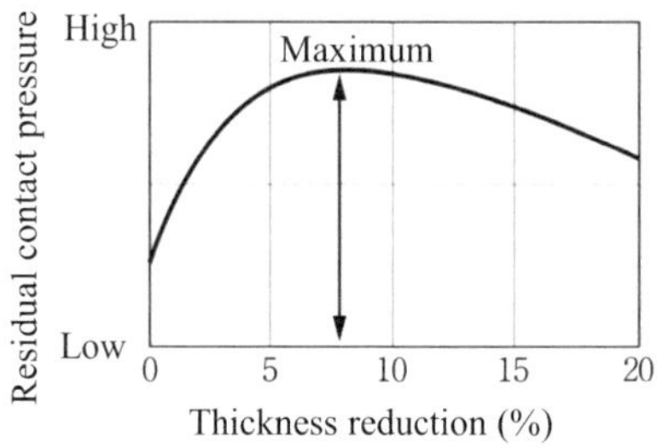

Fig. 8.5 Relationship between thickness reduction and residual contact pressure [2]

the inner diameter of the tube is expanded too much, the residual compressive pressure decreases, resulting in a decrease in fixation force. It is generally considered optimal for the thickness reduction rate Wt to be between 5 and 10%, and the inner diameter increase rate Wd to be between 1 and 1.2%. When a larger thickness reduction rate and inner diameter increase rate are used, the tube plate holes may undergo plastic deformation, resulting in a significant decrease in residual contact pressure.

The torque control method involves using electric, hydraulic, or pneumatic rotational drives and measuring and controlling the current or oil/air pressure flowing through the drive at the optimal tube expansion ratio. This ensures stable tube expansion rates during repeated operations.

When performing roller tube expansion on hundreds to thousands of tubes, such as in heat exchangers, it is time-consuming and challenging to measure dimensional changes in all the tubes. Therefore, a reasonable quality control method is to apply expansion under torque control, perform dimensional sampling inspections, and verify that the optimal tube expansion ratio has been achieved.

(4) **Factors affecting characteristics of joints**

Joints with sufficient fixing strength and high gas tightness are achieved through optimal elastic–plastic deformation. Various factors, such as the mechanical properties and dimensions of the tube material and plate material, as well as the specifications of the tube expander, influence the optimal joint conditions. The length of the joint affects the strength of the joint, with longer joints providing greater joint strength. It is possible to ensure the safety of a joint by designing it according to the thickness of the plate specified in pressure vessel construction standards. The residual contact pressure between the tube material and the plate hole, which generates fixing strength and gas tightness, is generally proportional to the thickness of the tube material. However, as the tube material becomes thicker, construction and availability may become difficult, necessitating the selection of an appropriate thickness. The clearance between the tube material and the plate hole must be set to an appropriate value to achieve a good joint. Based on numerous experiments, there are two theories regarding the size of the clearance: one suggests that the clearance should be approximately 1% of the outer diameter of the tube material, and the other suggests that it should be approximately 10% of the thickness of the tube material. For boiler tubes, a clearance of 0.7–1.0 mm is considered suitable. The forming or machining precision of the outer surface of the tube material and the plate hole also affects the joint strength. As the joint is a type of friction joint, factors that increase friction resistance contribute to increased joint strength. On the other hand, to ensure gas tightness, the contact surfaces need to be smooth and correctly cylindrical. Therefore, the plate holes are generally finished with a reamer after drilling. In some cases, grooves are machined in the plate holes to enhance fixing strength and gas tightness, especially when operating pressures are high. As the thickness of the tube material at the expanded section becomes thinner due to tube expansion, the tube material elongates in the axial direction. When securing multiple tubes to a plate, such as in a heat exchanger, attention must be paid to the sequence of tube expansion to prevent the plate from warping or the tube material from deflecting.

8.2.2 *Hydraulic Expansion Method*

(1) **Overview**

The hydraulic expansion method is a tube expansion technique, shown in Fig. 8.6, that applies ultra-high-pressure water directly to the inner surface of a tube, uniformly expanding the inner surface through pressure. The expander in the figure is called an "aqua mandrel". Similar to the roller expansion method, the hydraulic expansion method is a cold plastic deformation process that deforms the joining member (tube material) from the inside to achieve tight contact with the joined member (plate material). The required pressure for expansion varies depending on the tube material, outer diameter, and thickness, typically ranging from 100 to 400 MPa. The hydraulic expansion method utilizes a control unit to generate and regulate ultra-high-pressure water, an aqua mandrel inserted into the tube material to apply the ultra-high-pressure water, and a gun unit that holds the aqua mandrel and supplies the ultra-high-pressure water, as shown in Fig. 8.7 [2].

(2) **Advantages**

The hydraulic expansion method has the following advantages over the roller expansion method:

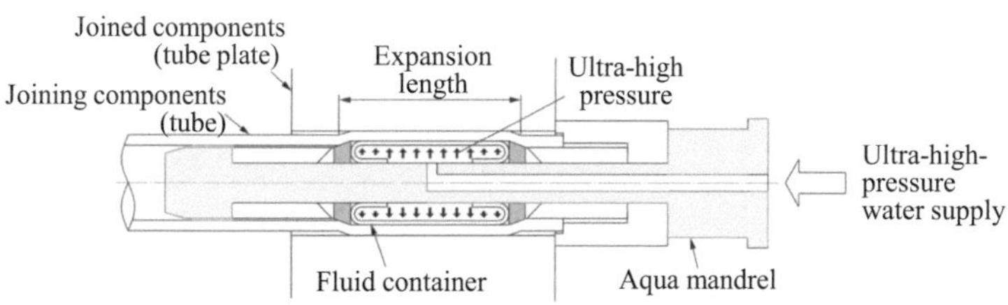

Fig. 8.6 Hydraulic expansion method using aqua mandrel

Fig. 8.7 Hydraulic expansion work using aqua mandrel [2]

(1) It expands the tube all at once without pressing it repeatedly with rollers, eliminating flaking.
(2) Since there is no rolling effect from the rollers, there is minimal reduction in thickness, and there is no elongation in the axial direction of the tube.
(3) It can expand even thick-walled tubes (tubes with a thickness exceeding 15% of the outer diameter).
(4) By extending the aqua mandrel, it is possible to expand widths longer than 500 mm in a single operation.
(5) Water is used as the expanding liquid, eliminating any contamination of the work environment.
(6) In the roller expansion method, the operator needs to withstand the rotational reaction force during expansion, which can be a significant burden. However, in the hydraulic expansion method, expansion is carried out by ultra-high-pressure water, eliminating the generation of rotational reaction forces and significantly reducing the burden on the operator during expansion.

(3) **Expansion conditions**

In the case of the hydraulic expansion method, it is possible to analyze the residual contact pressure after expansion using "the formula for a thick-walled cylinder subjected to internal pressure", and the required expansion pressure can be calculated based on the material, dimensions, and mechanical properties of the used components.

(1) Expansion pressure for plastic deformation of the tube: P_0 [MPa]

$$P_0 = \frac{\sigma_{st}}{2}\left(K^2 - 1\right) \tag{8.5}$$

(2) Minimum expansion pressure to generate residual stress in the tube and tube plate: P_{min} [MPa]

$$P_{min} = \frac{2\sigma_{st}}{\sqrt{3K^4 + 1}} \times \frac{\kappa^2 - 1}{\kappa^2(1 + \mu)} + \frac{\sigma_{st}\left(\kappa^2 - 1\right)}{2} \tag{8.6}$$

(3) Expansion pressure causing plastic deformation of the tube plate: P_{max} [MPa]

$$P_{max} = \sigma_{sp} \times \frac{\kappa^2 - 1}{\kappa^2 + 1} + \frac{\sigma_{st}\left(\kappa^2 - 1\right)}{2} \tag{8.7}$$

Here, σ_{st} [MPa] represents the yield stress of the tube material, σ_{sp} [MPa] represents the yield stress of the tube plate, v represents the Poisson's ratio of the tube plate, and K represents the radius ratio of the tube (K = outer diameter/inner diameter). The radius ratio of the tube plate, k, can be calculated from the pitch P of the tube plate holes and the hole diameter H.

$$\kappa = \frac{2P - H}{H} \tag{8.8}$$

To accurately calculate the expansion pressure, the actual yield stress (σ_{st} and σ_{sp}) of the tube material and tube plate is required. However, for simplicity, it is common to use the minimum specified yield stress values defined in the industrial standards of each country. By using these values, it is often recommended to expand at pressures approximately 20–30% higher than the calculated pressure to achieve the desired level of joint strength.

8.2.3 *Examples of Roller Expansion Method and Hydraulic Expansion Method*

Both the roller expansion method and the hydraulic expansion method are used for the joining of tube materials and tube plates in various types of multi-tube heat exchangers. Representative applications of these methods include:

(1) Condensers and feed water heaters in steam turbines for power generation.
(2) Heat exchangers for petroleum refining plants.
(3) Heat exchangers for petrochemical plants.
(4) Refrigeration equipment.
(5) Large-scale air conditioning equipment.

Figure 8.8 illustrates an example of a joint in a condenser [3]. The purpose is to achieve plastic joining of heat transfer tubes to the tube plate and maintain the bonding strength of the tube material as well as the airtightness between the outer surface of the tube material and the tube plate holes.

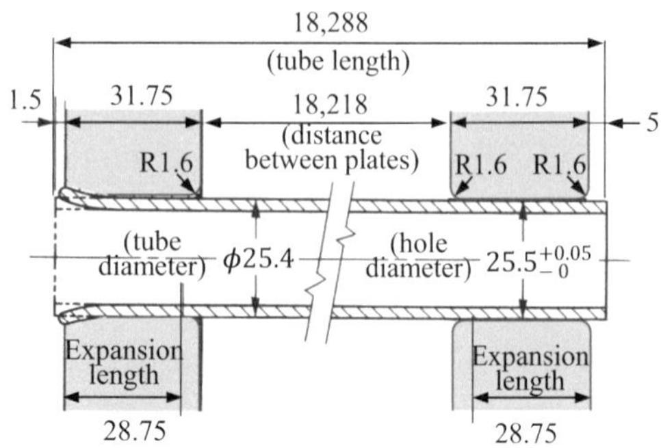

Fig. 8.8 Example of pipe joint in condenser [3]

8.2.4 *Mechanical Type Tube Joint*

(1) **Overview**

Stainless steel tubes used in various architectural piping applications, including water supply, hot water supply, chilled water, cooling water, and fire-fighting water, are defined by "Light gauge stainless steel tubes for ordinary piping" JIS G 3448. "Mechanical type" tube joints are commonly used for these tube joints.

The term "mechanical type" refers to a joining method where pipes and fittings are mechanically connected without the need for heating or welding. In this processing method, the tubes are usually cut or processed on-site and mechanically joined using fittings that incorporate seal members and mechanisms to prevent internal parts from falling out. These joints are classified into six types, as shown in Table 8.2 [4]. Among them, the "press-fit" and "expansion" types involve reducing or expanding the pipe diameter on the construction site before joining it with the fitting. This section introduces these two methods.

Light gauge stainless steel tubes for ordinary piping (JIS G 3448) were specifically developed for building facilities, including high-rise buildings, and were established in 1980. They are designed to be lightweight by using thinner walls compared

Table 8.2 Designation of mechanical tube joints [4]

Nominal and maximum pressure	Joining method	Designation
10 K (max.1.0 MPa) and 20 K (max. 2.0 MPa)	Press	10 K: 8 Su to 300 Su 20 K: 8 Su to 100 Su
	Expander	
	Nuts	
	Rolled thread	
	Insertion	
	Coupling	

Table 8.3 Designations and diameters of typical tubes for piping (excerpt)

"Su" pipe and tube[a]		"Su" pipe and tube[b]		Copper pipe and tube[c]	
Designation Su	Outer diameter (mm)	Designation A	Outer diameter (mm)	Designation A	Outer diameter (mm)
10	12.70	10	17.3	10	12.70
13	15.88	15	21.7	15	15.88
20	22.22	20	27.2	20	22.22
25	28.58	25	34.0	25	28.58
30	34.0	32	42.7	32	34.92
40	42.7	40	48.6	40	41.28
50	48.6	50	60.5	50	53.98
60	60.5	60	76.3	65	66.68

[a] JIS G3448 "Light gauge stainless steel tubes for ordinary piping"
[b] JIS G3459 "Stainless steel pipes"
[c] JIS H3300 "Copper and copper alloy seamless pipes and tubes"

to conventional stainless steel pipes, thus improving ease of installation. They are referred to as "Su" and are distinguished from other standards. The material is specified in four types: SUS 304, SUS 315 J1, SUS 315 J2, and SUS 316. A comparison of the outer diameters for each specification is shown in Table 8.3.

(2) Double press-type tube fittings

I will explain the double press-type tube fittings as an example of press-type tube fittings. Figure 8.9 illustrates the schematic of the fitting before and after tightening [5]. The tube and the fitting are made of SUS304 and the rubber ring is made of heat-resistant butyl rubber. Tubes marked with a line indicating the insertion length are inserted into the fittings. Figure 8.10 shows the tool used for tightening and the jointing situation [5]. The fitting's convex part (curl section) is aligned with the concave part (attached the rubber ring) on the dedicated tightening tool, and both sides are pressed simultaneously (double press). By reducing the section from the center of

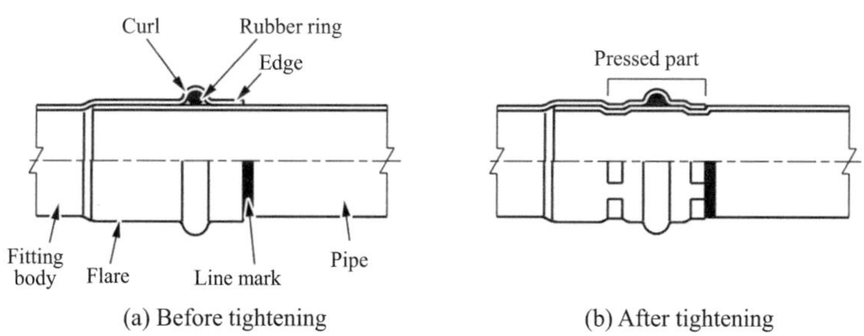

Fig. 8.9 Schematic of the double press type tube fitting before and after tightening

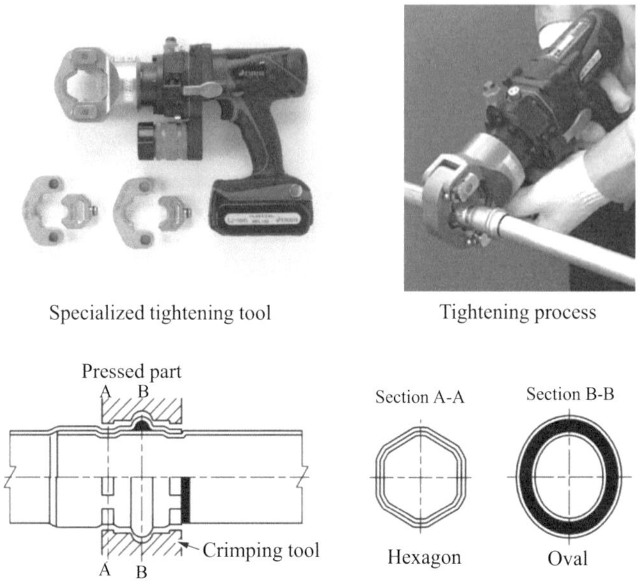

Fig. 8.10 Tools and details of double press-type tube fitting [5]

the flared (pipe insertion) zone to the fitting end into the hexagonal and elliptical cross-sectional shapes, the pipe and fitting undergo close contact deformation, resulting in sufficient joint strength.

The structure also ensures water-tightness through the compressive deformation of the rubber ring. The insertion depths of the pipe are shown in Fig. 8.11. *L1* and *L2* indicate the insertion depths, where *L1* represents the standard insertion depth. *L2* indicates the minimum insertion depth at which the water-tightness of the rubber ring is maintained, although it indicates insufficient insertion.

The notable feature of this fitting is that even if the pipe is under-inserted (*L2* situation), it can still exhibit sealing performance as long as the pipe passes through the rubber ring. Furthermore, even in cases of further under-insertion where the pipe does not pass through the rubber ring, it is difficult to remove the pipe once installed. These characteristics serve as measures against construction defects at the construction site.

(3) **Expanding type tube fitting**

An example of an expanding type tube fitting for JIS G3448 pipe is shown in Fig. 8.12 [5]. The fitting body and cap nut are made of SCS13 or SCS14. These are stainless steel castings, equivalent to SUS304 and SUS316, respectively. Additionally, the fastening confirmation ring is made of polyamide, and the sealing ring is made of

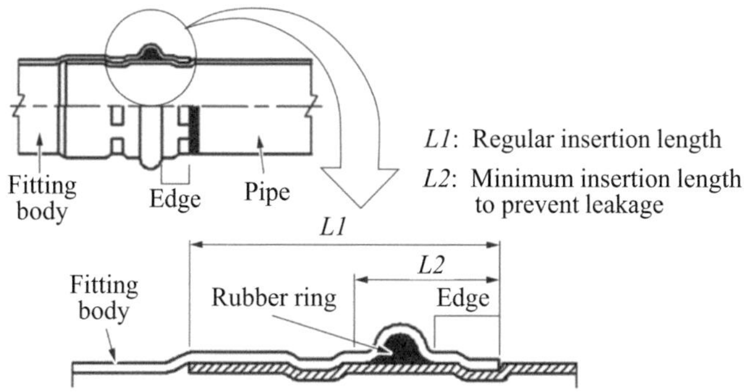

	13 Su	20 Su	25 Su	30 Su	40 Su	50 Su	60 Su
L1 (mm)	28	32	36	53	61	68	78
L2 (mm)	13	15	19	25	27	30	31

Fig. 8.11 Insertion length of double press-type tube fitting [5]

fluorine rubber. For expanding the tube to be connected to the fitting, a dedicated expanding tool as shown in Fig. 8.13 is used.

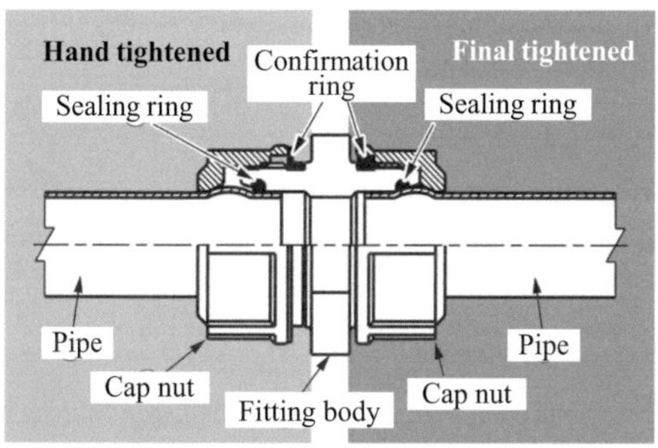

Fig. 8.12 Mechanism of expanding type tube fitting

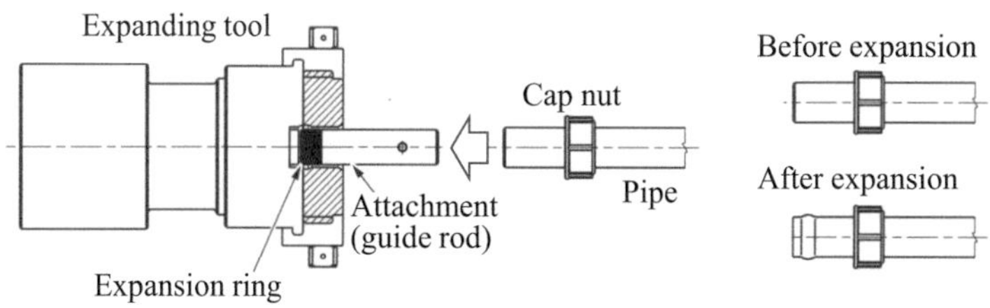

Fig. 8.13 Dedicated tool for expanding type tube fitting and formed shape of the fitting [5]

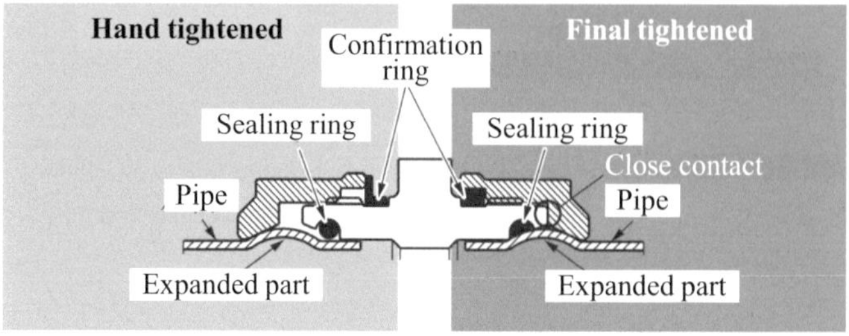

Fig. 8.14 Tightening states of expanding type tube fitting [5]

In the expanding process, first the deburred pipe is fitted with a cap nut, then the tube end is inserted into the expanding section at the back of guide rod of the dedicated tool. When the tool's operating switch is activated, the expanding process begins, and the tube is pushed outwards by the radial expansion caused by axial compressive deformation of the rubber fitted inside the attachment, forming an expanded shape at the tube end. During assembly, the expanded pipe is inserted into the fitting body, and the cap nut is hand tightened until it reaches the confirmation ring. The final tightening is performed using tools such as a pipe wrench to ensure a tight fit between the fitting body and the collar of cap nut. By tightening the cap nut, the expanded section of the tube is pressed against the body, achieving a proper compression on the rubber ring and ensuring a watertight seal and secure joint between the tube and fitting. Figure 8.14 illustrates the situation with the expanded tube inserted into the fitting and the cap nut tightened [5]. The quality of the joint (tightening) can be visually confirmed from the outside after installation, and if the tightening of the cap nut is insufficient, the confirmation ring will be visible. Also, if the cap nut is tightened without expanding the tube end or if the final tightening is forgotten after hand tightening, leaks may occur during hydraulic pressure testing, allowing for the detection of abnormalities.

(4) **Joint strength of tube fittings**

The performance standards that represent the joint strength of tube fittings are defined in the Japan Stainless Steel Association Standard "SAS 322:2016 Performance Standard for Tube Fittings of Stainless Steel Pipes for Ordinary Piping". This standard specifies 10 performance standards that must be met, including the fitting body, the air tightness of the barrel, negative pressure performance, water pressure performance, and other aspects.

8.3 Plastic Joining Techniques Primarily Used in Structural Assembly

8.3.1 Joining of Thin-Walled Materials

(1) **Seam fold crimping of can bodies** [6]

In the joining of can bodies, a method called seam fold crimping (lock seaming) is commonly employed, as shown in Fig. 8.15. In this method the seam is folded into a hook-like shape. As depicted in Fig. 8.16, the joining process involves aligning an L-shaped sheet bent by a press and an L-shaped sheet with a curled tip, and then winding them together using a press [6]. Typically, a single type of punch is sufficient for this joining process. In Fig. 8.16, the thicknesses of the left and right sheets may differ, and there may be a step on the sheet surface, but the joining method remains the same as that for can bodies. This joining technique is commonly referred to as seamed cans and is widely used for containers such as those for confectionery, nori (seaweed), and tea. It should be noted that welded bodies are used when tightness is required, such as for containing liquids.

Fig. 8.15 Seam fold crimping and seamed can body

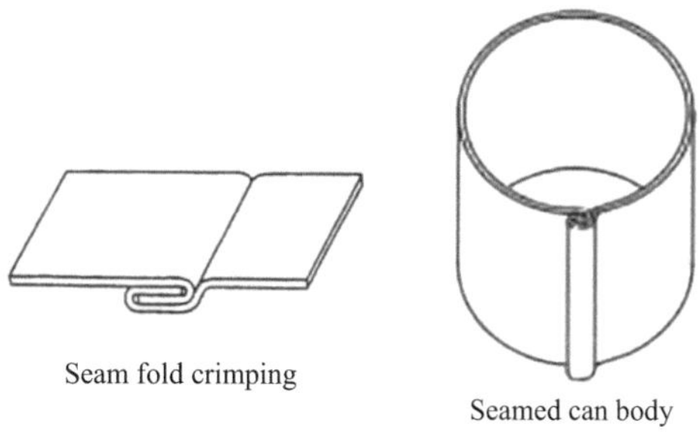

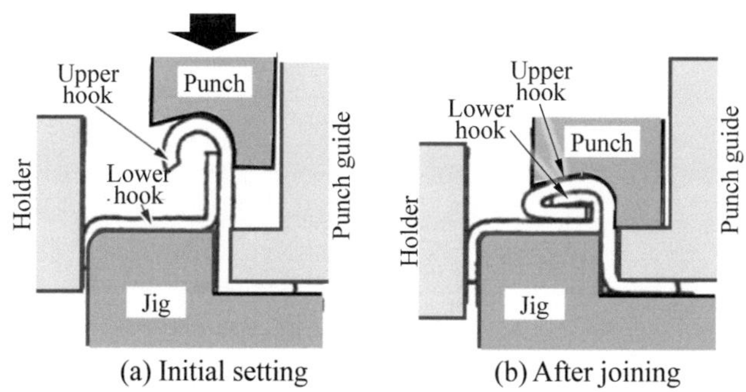

Fig. 8.16 Process of seam fold crimping

(2) **Joining of can body, bottom plate, and lid** [7]

The joining method used for the can body, bottom plate, and lid is called seaming. The joined assembly of the seamed can body, bottom plate, and lid is referred to as a three-piece can. The joined product of a deep-drawn and ironed can and a lid, is known as a two-piece can. The joining method is illustrated in Fig. 8.17. Typically, a two-step process (using two types of rolls) is employed to rotate and join the can body, bottom plate, and lid. The can body with bent ends and can lid are aligned, and the bending process that wraps both components is primarily performed by the first roll. The second roll is used to flatten the wrapped seam. This joining technique is widely used for various applications, ranging from beverage cans and canned foods to larger pail cans and drum cans. In liquid containers, sealing materials are often used at the seams to prevent leakage. Conventional containers, as shown in Fig. 8.18a, have a double seam (5-layer structure). However, for drum cans, international standards may require a triple seam (7-layer structure), as shown in Fig. 8.18b, to enhance drop resistance [8].

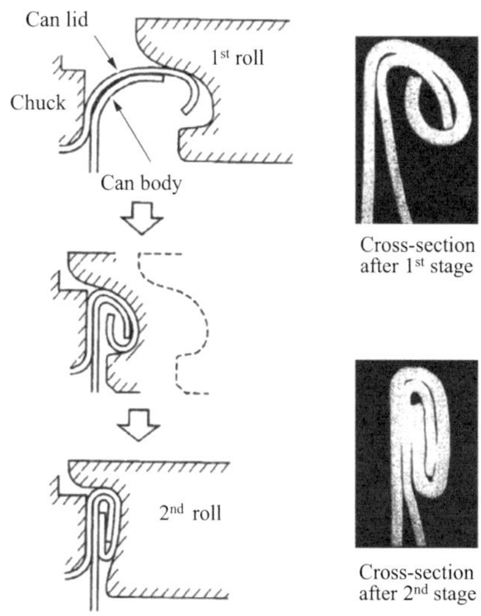

Fig. 8.17 Seaming processes and cross-sections of seams

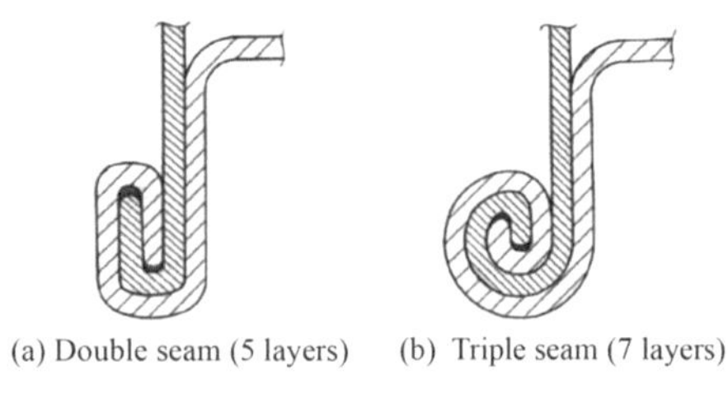

Fig. 8.18 Cross-sections of double and triple seams

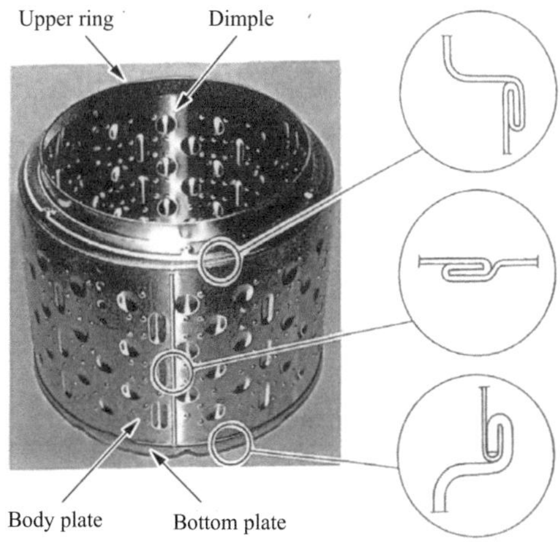

Fig. 8.19 Joining methods applied to washing machine drum [6]

(3) **Stainless steel washing machine drum** [6]

Figure 8.19 illustrates an overview of a stainless steel washing machine drum.

The manufacturing process of the stainless steel drum involves initially perforating numerous holes for drainage purposes. Additionally, dimples and other surface profiles are formed on a stainless steel sheet, which is then bent into a cylindrical shape. The ends are folded and hooked to create the body plate. Subsequently, the pre-pressed bottom plate and top ring are seamed to the body plate. It should be noted that the top ring is used to enhance the rigidity of the washing machine drum and may not be used in all cases.

8.3.2 Plastic Flow Joining Method

(1) **Principles of plastic flow joining** [9]

The plastic flow joining method was developed in the early 1980s. The principle is illustrated in Fig. 8.20. Firstly, when joining two components, a groove (referred to as a bonding groove) is machined on the hard component A (Figure (a)). Component A and the soft component B are positioned in a fixture, and a portion of the end face of component B is subjected to pressure using a punch [9].

This causes plastic flow in the groove of component A, resulting in mechanical interlocking and bonding. In the case of both components A and B being hard, an additional soft component C is introduced. After positioning all three components in the fixture, pressure is applied to the intermediate soft component C, causing plastic

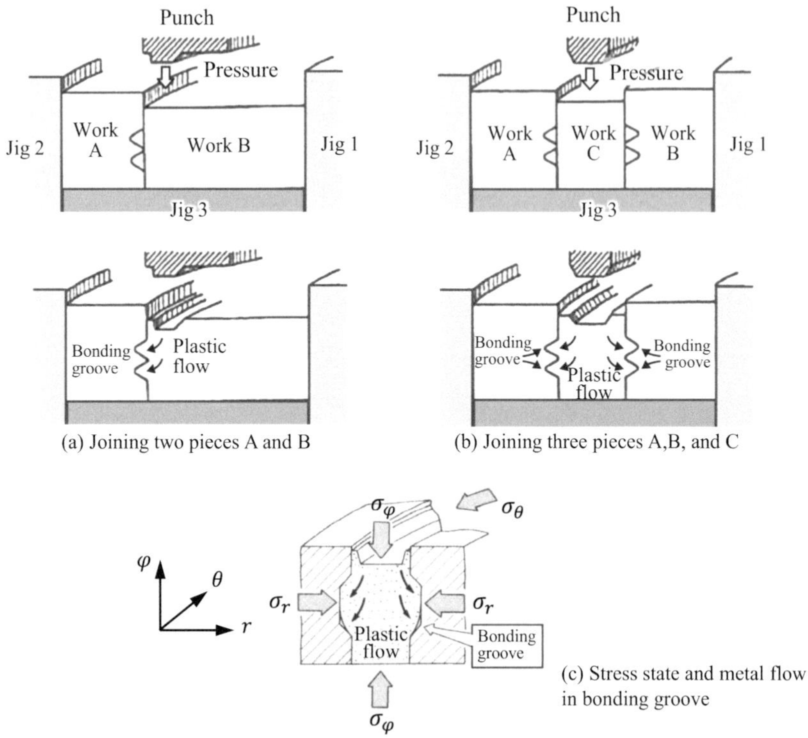

Fig. 8.20 Mechanism of plastic flow bonding

flow in the grooves of hard components A and B for bonding (Figure (b)). Component C can be a thin-walled cylinder or a bent thin sheet.

Plastic flow joining is a method of bonding by compressing the material axially and causing it to flow radially (Figure (c)). The material is in a compressive stress state as a whole due to the enclosed condition. In this state, axial compression increases the hydrostatic pressure, but its contribution to yielding is small. Furthermore, even after plastic flow begins, the friction between the tools and the material remains high. As a result, the axial compressive stress $|\sigma_{\varphi}|$ corresponding to the processing force can reach 4 to 6 times the flow stress.

The schematic diagram in Fig. 8.20 shows a cross-sectional view with components A and B being cylindrical parts. If component A is shorter than component B along the axial direction, it represents the bonding between a hollow shaft and a disk. In the case of bonding between a solid shaft and a disk, fixture 1 is not required. Various combinations of components can be bonded based on their shapes and dimensions. Additionally, depending on the shape and dimensions of the components, restraint of radial deformation and support for axial bonding load are necessary.

The distinctive features of this technique are as follows:

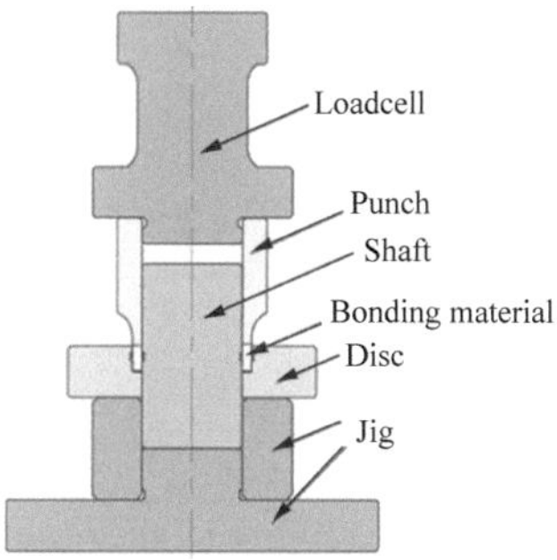

Fig. 8.21 Testing machine for plastic flow joining of three components

1. It allows easy bonding of dissimilar materials or hard brittle materials, with fewer constraints on material combinations.
2. There is no relative movement between the components after fitting (assembly) for pressure bonding, resulting in high axial positioning accuracy.
3. The fitting between components can be achieved with gap fitting or intermediate fitting, leading to high productivity.

(2) **Basic data on plastic flow joining**

Basic characteristics of the shaft and disk plastic flow joining are demonstrated through fundamental experiments shown in Fig. 8.21 [10]. This refers to the case of three-component bonding in Fig. 8.20b, specifically when there is no fixture 1 and a solid shaft is used. A heat-treated steel (hardness approximately HRC 40) is used for a 10 mm thick and 50 mm outer diameter disk, and SKD 11 (hardness approximately HRC 60) is used for a 20 mm diameter shaft, S30C is used as the soft bonding material.

Figure 8.22 shows the punch pressure, punch stroke, and material flow state into the bonding groove based on the applied pressure load [10]. In plastic flow joining, the punch penetration (press stroke) is small, typically less than 1 mm, as bonding is performed after fitting the components together. As the punch face pressure (load) increases, the filling ratio into the two grooves (material flow area into the groove/groove area) monotonically increases. It reaches approximately 57% at a punch face pressure of 1200 MPa and approximately 95% at 2000 MPa. Although the tensile strength of the bonding material S30C is approximately 500 MPa, in the case of plastic flow joining, the high constraint between the disk and shaft components indicates that bonding occurs under very high hydrostatic pressure.

Figure 8.23 presents the results of four types of bonding grooves (including the case without a groove) showing the influence of the bonding load (punch face pressure) on the bonding strength. From these results: **(1)** The bonding strength increases proportionally with the bonding load, regardless of the presence of a bonding groove, and approximately 20% of the bonding load (100–180 kN) can be obtained as the bonding strength. **(2)** In the absence of a bonding groove, the bonding strength

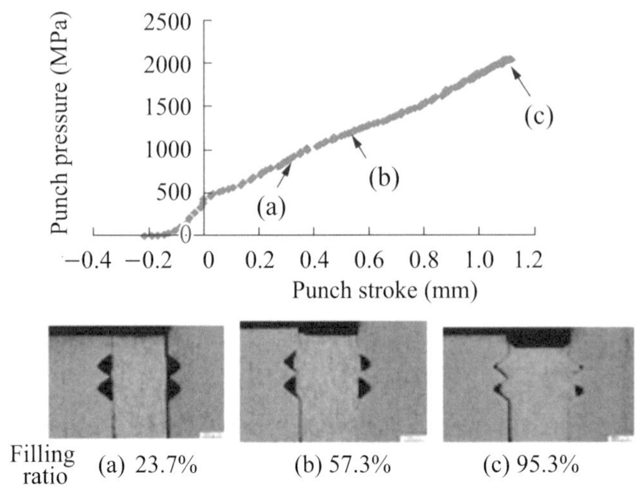

Fig. 8.22 Punch pressure, punch stroke, and filling ratio in plastic flow joining

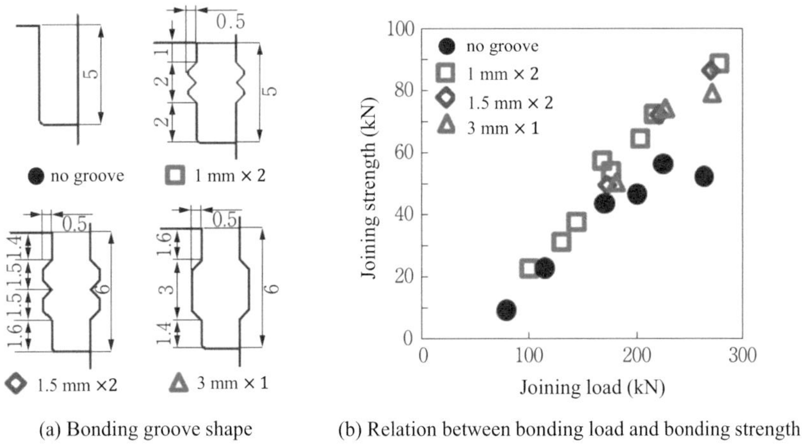

Fig. 8.23 Effect of groove shape and joining load on joining strength [10]

remains nearly constant at 50 kN, even with an increase in the bonding load beyond 200 kN. **(3)** With a bonding groove, the bonding strength increases (up to 90 kN) along with an increase in the bonding load (up to 300 kN), demonstrating the effect of the bonding groove. **(4)** The influence of the shape of the bonding groove is not significant among the four types examined. While the frictional force can be expected to contribute to the bonding strength in the range of 100–180 kN of bonding load, the anchor effect from the bonding groove is necessary to achieve higher strength.

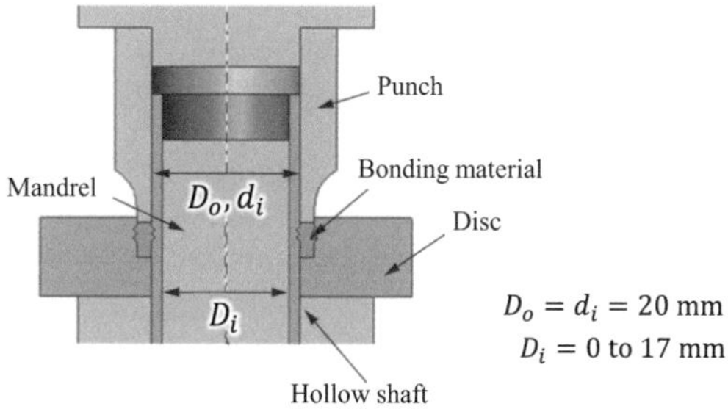

Fig. 8.24 Plastic flow joining of hollow shaft and disc [11]

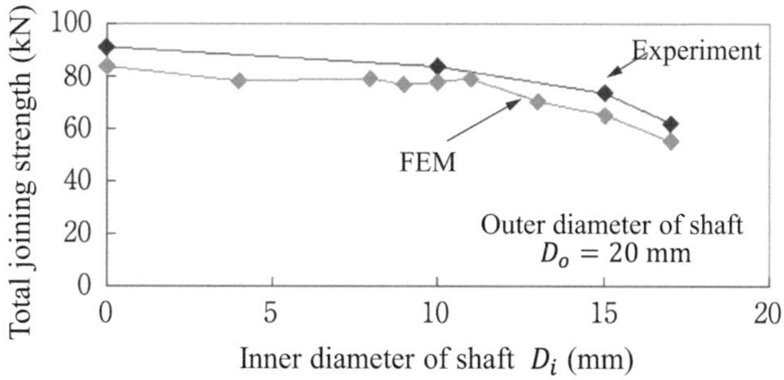

Fig. 8.25 Effect of inner diameter of hollow shaft on total joining strength [11]

For hollow shafts, as shown in Fig. 8.24, plastic flow joining with a disk is performed by using a mandrel to prevent deformation inside the hollow shaft and promote material flow into the bonding groove [11]. In this case, the outer diameter do and inner diameter di of the disk are kept constant at 50 mm and at 20 mm, respectively, and the hollow shaft with outer diameter Do of 20 mm varied from an inner diameter Di of 0 mm (solid shaft) to 17 mm (with a wall thickness of 1.5 mm).

Figure 8.25 shows the experimental results (four conditions) and finite element analysis results (nine conditions) of the relationship between the inner diameter of the hollow shaft and the bonding strength. Even with an inner diameter Di of 15 mm for an outer diameter Do of 20 mm, the hollow shaft still has approximately 80% of the strength of the solid shaft, indicating the applicability of this bonding method to hollow shafts. Since the experimental results and analysis results are nearly identical, the breakdown of the two types of strengths, which is difficult to determine in experiments, was estimated from the analysis results.

Figure 8.26 presents the analysis results of the shear strength of the bonding groove section and the frictional strength from the tightening of the cylinder section on the bonding strength of a hollow shaft. According to these results, the tightening strength of the cylinder section decreases as the hollow shaft becomes thinner, while

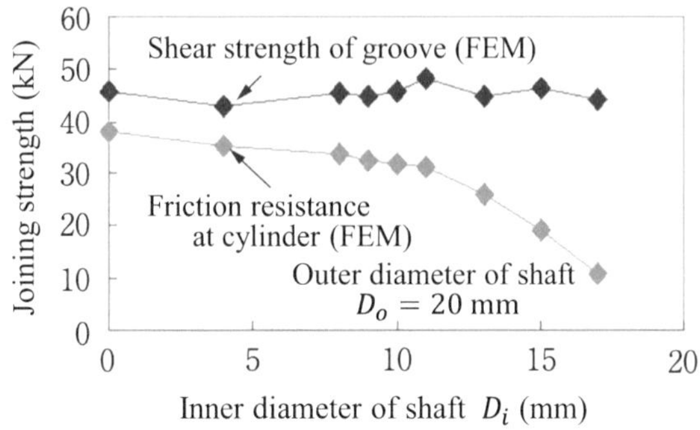

Fig. 8.26 Effect of inner diameter of hollow shaft on joining strength (FEM) [11]

the bonding strength of the groove section remains nearly constant regardless of the thickness of the material. This indicates that the anchor effect from the bonding groove is particularly important for thin-walled materials.

(3) **Examples of applications of plastic flow joining technology**

The following are some examples of the application of plastic flow joining technology. Although the components are not tubular, they include hollow shapes, making plastic flow joining an effective technique for tubular products.

Figure 8.27 shows a part of a flywheel component for outboard motors [9]. Previously, a cold forged center piece (with holes) and disc-shaped wheel were joined with nine rivets, and the hollow wheel was joined to the ring gear with six bolts. However, these connections were converted to plastic flow joining, reducing the overall length from 46 to 42 mm. Applying a 210 tonf load results in a pulling out strength of 15 tonf and a torsional strength of 550 kgf-m.

Figure 8.28 depicts a gasoline injector for automobiles [12]. Plastic flow joining is applied at four locations: between the core and yoke (hollow shape), between the core and nozzle, and others. These applications were converted from conventional

Fig. 8.27 Flywheel component for outboard motors

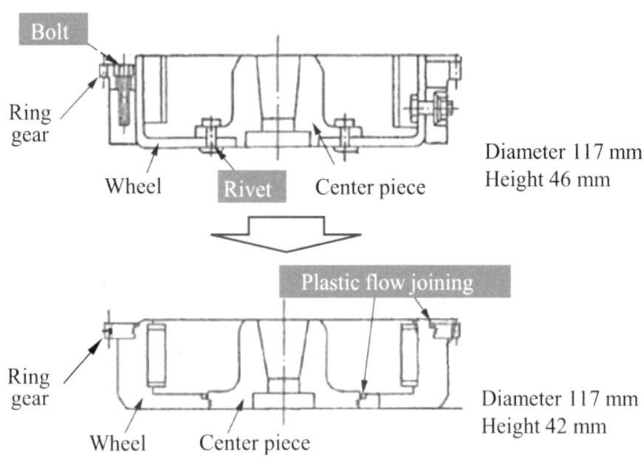

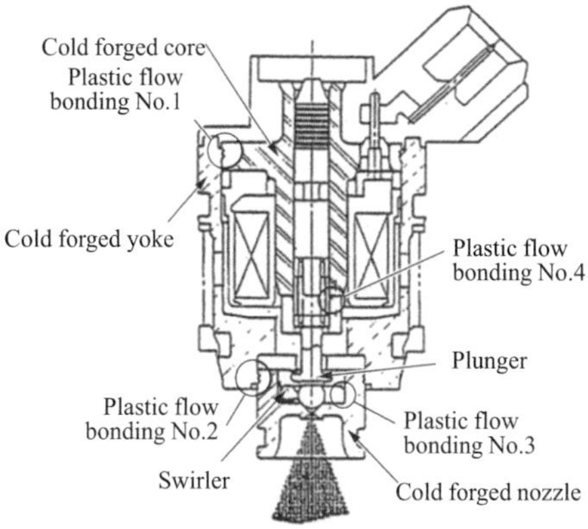

Fig. 8.28 Example of plastic flow joining on gasoline injector for automobiles [12]

laser welding methods, resulting in improved precision, productivity, and reduced equipment investment.

Figure 8.29 illustrates an attempt to use plastic flow joining for a two-stage hollow gear [11]. Normally, space is required to avoid interference with the larger gear during the cutting process of the small gear. However, by performing plastic flow joining after the cutting process of the thin-walled small gear, space can be saved, enabling space-efficient design. As mentioned in the principle of plastic flow joining, it is also possible to join hard brittle materials [13], as demonstrated by the assembly of an actuator housing for the variable valve timing mechanism of an automobile engine, which involves joining an aluminum die-cast part and a steel shaft [14]. This method has been put into practical use, achieving benefits such as 18% reduction in weight and 34% cost reduction compared to conventional bolted connections.

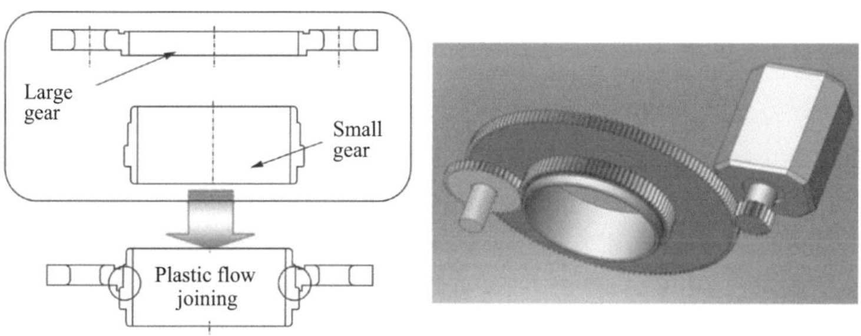

Fig. 8.29 Example of plastic flow joining on two-stage gear [11]

8.4 Other Joining Methods

Flaring the tube end using a press to create stepped shapes, necking down the tube end to reduce its diameter (refer to Sect. 5.2), and rotary swaging (refer to Sect. 6. 3) are also used in tube joining processes. Not only metal tools but also rubber tools for expansion, as shown in Fig. 8.30, can be used [15].

These techniques are applied to joining in heat exchangers, piping, and other applications [15]. Electromagnetic forming, which utilizes the magnetic field generated by passing a large current through an electromagnetic coil, is also used in joining (Fig. 8.31).

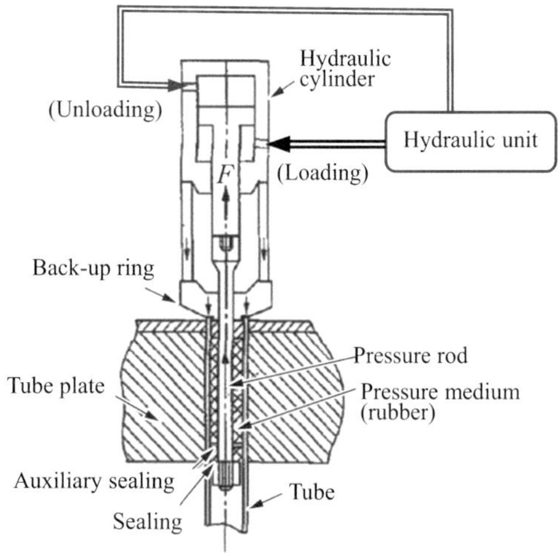

Fig. 8.30 Expanding tool using rubber for expansion

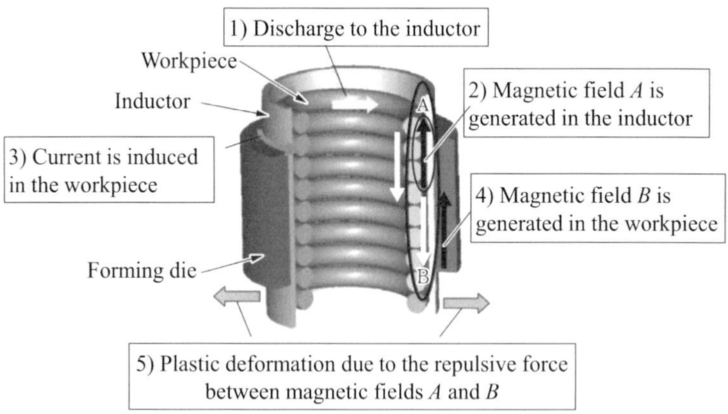

Fig. 8.31 Electromagnetic forming of tubes [20]

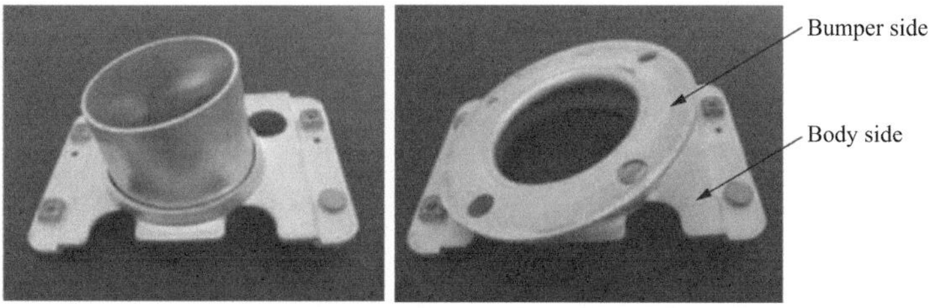

Fig. 8.32 Example of electromagnetic tube expansion of aluminum bumper stays [21]

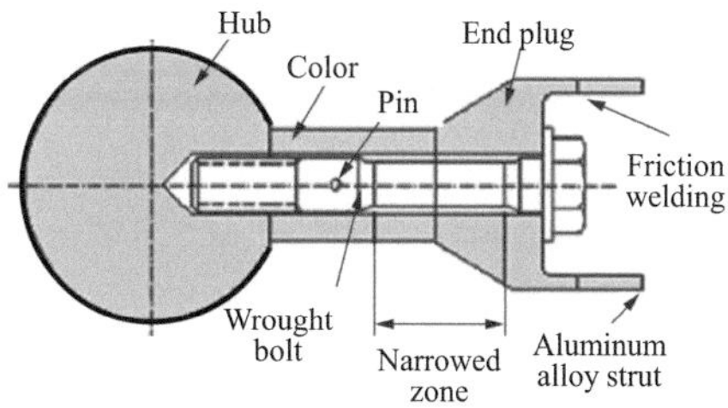

Fig. 8.33 Example of friction welding on aluminum alloy members for seismic truss [23]

Practical machines manufactured in the United States have been available since the 1960s, and research on circular tubes was conducted in Japan at that time [16]. Subsequent research has been conducted [17, 18], and it has been utilized as a joining method for mass-produced aluminum alloy parts in Japan since the mid-2000s [19]. In electromagnetic forming, expanding can be achieved by placing the electromagnetic coil on the inside of the tube, and contracting can be achieved by placing it on the outside. Figure 8.32 shows an example of applying electromagnetic tube expansion to the forming of aluminum bumper stays for automobiles.

Friction welding is a method of joining where the materials to be joined are rubbed together at high speed, softening the materials with the resulting frictional heat while applying pressure.[1] JIS Z 3607:2016 specifies friction welding for metallic materials. For circular tubes, research on friction welding between aluminum alloy tubes and carbon steel tubes has been conducted [22]. An example of recent practical application of friction welding is the seismic truss members made of aluminum alloy (Fig. 8.33) [23].

[1] Japan Friction Joining Association: http://www.jfja.or.jp (as of 2019 February).

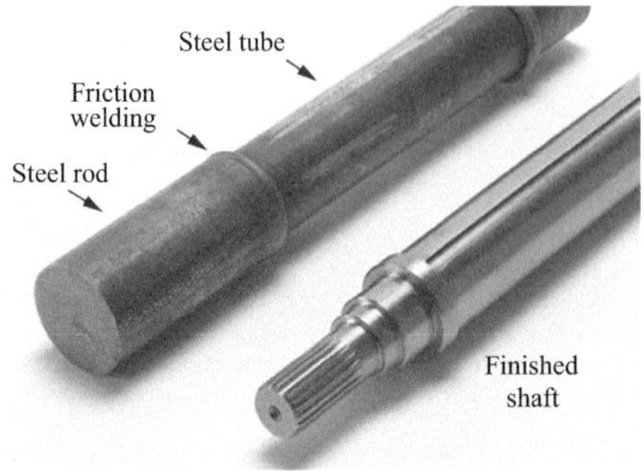

Fig. 8.34 Lightweight steel shaft by friction welding[2]

In 2000, the Building Standards Act of Japan was revised, allowing the use of aluminum alloy for the structural parts similar to steel. Furthermore, a lightweight shaft has been developed by friction welding solid rods to both ends of a hollow tube and machining the welded ends (Fig. 8.34).

Figure 8.35 shows an example of the application of shaving joining in the assembly of a camshaft. In shaving joining, the hard member (shaft roulette in the figure) is joined to the soft member (cam inner piece in the figure) while shaving a small

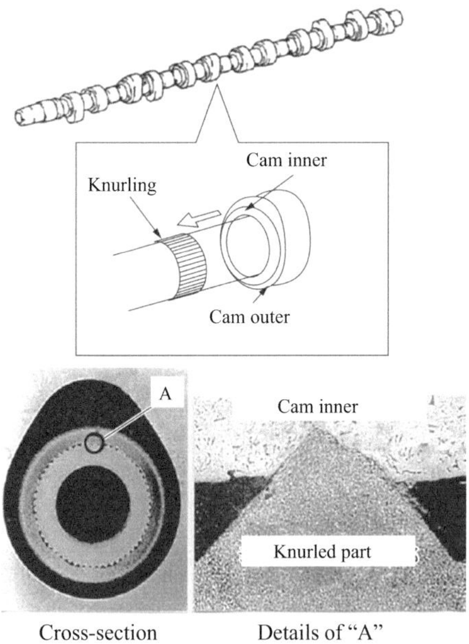

Fig. 8.35 Example of shaving joining on cam shaft assembly [24]

[2] JST Co. Ltd. https://www.jstco.jp/service/lightweight_shaft/ (as of 2025 April).

amount of material, resulting in high adhesion on both sides of the joint and a joint formed with fresh surfaces.

References

1. Sugino Machine Ltd.: Tube expanders and Accessories Catalog, vol. 18, (2015)
2. Sugino Machine Ltd.: Tube expanders and Accessories Catalog, vol. 2, (2015)
3. Expander Forming Research Group (ed.): Comprehensive guide to expander forming technology, vol. 139. Maruzen Publishing, (1966)
4. Japan Stainless Steel Association (ed.): Guide to stainless steel piping, (2019)
5. Benkan Corporation.: Double press-type tube fitting catalog, (2017)
6. Murakami, H., et.al.: J. JSTP, **34**(391), 905–909 (1993)
7. Ota, T., Taura, Y., Otsuka, M., Tamiya, S.: Mitsubishi heavy industries technical review, vol. 35, no. 1, pp. 64–67 (1998)
8. Association, J.S.D.: Journal Hibiki **34**, 6 (2012)
9. Kanamaru, H., Shoji, A., et.al.: Hitachi Rev, **64**(2), 147–152 (1982)
10. Murakami, H., Takada, M., Nishikawa, S., Kanamaru, H., Imura, T.: In: Proceedings of the 2009 spring meeting, JSTP, pp. 187–188 (2009)
11. Murakami, H.: In: Proceedings of the 322nd symposium on the technology of plasticity, JSTP, pp. 1–8 (2017)
12. Muarakami, H., Otsu, E., Kanamarau, H.: In: Proceedings of the 1997 spring meeting, JSTP, pp. 237–238 (1997)
13. Murakami, H., Kawame, N., Suzuki, Y., Watabe, M.: In: Proceedings of the 65th joint conference on the technology of plasticity, JSTP, pp. 81–82 (2014)
14. Nikkei, B.P. (ed.): Nikkei Monozukuri (The solutions magazine for design and manufacturing), August issue, pp. 37–39. Nikkei Business Publications, (2014)
15. Yoshitomi, Y., Kanbara, H., Omura, K., Takada, T. J. JSTP, **28**(322), 1128–1132 (1987)
16. Teramoto, J., Kiyota., K.: J. Jpn. Soc. Precis. Eng., 38(3), 303–307 (1972)
17. Murata, M., Negishi, H., Suzuki, H.: J. JSTP., **25**(283), 702–708 (1984)
18. Suzuki, H.: J. JSTP., **30**(339), 489–496 (1989)
19. Tsuyoshi, T., Hashimoto, S., Hashimoto, T.: R&D kobe steel engineering reports, vol. 59, no. 2, pp. 17–21 (2009)
20. Tsuyoshi, T., Hashimoto, S.: R&D kobe steel engineering reports, vol. 62, no. 1, p. 96 (2012)
21. Goto, T, Imamura.Y., Kaitoku, K., Mizugaki, T.: Preprints of autumn national meeting of JWS, (2012)
22. Kawai, G., Ogawa, K., Tokisue, H.: J. Jpn. Inst. Light. Metals, **49**(11), 559–563 (1999)
23. Hiyama.Y., Okubo, S.: J. Jpn. Inst. Light. Metals, 60(2), 93–99 (2010)
24. Egami, Y., Taniji, T.: J. JSTP., **52**(603), 439–442 (2011)

Chapter 9
Distinctive Processing Examples

9.1 Other Processing

In this section, we will introduce distinctive processing examples that have not already been discussed, as well as processing examples in the research stage.

9.1.1 Tube Forging

In the case of secondary forming of thick-walled tubes, because thick-walled tubes require higher energy for processing, forging is sometimes used. When forging thick-walled tubes, there are two main methods: compressing the side wall of the tube and compressing it in the axial direction. When compressing the side wall of a thick-walled tube, variable thickness can be achieved by applying multiple dies with surfaces inclined to the tube axis, as shown in Fig. 9.1 [1]. As shown in the figure, a mandrel is inserted inside the tube to improve the shape accuracy. On the other hand, when compressing a thick-walled tube in the axial direction, a large thickness increase, approximately 2.2 times the initial thickness, can be achieved. Figure 9.2 shows this process, inserting a core bar into a tube and completely constraining the outer circumference of the tube. This enables the application to pipe fittings [2]. It is also possible to create stepped shapes by providing clearance between the outer circumference of the tube and the die without complete constraint.

Y. Kuriyama et al., *Tube Forming*, https://doi.org/10.1007/978-981-95-3421-0_9

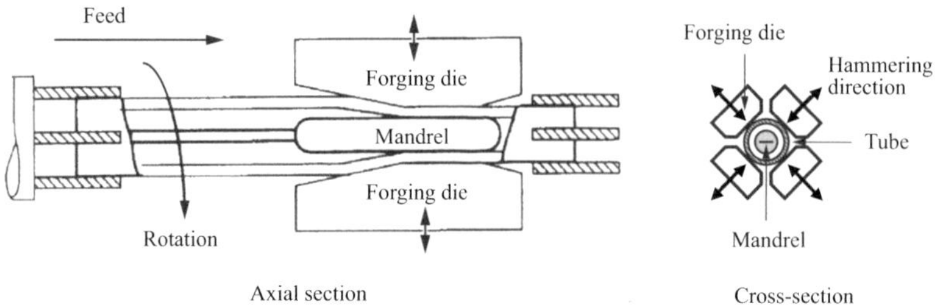

Fig. 9.1 Schematic of tube forging [1]

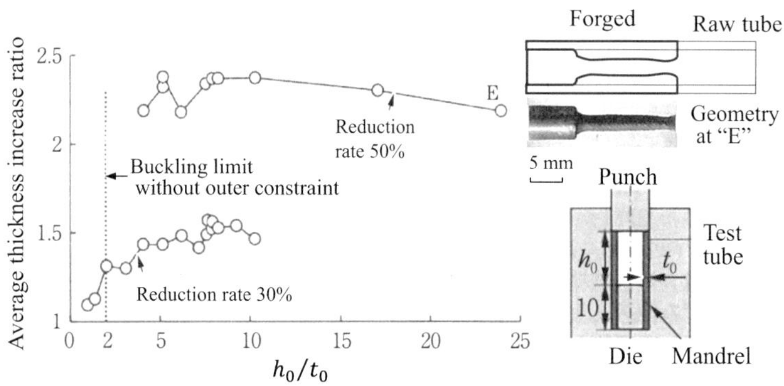

Fig. 9.2 Axial compression of thick-walled tube under outer constraint [2]

9.1.2 Crushing

There are many examples of tube crushing in various applications such as the front fork of bicycles or mounting parts for cargo racks. Some important considerations for the process include avoiding rapid changes in cross-sectional shape, predicting the occurrence of plastic hinges to prevent cracking, and considering the position of welding joints [3, 4]. The crushing of tubes shown in Fig. 9.3a–c involves inserting rubber or applying internal pressure inside the tube and pushing a punch to shape it. In some cases, as shown in Figure (d), oval shaping is achieved without inserting

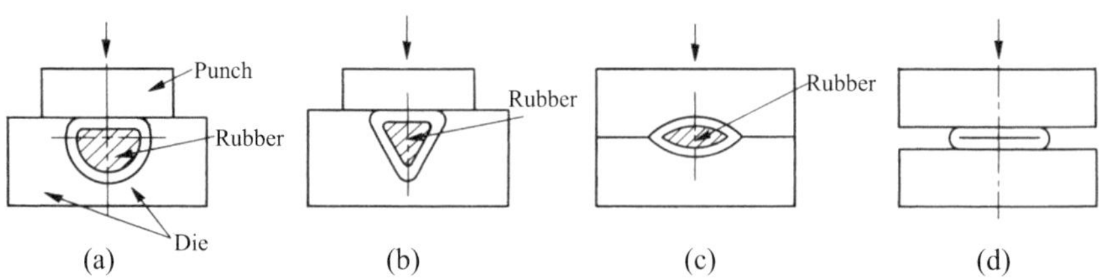

Fig. 9.3 Different types of crushing [5]

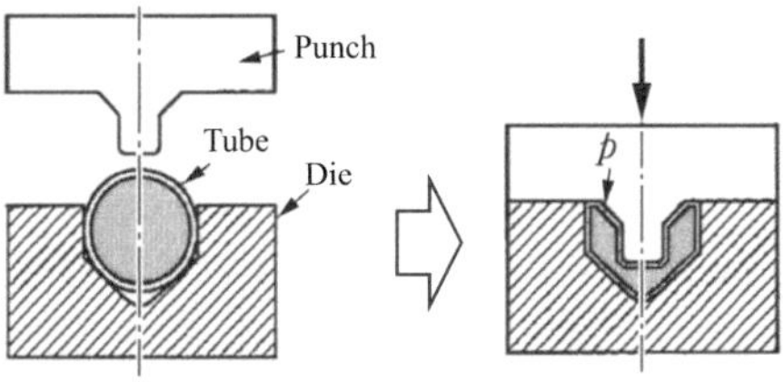

Fig. 9.4 Sealed fluid forming [6]

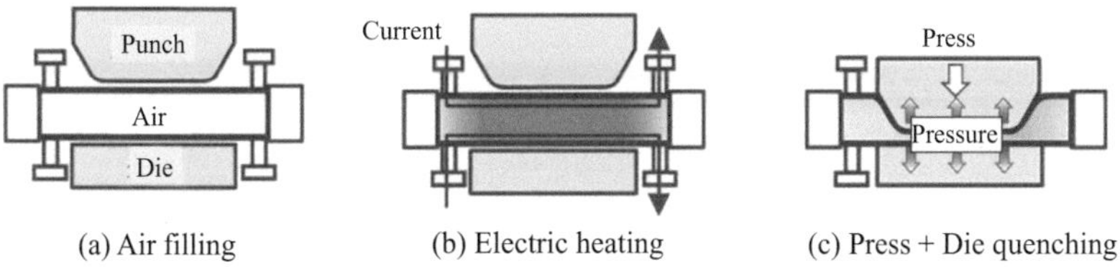

Fig. 9.5 Tube forming using electric heating [9]

anything inside the tube [5]. Additionally, there is a technique called sealed fluid forming, shown in Fig. 9.4, where liquid is sealed inside the tube, and the internal pressure p is increased by pushing the punch or moving the die, causing deformation conforming to the shape of the die or punch [6]. This method is applied to automotive components such as front arms [7, 8]. Furthermore, for steel tubes, as shown in Fig. 9.5, electric heating is used to heat the tube, enabling easier press forming [9].

9.1.3 Non-circular Shape Forming [10]

Non-circular shape forming refers to the process of shaping a portion of a tube material into a non-circular cross-section. An example of the forming method is shown in Fig. 9.6. The processing steps are as follows:

(a) After advancing the punch, squeeze the tube by advancing the die.
(b) Ironing the tube by holding the outer die and retracting the inner punch.
(c) After withdrawing the punch, retract the die to complete the forming process.

Because this process involves ironing, products have high precision and low surface roughness, but it may require multiple processing steps due to the high degree of deformation. By combining different die shapes and punch shapes, it is possible to shape various non-circular cross-sectional shapes shown in Fig. 9.7.

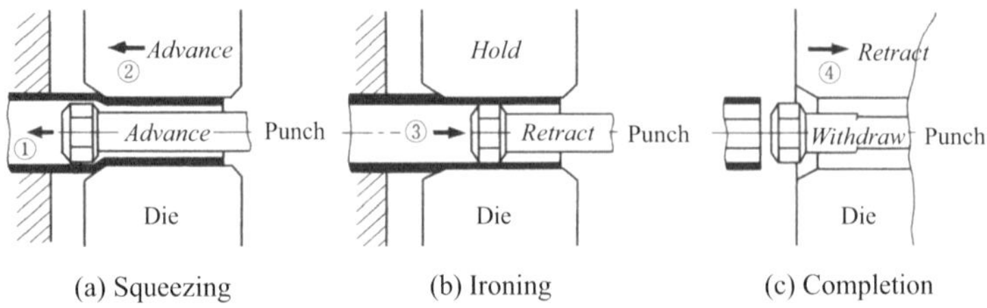

Fig. 9.6 Non-circular shape forming [10]

Fig. 9.7 Examples of non-circular cross-sectional shapes

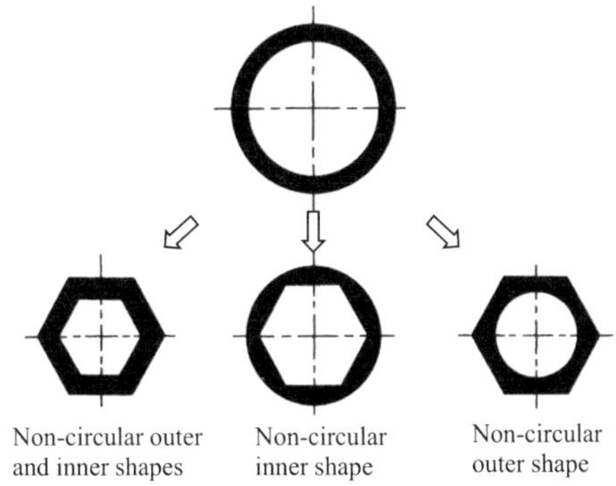

An example in Fig. 9.8 shows a case where there is a circular die hole and a shaped inner punch. The die clamps the outer side of the tube, while the punch is withdrawn to form the inner side into the desired shape.

Large changes in cross-sectional shape result in significant variations in wall thickness. In such cases, the inner punch is given a step, shown as "B" in the figure, similar to that used in broaching. Specialized machines are often required due to the need for complex operations. Hydraulic systems that are suitable for ironing are commonly adopted for the forming process. Examples of formed products are shown in Fig. 9.9.

Fig. 9.8 Circular hole die and punch with different hexagonal shapes [11]

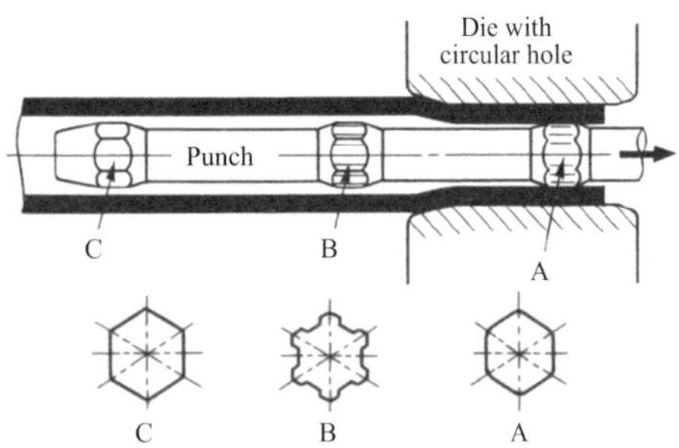

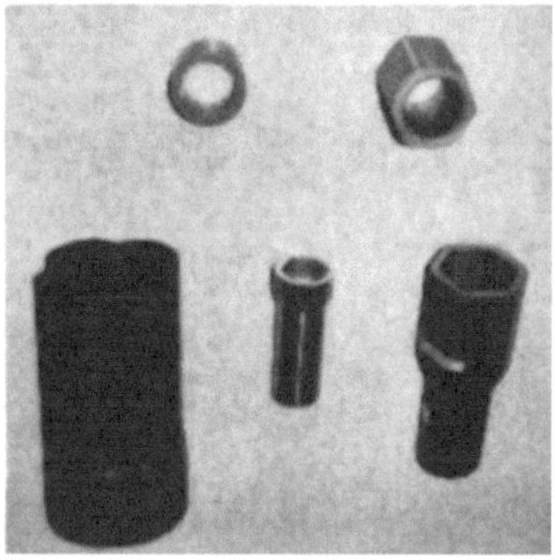

Fig. 9.9 Examples of non-circular shape formed products

9.1.4 Butted Tubes

Some bicycles parts are required to have the conflicting properties of "lightness" and "strength". Therefore, there are special tubes used as structural components. One example is the butted tube, which has varying wall thickness along its axis [11]. The butted tube used in bicycles has thicker walls at both ends, where strength is necessary, and thinner walls in the middle to achieve light weight. In general, the tubes used in the front triangle and fork stem of sport bicycles have a constant outer diameter but varying wall thickness along their length, even though this variation is not visible externally. The manufacturing methods are shown in Fig. 9.10 [12].

Figure (a) demonstrates a method where a portion is reduced in thickness by rolling, and then the entire tube is drawn to achieve a uniform outer diameter. Figure (b) shows a method involving the insertion and extraction of a stepped mandrel. Figure (c) involves constraining the outer side with a die while inserting a stepped mandrel and upsetting the smaller diameter side. Figure (d) illustrates a method where a portion is bulged to reduce the wall thickness, followed by drawing the entire tube to achieve a uniform outer diameter. Figure (e) demonstrates a method where the mandrel is inserted and extracted during the drawing process. These processes include drawing, rolling, bulging, and upsetting, which are combined to achieve the desired results. In the case of the drawing process shown in Fig. 9.11, by moving the drawing plug and synchronizing it with the die operation, it is possible to produce tubes with varying wall thickness at any desired location.[1]

[1] Sango Co., Ltd.: http://www.sango.jp/products_technology/processing/pull_out/l (as of 2019 February).

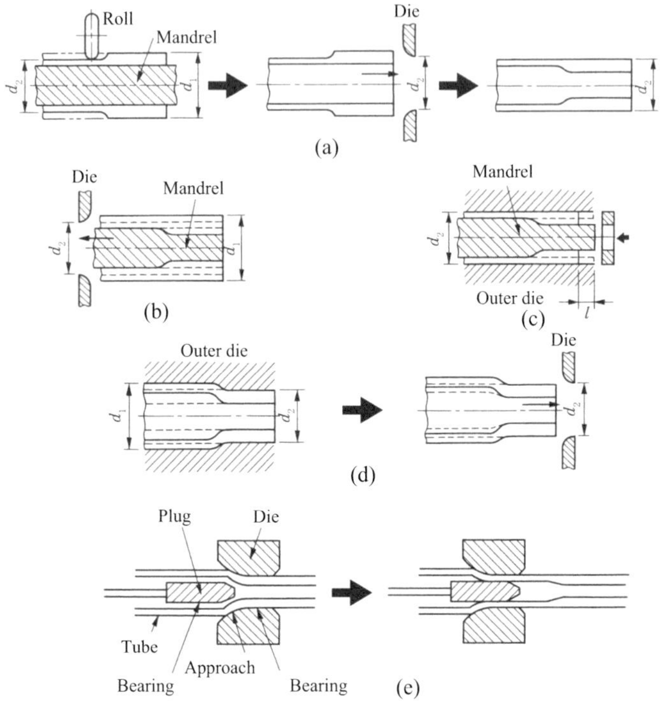

Fig. 9.10 Typical manufacturing methods of butted tubes [11]

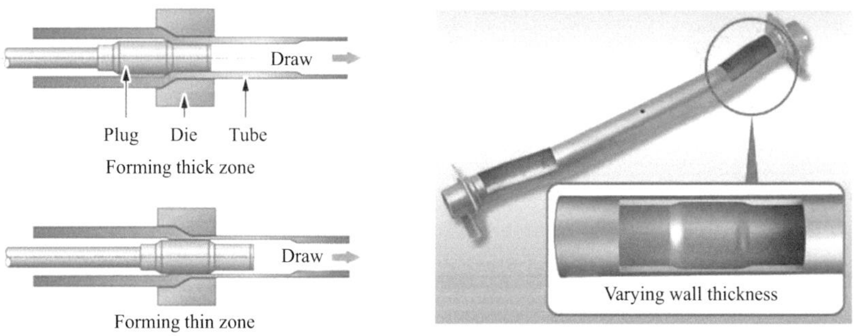

Fig. 9.11 Forming process of butted tube by drawing (see Footnote 1)

9.1.5 Thickening Method of Tube Using Laser

When performing bending or bulging processes on tube materials, the wall thickness of the tube becomes thinner, resulting in decreased strength and potential fracture. By previously thickening the thinning areas, such phenomena can be mitigated. For this purpose, it is possible to achieve a high thickening rate by localized heating on

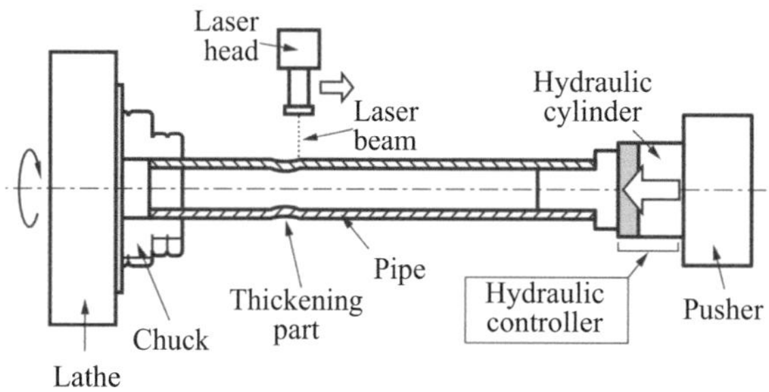

Fig. 9.12 Thickening process of thin-walled tube by localized heating using laser [13]

the tube material using a laser and subsequently compressing the tube, as shown in Fig. 9.12 [13].

9.1.6 *Spiral Grooved Tubes and Spiral Tapered Tubes*

Tubes with spiral grooves on the inner surface contribute to energy savings in heat exchangers. By extrusion using a mandrel with spiral grooves, as shown in Fig. 9.13, circular tubes with twisted protrusions on the inner wall can be obtained [14].

A new roll forming apparatus has been developed that combines bending in both directions with twisting, enabling the formation of spiral tubes. This roll forming apparatus imparts in-plane bending to a strip sheet by the in-plane bending roll, as shown in Fig. 9.14, and out-of-plane bending is induced by displacing the forming rolls vertically. By further tilting the forming rolls, twisting can be applied, and the strip sheet bent along the arc can be twisted, enabling the production of long-length tapered spiral tubes, as shown in Fig. 9.15 [15].

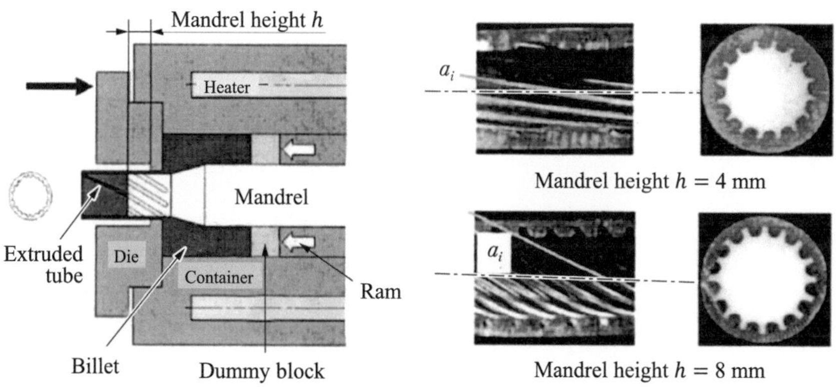

Fig. 9.13 Extruded tube with inner spiral groove [14]

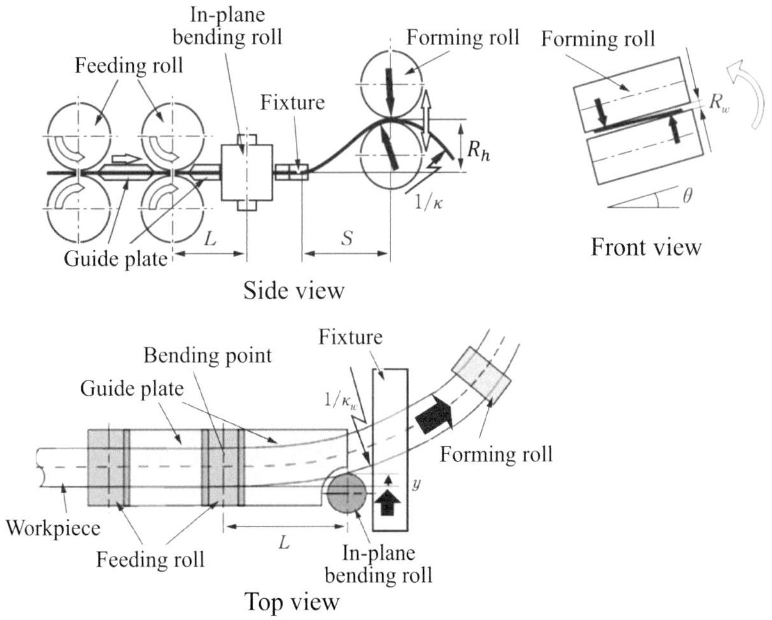

Fig. 9.14 Roll forming machine for tapered spiral tube [15]

Fig. 9.15 Tapered spiral tube [15]

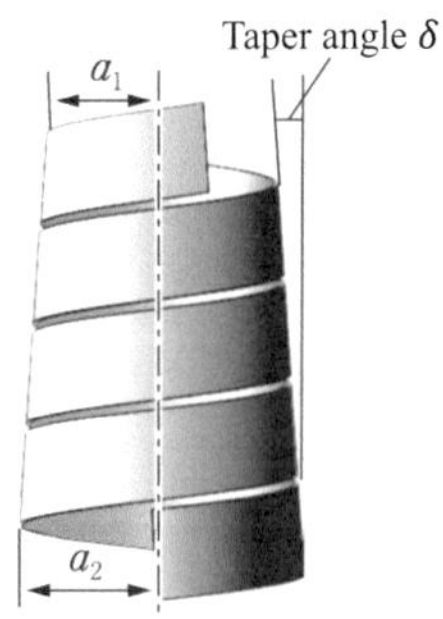

9.2 Microtube Forming

9.2.1 Applications and Differences from Macroscale Tubes

In recent years, the manufacturing of small and highly precise components and parts has led to miniaturization and enhanced performance of equipment used in the electronics, electrical, telecommunications, medical, bio, and chemical fields. Particularly in the medical, chemical, and bio fields, there is a high demand for microtube forming that deals with metal microtubes capable of transporting liquids such as chemicals or gases. Conventional macroscale tubes have been used as piping for gas

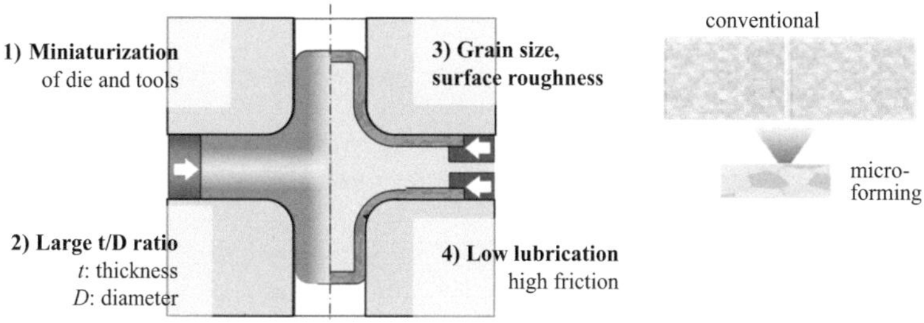

Fig. 9.16 Size effect in microtube forming

and fluid transportation, as well as component parts for lightweighting and increased rigidity in various structures, drawing attention as components for transportation equipment such as automobiles. However, the applications of microtubes are not limited to the advantages of lightweighting and increased rigidity for component parts. They are often utilized for specialized purposes unique to microcomponents. They have been applied to semiconductor inspection contact probes, electrodes for discharge machining, micronozzles, optical fibers, lasers, micro heat exchangers, medical devices, gas sensors, and other applications, with some of them reaching practical use. While there is no definitive definition for the dimensions of microtubes, generally speaking, those with an outer diameter of 1 mm or less are considered as microtubes in this context.

When manufacturing or secondary processing microtubes, it is important to consider the size effects as the dimensions significantly reduce compared to the traditional macroscale tube-forming techniques. Figure 9.16 summarizes the main size effects that need to be taken into account as an example of performing tube hydroforming at the microscale. In microtube forming, miniature dies and tools are required. As a result, it becomes challenging to process the dies and tools themselves, and high precision is also required. Additionally, for microtubes, the inner diameter also becomes extremely small, which can make the insertion and handling of dies and tools very difficult.

Another size effect to consider is the increasing ratio of thickness t to diameter D as the tube size reduces at the microscale. Figure 9.17 presents the manufacturing range of microtubes from various companies providing cold-drawn tubes, organized by the ratio of thickness to outer diameter t/D.[2,3] Generally, as microtubes with an outer diameter of 1 mm or less are considered, the achievable t/D ratio increases, resulting in the handling of relatively thicker-walled tubes compared to the macroscale. For example, in the case of microtube hydroforming, it can be observed that a larger internal pressure is required as the t/D ratio increases.

It is known that lubrication effects decrease at the microscale. This is because the confinement effect of lubricants diminishes with the reduction of dimensions [16].

[2] Nippon Tokushukan MFG. Co. Ltd.: http://www.nittoku.com/ (as of 2019 February).

[3] Ohbakiko Co. Ltd.: http://www.ohbakiko.co.jp/ (as of 2019 February).

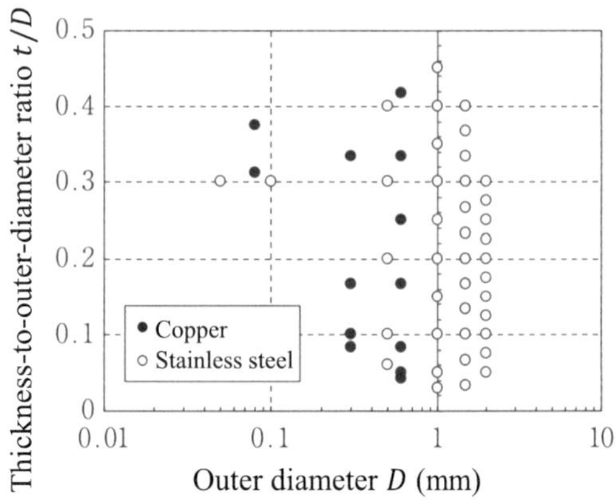

Fig. 9.17 Available range of cold-drawn micro tube

Therefore, in microtube forming, there is a general tendency for the friction coefficient to increase. Furthermore, at the microscale, it is known that the ratio of grain size and surface roughness to thickness increases. For example, when the material dimensions reduce to 1/100, the grain size and surface roughness do not decrease to 1/100 [17]. As a result, the grain size and surface roughness have a significant influence on the deformation behavior and formability of the material itself. In fact, the decrease in strength and ductility at the microscale is well-known [18]. Thus, considering the aforementioned size effects is crucial in microtube forming.

9.2.2 Primary Processing

The primary processing techniques for manufacturing metal microtubes can be broadly classified into two categories: plastic deformation methods, represented by drawing processes, and physical/chemical methods that utilize bottom-up approaches where atoms are stacked one by one or etching.

Drawing processes, which are a representative plastic deformation method, constitute the most common method for tube miniaturization. It involves passing the tube through a die to obtain a material with the same cross-sectional shape as the die exit. It is suitable for the production of long-length tubes due to its continuous and high-speed production capabilities. Here are examples of drawing processes: drawing without a plug, where the thickness increases as the diameter reduces; fixed plug drawing, which ensures good surfaces on both the inside and outside of the tube; floating plug drawing, which enables the drawing of thin tubes; and mandrel drawing, which reduces the wall thickness and is commonly employed for small-diameter or thin-walled tubes [19].

However, these methods are intended for tubes with sizes in the millimeter range, and it becomes challenging to insert plugs or mandrels into extremely small-diameter tubes like metal microtubes. Therefore, for tubes with an inner diameter smaller than

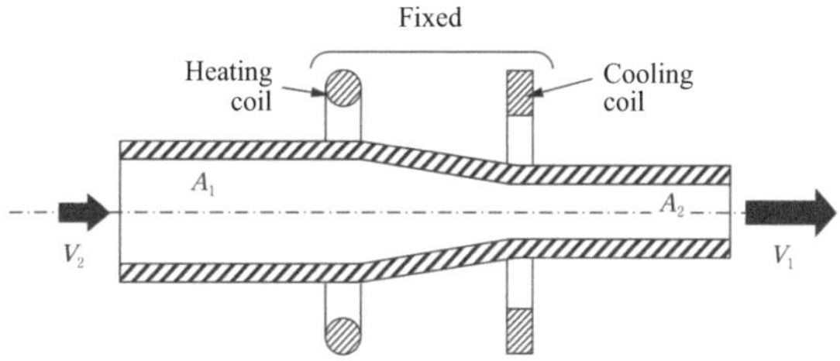

Fig. 9.18 Schematic of die-less drawing

0.5 mm, empty drawing is usually employed. In the case of empty drawing, the wall thickness increases, and eventually, the inner diameter becomes blocked, limiting the production of tiny microtubes. To address these issues, alternative techniques such as soft mandrel drawing that uses a softer material for the mandrel [20] and fluid mandrel drawing that fills the material's interior with liquid [21] have been developed. The main materials for metal microtubes include commercially available copper pipes with an outer diameter of 80 μm and an inner diameter of 30 μm, and stainless steel pipes with an outer diameter of 50 μm and an inner diameter of 20 μm. Additionally, microtubes made of various metal materials such as brass, nickel, cobalt, titanium, tantalum nickel silver, platinum, tungsten, Inconel, and permalloy are available in the market.

Regarding the narrowing of microtubes, a significant challenge lies in the presence of a die placed externally or tools inserted inside the tube. Therefore, a method called die-less drawing is employed to achieve tube reduction without the use of dies or tools [22]. As shown in Fig. 9.18, die-less drawing is a drawing process where the raw tube is locally heated and pulled while being moved, allowing for tube reduction without the need for dies or tools.

It enables the fabrication of microtubes with an outer diameter below 200 μm and is also effective in producing hollow rectangular microtubes with a single rib as shown in Fig. 9.19 [23]. In addition to the method of fabricating metal microtubes by reducing the diameter of raw tubes, a technique has been developed to create metal microtubes by bending and shaping sheet metal through press bending (see Footnote 3). This method, known as progressive press bending, offers advantages such as low cost, high speed, and mass production capabilities. It is commonly used in the production of stainless steel injection needles for insulin delivery. The tip of the needle has an outer diameter of 200 μm, an inner diameter of 50 μm, and a tapered shape that becomes thicker towards the base. The tapered shape reduces pain during injection and helps minimize the increase in injection resistance. To achieve the tapered shape, fan-shaped sheet metal is rounded using progressive press bending dies, and then the seams are welded to form the tube shape.

Furthermore, other methods for fabricating microtubes at the microscale using bottom-up approaches include electroforming applying the plating technique, which produces nickel microtubes with an outer diameter of 20 μm and an inner diameter

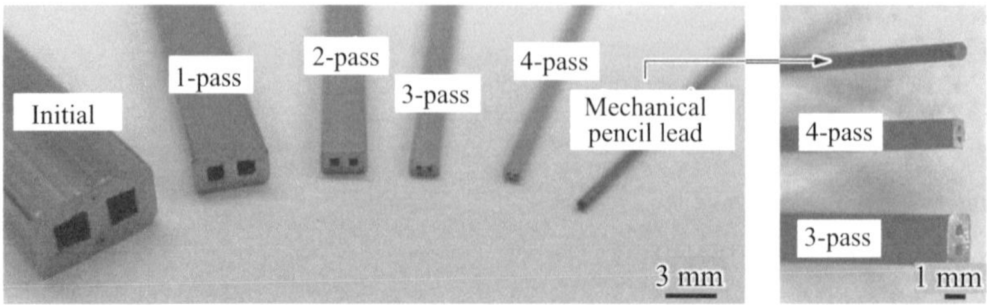

Fig. 9.19 Hollow rectangular microtube formed by die-less drawing [23]

of 10 μm.[4] Additionally, titanium microtubes with an outer diameter of 60 μm and an inner diameter of 30 μm have been fabricated using the sputtering technique [24].[5]

9.2.3 Secondary Processing

In this section, examples of secondary processing in microtube forming are introduced.

(1) **Flaring**

Microtube flaring is a process of expanding the tube by pressing a conical die against the end of the microtube [25]. Microtubes produced by drawing processes often employ drawing without the use of plugs or mandrels, resulting in a higher surface roughness on the inner surface compared to the outer surface in contact with the die. Particularly at the microscale, the surface roughness becomes relatively more significant compared to the tube wall thickness. In the case of flaring with an outer diameter of 500 μm and wall thickness of 50 μm, the conical die contacts the inner surface of the tube, making the surface roughness of the tube inner, the surface condition of the conical die, and the lubrication state have a significant influence on the flaring limit (Fig. 9.20) [26]. Therefore, surface treatment of the die becomes important compared to the macroscale.

(2) **Spinning**

In microtube forming, there is a demand for processing methods that require fewer dies and tools, which are difficult to manufacture. Therefore, flaring by spinning, as shown in Fig. 9.21, has been proposed. Micro-spinning is a form of incremental forming, where the microtube is rotated, and a rod-like tool is pressed against it to expand the tube. Since dedicated dies are not required, various angles of flaring can

[4] Terumo Corporation: http://www.terumo.co.jp/ (as of 2019 February).

[5] Luzcom Co. Ltd., http://www.luzcom.jp/index.html (as of 2019 February).

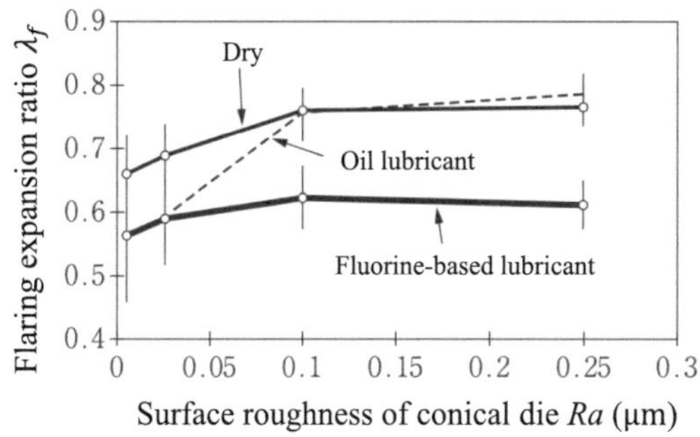

Fig. 9.20 Effect of surface roughness of tool and lubrication on flaring expansion limit [26]

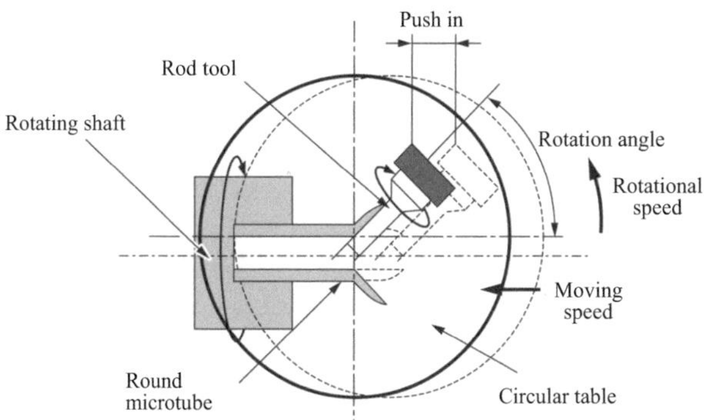

Fig. 9.21 Flaring of tube by micro-spinning [27]

be achieved using only the rod-like tool [27]. Furthermore, by sequentially contacting and forming the tube rather than having the die contact the entire inner surface, even low ductility microtubes can achieve significant expansion ratios. There are also cases where spinning is applied to create a bellows shape by introducing a corrugated shape to the tube body [28].

(3) **Die-less forming**

Die-less forming is a processing method in microtube forming that completely eliminates the need for dies and tools. By applying the die-less drawing introduced as the primary processing method for microtubes, it is possible to fabricate micro-tapered tubes. The reduction ratio of the cross-section in die-less drawing is determined by the ratio of the drawing speed to the feeding speed. Therefore, by fixing the feeding speed and varying the drawing speed, it is possible to change the reduction ratio during the drawing process and fabricate tapered tubes without using any dies or tools. Additionally, by locally heating one end of the microtube and compressing the other end, it is possible to control the location of buckling that occurs at the locally heated region. In this way, continuous buckling deformation is induced, and die-less bellows forming, which forms a bellows shape on a microtube with an outer diameter

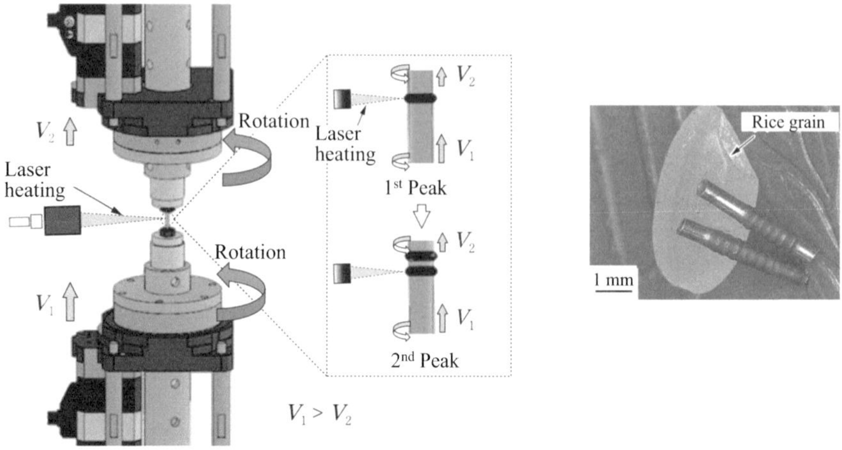

Fig. 9.22 Schematic of laser die-less forming and micro-bellows [30]

of 500 μm and a wall thickness of 100 μm, has been proposed. In this processing method, a laser heating method is adopted to achieve localized heating (Fig. 9.22).

(4) **Swaging process**

Micro-swaging is a forming process where equally divided dies are rotated and a microtube is inserted inside, and then it is compressed by striking from the outside. It is a particularly suitable method for producing tapered tubes. In the tapered section, compressive stresses in the circumferential direction are applied, making it prone to thickening or wrinkling, requiring careful selection of the taper angle. By gradually increasing the taper angle, it is possible to achieve a significant taper angle without causing wrinkles. The swaging process has been reported as an example of diameter reduction, where a microtube with an outer diameter of 618 μm and an inner diameter of 490 μm was swaged to an outer diameter of 230 μm and an inner diameter of 90 μm in the case of SUS 304 stainless steel tube (Fig. 9.23) [31].

Fig. 9.23 Micro-tapered tube by micro-swaging [31]

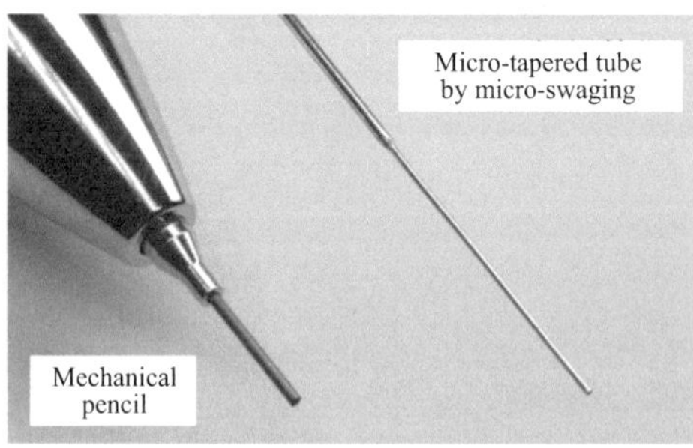

(5) Hydroforming process

Microtube hydroforming is a forming process where internal pressure is applied to expand the microtube, and by placing a die on the outside of the tube, it is possible to form micro hollow parts with shapes corresponding to the die. One of the challenges in microtube hydroforming is the increased internal pressure required for forming compared to the macroscale. Generally, the maximum internal pressure p^* of an infinitely long thin-walled cylindrical tube is given by the following equation:

$$p^* = C\left(\frac{t}{D}\right)\frac{4}{\sqrt{3}^{n+1}}\left(\frac{n}{\mathrm{e}}\right)^n \tag{9.1}$$

Here, C represents the strength coefficient, and n represents the work-hardening exponent. As mentioned earlier, as the dimensions of the tube decrease at the microscale, the ratio of wall thickness t to outer diameter D generally increases. Therefore, at the microscale, the required internal pressure for forming becomes significantly higher. Additionally, at the macroscale, a hole is drilled in the axial feed punch to supply the pressure medium, but in the case of microtubes with a diameter of less than 1 mm, it becomes extremely challenging to create a small deep hole for supplying the pressure medium in the already thin axial punch. As a solution, methods such as providing grooves on the side of the axial punch for supplying the pressure medium without leakage have been proposed. Successful examples include the hydroforming of highly delicate microtubes with an outer diameter of 500 μm and a wall thickness of 100 μm, achieving a cross-shaped branch expansion as shown in Fig. 9.24 [32]. In addition, a method has been proposed where the die is allowed to float without being fixed, enabling the supply of pressure medium inside the tube without leakage [33].

(6) Simulation

In the microscale, the number of grains present in the thickness direction becomes extremely small, and as a result, the individual characteristics of each grain start to influence the deformation behavior and formability in microtube forming. Therefore, there have been reports of applying crystalline plasticity theory considering the slip deformation of individual grains to finite element analysis. An example includes the crystalline plasticity finite element analysis of tube hydroforming for a microtube with an outer diameter of 800 μm and a wall thickness of 40 μm, where the number of grains in the thickness direction is around 2–3. This analysis elucidated the occurrence behavior of necking caused by the heterogeneous deformation of grains during expansion [34].

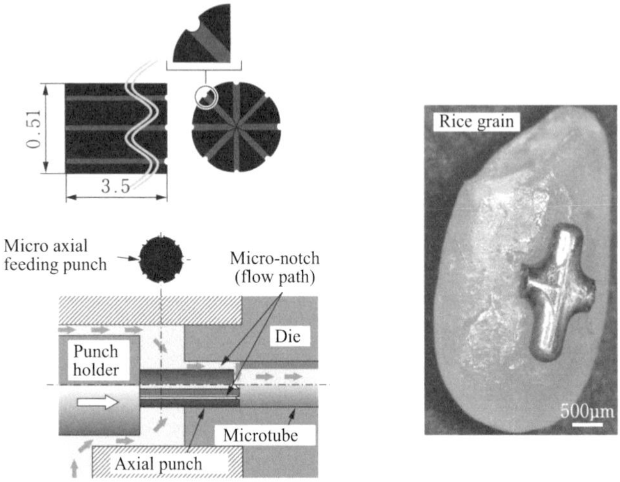

Fig. 9.24 Tool and die structure of microtube hydroforming and cross-expansion [32]

9.3 Process Design Case Studies

As mentioned in Chap. 1, tube forming is a holistic secondary processing technique that uses tubes as the material. The appropriate method and intermediate shape design depend on the shape and material of the product being produced. In some cases, drawings of intermediate shapes remain, but information about the forming methods used to achieve those shapes is rarely preserved. Furthermore, the selection of the intermediate shape and methods is a thought process based on the experience of the process designer, and the reasons behind the resulting choices are often unknown. While the previous edition included examples of processing, it was difficult to reproduce them due to a lack of information on the process involved.

Therefore, the Tube Forming Research Committee worked on the technology transfer of process design. As described later, we conducted interviews with process designers and developed a method to extract important perspectives and design rationale by documenting them in graph format. Here, we present three case studies selected from the interviews. Please note that the content of these case studies may differ from previous editions or other sources, but we consider the information obtained through interviews with the designers themselves as accurate.

9.3.1 Torsion Spring Automotive Suspension Component

Figure 9.25 shows a drawing of the target component. This component is a suspension part used as a torsion spring in automobiles. According to the skilled engineer who actually carried out the process design, the original method involved machining to

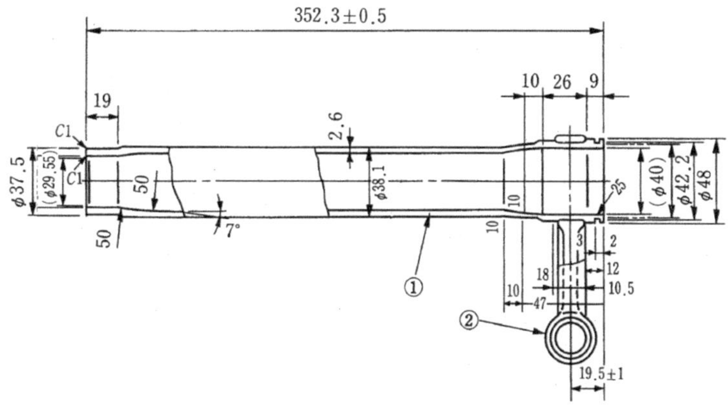

Fig. 9.25 Torsion spring of automobile suspension

create the shape, followed by heat treatment of high-carbon steel with added boron to have the desired strength. The arm was subsequently welded to the torsion spring body, which necessitated additional processes to realign the deformation caused by welding. As a result, this original process incurred high processing and material costs. The goal of the process design was to shape the component through plastic deformation, ensure strength through work-hardening, and eliminate welding and realignment processes through plastic bonding.

The results of the interviews were described in a graph and clustered into four main clusters. Through interpretation, it was revealed that the skilled designer considered (1) product performance, (2) crimping, (3) raw tube specifications, and (4) post-processing ductility as the most important factors in the process design.

Item (1) refers to ensuring the performance of the required torsion spring, with the design of pipe diameter and wall thickness distribution being the main considerations. Item (2) involved not only the shape of the serrations but also the design of the die to accommodate elastic deformation and leave compressive residual stress similar to autofrettage treatment. Item (3) specified seam heat treatment to ensure strength by work hardening through significant plastic deformation and to achieve the required shape and wall thickness. Item (4) focuses not only on processing to achieve the desired shape but also on ensuring the ductility of the processed product.

The intermediate process drawings are shown in Fig. 9.26. The circled areas in the drawing indicate important points in this process. Particularly in the second step, it is crucial to create the diameter and wall thickness (Figure b) and to form the serration shape, and the die design for autofrettage effect during crimping is essential.

Determining the dimensions of the tube involves finding a balance between achieving sufficient plastic deformation for strength through work-hardening and avoiding excessive deformation that can lead to defects, such as cracks and wrinkles. Based on his experience, a reduction rate of 40% has been determined to achieve this balance. For the original process where strength was achieved through heat treatment, tube dimensions were an outer diameter of 38.1 mm and wall thickness of

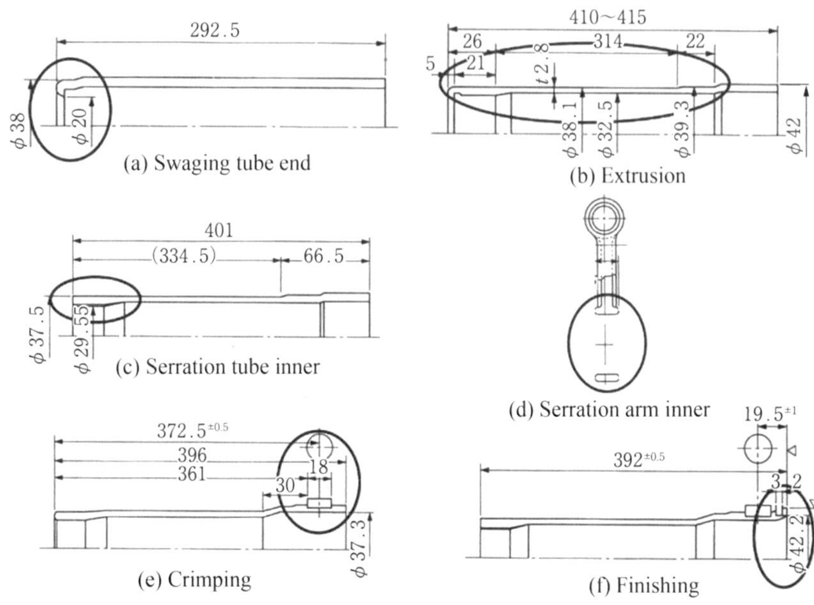

Fig. 9.26 Intermediate process drawings

2.6 mm. The selected tube dimensions for the new process were an outer diameter of 42.7 mm (two sizes larger than the original) and a wall thickness of 4.0 mm to achieve a 40% reduction rate. In the second process, which is the extrusion process (Figure (b)), confirming an overall reduction rate of 40%. Additionally, it is observed that the cold working of the tube resulted in an increase in yield stress, from 320 MPa for the raw tube to 830 MPa for the formed part, leading to the elimination of heat treatment.

9.3.2 Vacuum Valve Body [35]

Figure 9.27 shows a drawing and photograph of a vacuum valve body. This component is part of a device used for creating a vacuum and was previously manufactured by welding together four separate parts. It was made of aluminum (A6030), which presents difficulties in welding, as welding pinholes tend to occur, and the gas trapped in these pinholes becomes problematic during vacuum operation. Therefore, the aim was to achieve one-piece forming without the need for welding. Although its complex shape may not appear readily suitable for one-piece forming, the experienced engineer determined that one-piece forming could be achieved by extending branch pipes using T-forming hydroforming and reducing the diameter at both ends of the main pipe through necking, with the additional processing of three tube ends.

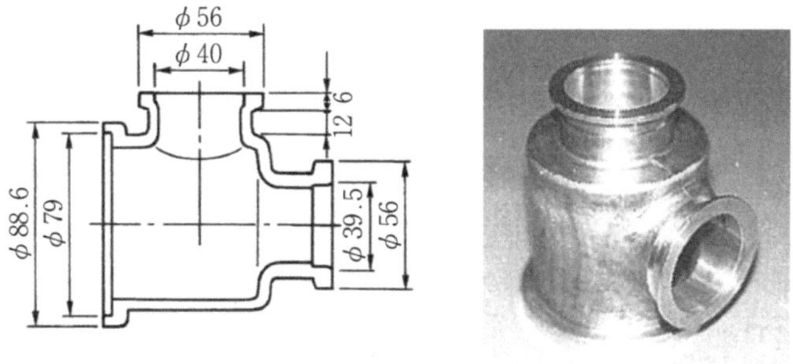

Fig. 9.27 Vacuum valve body [35]

Figure 9.28 illustrates the processing stages for the vacuum valve body. The sequence of the first process, necking, and the second process, bulging, was determined to perform necking before bulging because it would be difficult to neck the tube if there was already a bulged dome. Additionally, appropriate processing conditions were selected to ensure that the bulged dome does not become too thin. The burring process for the dome section involves drilling pilot holes, and as the conventional burring tool cannot fit, a detachable burring tool is used for processing.

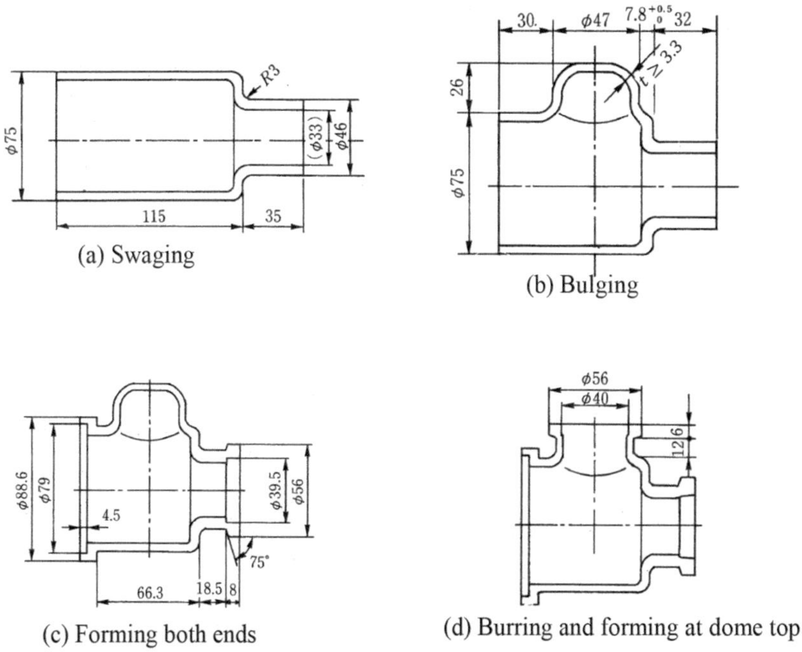

Fig. 9.28 Processing stages of vacuum valve body [35]

9.3.3 Front Lower Arm

Figure 9.29 shows the processing stages and product photo of a front lower arm. By performing necking at both ends of the raw tube, followed by bending and further sealed fluid forming, it is made possible to shape a high-strength steel tube with a tensile strength of 780 MPa. Among these processes, we interviewed a skilled engineer of a bending machine manufacturer regarding the bending process. This engineer, in addition to process design, also developed the new bending machine required for this process.

Using a mandrel would be ideal for the demanding bending process, but it is not possible due to the narrowing of both ends. As the section distortion is allowed for in the subsequent process of liquid forming, the bending die shape was carefully designed to enhance external restraint during bending. However, the different bending radii at two locations made die design challenging. Bending with strong external restraint is achieved through the utilization of a bending machine called a push rotary bender (Fig. 9.30), where a pusher rail is pushed against a forming shoe while bending. Reducing the eccentricity of the hydraulic cylinder in this bending method reduces defects such as wrinkles.

However, if the eccentricity of the hydraulic cylinder is reduced to zero, the bending moment arm becomes zero, resulting in an inability to bend. Therefore, a PRB bending machine (Fig. 9.31) was developed, which combines a rotary draw bender and a push rotary bender. This bending machine utilizes the rotation of the rotary draw bender to provide driving force, and it incorporates a new cylinder that applies strong external restraint by pressing the pressure die. This new bending machine allows adjustment of external restraint not through die shape but by adjusting the pressing load, making it more adaptable to variations in raw tubes and other

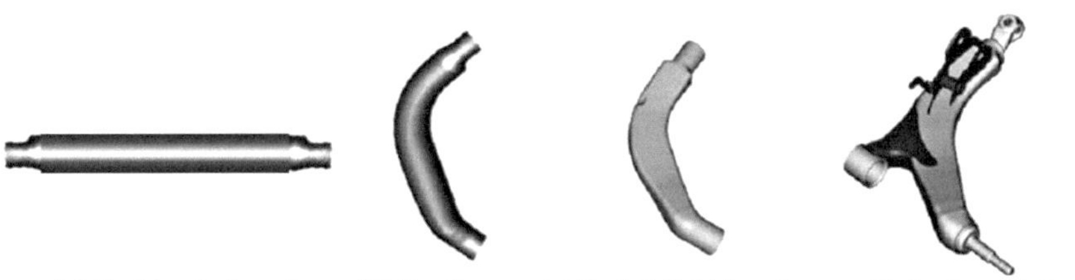

(a) Necking both ends (b) Bending (c) Sealed fluid forming (d) Cutting, joining

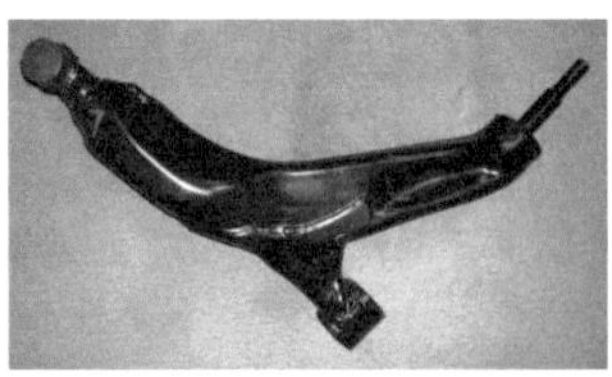

Fig. 9.29 Processing stages and product photo of front lower arm [36]

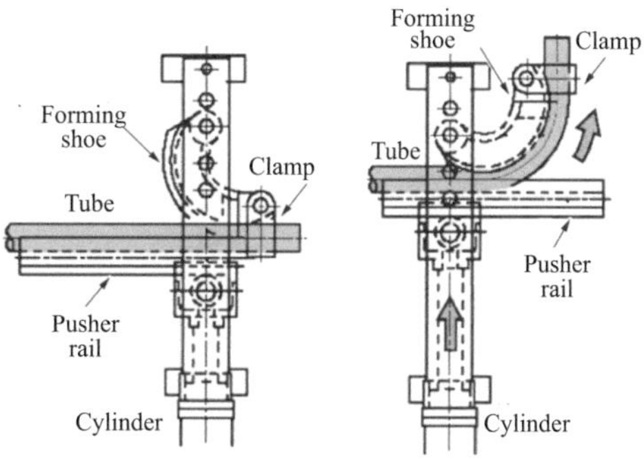

Fig. 9.30 Push rotary bender

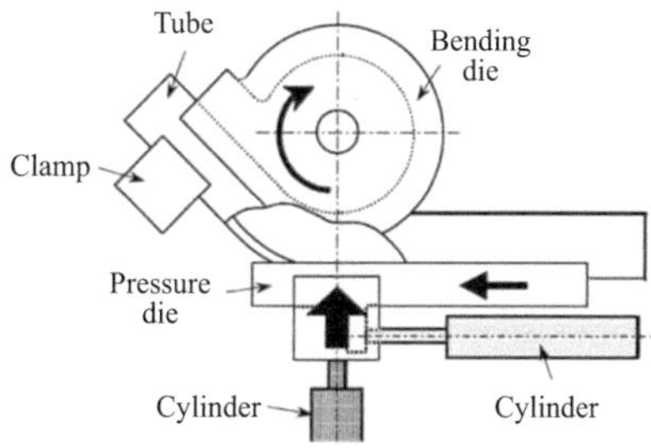

Fig. 9.31 PRB bending machine

factors. Furthermore, this bending method can be interpreted as facilitating bending in the axial direction by compressing in the circumferential direction and inducing plastic deformation in the cross-section.

9.3.4 Technology Transfer Through Graph Representation of Process Design

Plastic forming is an excellent processing method that has higher yield compared to machining and requires less energy compared to 3D printing. However, while machining and 3D printing can be carried out based on product drawings, plastic forming, particularly tube forming, requires the design of intermediate shapes and the selection of appropriate forming methods. This process design relies on a thought process based on the experience of skilled process designers, and the transfer of this information is not always efficient. Therefore, during the revision of this book, we developed a method for transferring the skills and technology of process design engineers and archived this knowledge. The process design examples in this section were extracted using this method. Regarding the extraction of design knowledge, a study was conducted on graph-based representation using drawings, focusing on machine tools [37, 38]. However, in process design, it is crucial to extract information that does not appear in the drawings, requiring a different approach. Furthermore, in this study, the nodes were not separated into "entities", "attributes", and "relationships", and it was difficult to understand the design rationale, and clustering did not result in meaningful extraction of information.

On the other hand, research on knowledge extraction that is not directly apparent in documents has been conducted, and studies on semantic analysis using natural language processing have made progress. Latent Semantic Analysis (LSA) [39] is a method that applies singular value decomposition to vast amounts of text data to extract topics by eliminating small singular values. Latent Dirichlet Allocation (LDA) [40] is a topic extraction method that improves efficiency by Bayesian estimation of the distribution parameters and is used for analyzing internet information, among other things. KeyGraph [41] is a topic analysis method that utilizes graphs, but since it lacks structure, it is difficult to extract design principles or important meanings, although it can indicate the proximity of items. Performing semantic analysis using these natural language processing techniques requires a vast amount of data, and the information obtained from interviews with process designers is insufficient for such analysis. Grounded Theory [42] has been developed as a method to extract knowledge from individuals with certain knowledge and experience and is used in qualitative research in the field of social sciences. Although this method is considered effective for interviews with process designers, it has challenges such as results being influenced by the analyst's subjectivity and the lack of understanding the basis of the designer's judgment. Therefore, to address these challenges, we have developed a method [43–46] that allows for semantic extraction even with a moderate amount of information obtained from interviews with process designers, enables tracing of design principles through graph representation of the extracted information, and ultimately facilitates semantic extraction.

The flow of the developed method for documenting and archiving process design knowledge in a transferable format is shown in Fig. 9.32. Since process design is a thought process, it is verbalized through interviews and then structured through graph description, connecting nodes to reveal the design principles (the reasons behind the design) and allowing the extraction of important considerations through clustering without bias from the analyst.

When conducting an interview for a single process design case, approximately 100 statements can be obtained after about half a day of interviewing. Since it is an

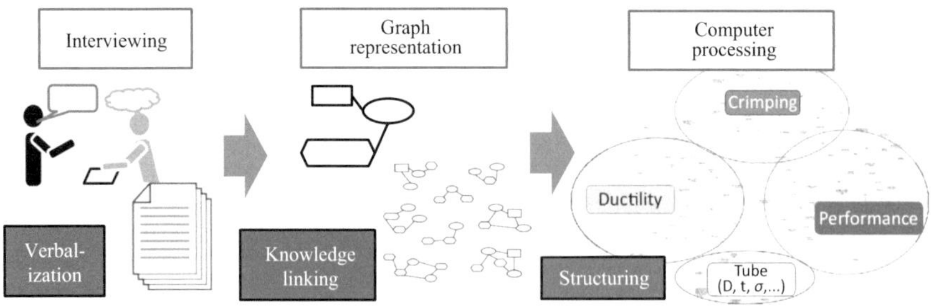

Fig. 9.32 Documenting and archiving process design knowledge in transferable format

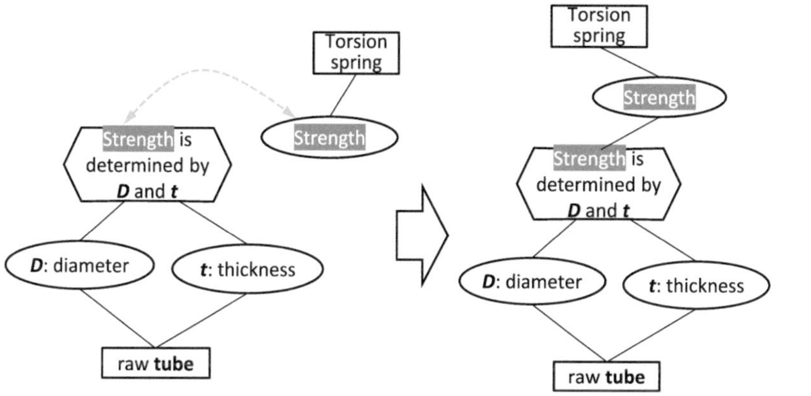

Fig. 9.33 Partial graph representation of statements and integration of partial graphs

interview, the topics may change based on the questions and answers, and the statements may not be systematic. However, it is possible to verbalize the thought process. Each individual statement (sentence) is described in graph format using three types of nodes: entity (rectangle), attribute (ellipse), and relationship (hexagon). These partial graphs are then combined based on the keywords they contain (Fig. 9.33).

When these partial graphs are combined to create the overall graph, areas with fewer nodes and sparse edges appear, indicating missing information. Skilled process designers may naturally assume that information without explicit mention is present in those areas.

By repeating the interviews focusing on such areas 2–3 times, it can be observed that the growth in the number of nodes obtained by an interview decreases, indicating that almost all information has been extracted. Next, when the obtained overall graph is subjected to clustering using a computer, in the case of a torsion spring automobile suspension part, four clusters appear. The interviewing engineer can interpret the meaning of each cluster from the information of the nodes it includes.

In the case of Fig. 9.32, it can be deduced that “product performance”, “post-forming ductility”, “crimping (plastic bonding)”, and “raw tube (dimensions and properties)” are considered important for process design. This method can be used not only for tube-forming process design but also for knowledge extraction and structuring in various fields of plastic processing and other disciplines.

References

1. Domblesky, J.P., Shivpuri, R., Painter, B.: Application of the finite-element method to the radial forging of large diameter tubes. J. Mater. Process. Technol. **49**, 57–74 (1995)
2. Hasegawa, S., Kitamura, K., Hasegawa, K., Mori, K.: In: Proceedings of the 65th joint conference on the technology of plasticity, JSTP, pp. 255–256 (2014)

3. Johnson, W.: Impact strength of materials, vol. 178. Edward Arnold Limited, (1972)
4. Kitazawa, K., Kobayashi, M., et al.: In: Proceedings of the 37th joint conference on the technology of plasticity, JSTP, pp. 399–402 (1986)
5. Saito, Y., Takahashi, A., Kato, K.: In: Proceedings of the 32nd joint conference on the technology of plasticity, JSTP, pp. 315–320 (1981)
6. Manabe, K., Fuchizawa, S.: J. JSTP, **52**(600), 36–41 (2011)
7. Nikkei, B.P. (ed.): Nikkei Monozukuri (The solutions magazine for design and manufacturing). Nikkei Business Publications, May issue (2005)
8. Ueno, Y., Kishimoto, A., Okada, M., Yumimochi, T., Uwai, K., Suzuki, T.: J. JSTP, **48**(563), 1055–1059 (2007)
9. Maeno, T.: J. Mater. Process. Technol. Cent., **55**(11), 16–21 (2014)
10. Nakamura, M.: Pipe forming techniques, pp. 83–92, Nikkan Kogyo Shimbun, (1982)
11. Shimabayashi.: In: Proceedings of the 31st Joint conference on the technology of plasticity, JSTP, pp. 275–278 (1980)
12. Takagi, R.: J. JSTP., **24**(274), 1101–1108 (1983)
13. Matsumoto, Unno, Y., Osakada, K. Shiomi, M.: J. JSTP, **46**(530), 239–243 (2005)
14. Murata, M., Kuboki, T.: J. JSTP, **55**(637), pp. 127–131 (2014)
15. Koyama, H., Saito, T., Kaneko, R.: In: Proceedings of the 66th joint conference on the technology of plasticity, JSTP, pp. 331–332 (2015)
16. Shimizu, T., Yang, M., Manabe, K.: J. JSTP, **54**(626), 235–239 (2013)
17. Furushima, T., Tsunezaki, H., Manabe, K., Alexandrov, S.: Int. J. Mach. Tools Manuf **76**, 34–48 (2014)
18. Engel, U., Eckstein, R.: J. Mater. Process. Technol. **125–126**, 35–44 (2002)
19. The Japan Society for Technology of Plasticity (ed.): Drawing, 22. Corona Publishing, (1990)
20. Yoshida, K., Furuya, H.: J. Mater. Process. Technol. **153–154**, 145–150 (2004)
21. Yoshida, K., Onitsuka, Y., Yamashita, S.: In: 2nd ASMP, CD-ROM (2009)
22. Furushima, T., Manabe, K.: J. Mater. Process. Technol. **191**, 59–63 (2007)
23. Furushima, T., Shirasaki, A., Manabe, K.: J. JSTP, **52**(602), 380–384 (2011)
24. Tsuchiya, K., Nakanishi, N., et al.: Trans. Jpn. Soc. Mech. Engineers. C, **71**(702), 603–609 (2005)
25. Mirzai, M.A, Manabe, K., Mabuchi, T.: J. Mater. Process. Technol., **201**, 214–219 (2008)
26. Mirzai, M.A., Manabe, K.: J. Solid Mech. Mater. Eng., **3**(2), 375–386 (2009)
27. Koyama, H., Yoshino, K.: In: Proceedings of the 56th joint conference on the technology of plasticity, JSTP, pp. 645–646 (2005)
28. Takahashi, T., Koyama, H.: In: Proceedings of the 64th joint conference on the technology of plasticity, JSTP, pp. 275–276 (2013)
29. Furushima, T., Esaka, Y., Manabe, K.: In: Proc int conf mater process technol, pp. 68–71 (2010)
30. Furushima, T., Kitamura, R., et al.: In: Proceedings of the 65th joint conference on the technology of plasticity, JSTP, pp. 237–238 (2014)
31. Tada, M.: Proceedings of the 2015, spring meeting. JSTP, 81–82 (2015)
32. Mori, S., Sato, H., Itai, K., Manabe, K.: J. JSTP **58–672**, 72–77 (2017)
33. Ngaile, G., Lowrie, J.: J. Micro Nano-Manuf. **2–4**, 041004 (2014)
34. Zhuanga, W., Wang, S., Lin, J., Balint, D., Hartl, C.: Eur. J. Mech. A/Solids. **31**, 67–76 (2012)
35. Nakamura, M., Maruyama, K., et al.: Pipe forming techniques, 2nd edn, pp. 3101–3103. Nikkan Kogyo Shimbun, (1998)
36. Ueno, Y., Kishimoto, A., Okada. M., Yuminamochi, T., Uwai, A., Suzuki, K.: J. JSTP, **48**(563), 1055–1059 (2007)
37. Ito, Y., Shinno, H., Nakanishi, S.: Annals of the CIRP **38–1**, 141–144 (1989)
38. Ihara, T., Ogawa, M., Ito. Y.: J. Jpn. Soc. Precis. Eng., **59**(3), 67–72 (1993)
39. Deerwester, S., Dumais, S.T., Furnas, G.W., Landauer, T.K., Harshman, R.: J. Am. Soci. Inf. Sci. **41–6**, 391–407 (1990)
40. Blei, D.M., Ng, A.Y., Jordan, M.I.: J. Mach. Learn. Res. **3**, 993–1022 (2003)
41. Osawa, Y., Benson, N.E., Yachida, M.: IEICE Transactions, D-1, 82–2, 391–400 (1999)

42. Glaser, B.G., Strauss, A.L., Anselm, L.: The discovery of grounded theory: Strategies for qualitative research. Transaction Publishers, (2009)
43. Kuriyama, Y., Kondo, S., Takamoto, H., Shirayori, A., Yoshida, Y.: Proceedings of the 2015 Spring Meeting. JSTP, 157–158 (2015)
44. Kondo, S., Takamoto, H., Kuriyama, Y., Shirayori, A., Yoshida.: Proceedings of the 2015 Spring Meeting, JSTP,159–160 (2015)
45. Kondo, S., Takamoto, H., Masui, K., Kuriyama, Y., Shirayori, A.: Trans JSME, **82**, 842 (2016)
46. Iwasaki, K., Kuriyama, Y., Kondoh, S., Shirayori, A.: Structuring engineers' implicit knowledge of forming process design by using a graph model. In: Proc 11th CIRP conf intelligent computation in manufacturing engineering, (2017)

MIX
Papier aus verantwortungsvollen Quellen
Paper from responsible sources
FSC® C105338

If you have any concerns about our products,
you can contact us on
ProductSafety@springernature.com

In case Publisher is established outside the EU,
the EU authorized representative is:
Springer Nature Customer Service Center GmbH
Europaplatz 3, 69115 Heidelberg, Germany

Printed by Libri Plureos GmbH
in Hamburg, Germany